SOLID STATE PHYSICS

SOLID STATE PHYSICS

For the Students of B.Sc. (Pass and Honours)

R.K. PURI
Ph.D.
Professor, Department of Physics
Indian Institute of Technology Delhi
New Delhi

V.K. BABBAR
Ph.D.
Lecturer, Department of Physics
Guru Nanak Dev University, Amritsar
Punjab

S. CHAND
PUBLISHING

S Chand And Company Limited

(ISO 9001 Certified Company)

S Chand And Company Limited
(ISO 9001 Certified Company)
Head Office: D-92, Sector–2, Noida – 201301, U.P. (India), Ph. 91-120-4682700
Registered Office: A-27, 2nd Floor, Mohan Co-operative Industrial Estate, New Delhi – 110 044, Phone: 011-49731800
www.**schandpublishing.com**; e-mail: **info@schandpublishing.com**

Marketing Offices:

Chennai	:	Ph: 23632120; chennai@schandpublishing.com
Guwahati	:	Ph: 2738811, 2735640; guwahati@schandpublishing.com
Hyderabad	:	Ph: 40186018; hyderabad@schandpublishing.com
Jalandhar	:	Ph: 4645630; jalandhar@schandpublishing.com
Kolkata	:	Ph: 23357458, 23353914; kolkata@schandpublishing.com
Lucknow	:	Ph: 4003633; lucknow@schandpublishing.com
Mumbai	:	Ph: 25000297; mumbai@schandpublishing.com
Patna	:	Ph: 4011400; patna@schandpublishing.com

First Edition 1997
Subsequent Editions and Reprints 2001, 2003, 2004, 2006, 2007, 2008, 2009 (Twice), 2010, 2011, 2012, 2013, 2014, 2015, 2016 (Twice), 2017 (Twice), 2018, 2019, 2020 (Twice), 2021 (Thrice), 2022 (Twice), 2024 (Twice)

Reprint 2025

ISBN: 978-81-219-1476-5 **Product Code:** H6SSP68PHYS10ENAC0XO

PRINTED IN INDIA

By Vikas Publishing House Private Limited, Plot 20/4, Site-IV, Industrial Area Sahibabad, Ghaziabad – 201 010 and Published by S Chand And Company Limited, A-27, 2nd Floor, Mohan Co-operative Industrial Estate, New Delhi – 110 044.

PREFACE TO THE THIRD EDITION

The present edition is brought up to incorporate the useful suggestions from a number of readers and teachers for the benefit of students. Keeping in view the present style of University Question Papers, a number of Very Short, Short and Long questions are given at the end of each chapter. The manuscript has been thoroughly revised and corrected to remove the errors which crept into earlier editions. We hope the present improved edition would serve the students in a better way. However, the comments and suggestions from readers for further improvement of the book will be gratefully acknowledged.

R.K. PURI
V.K. BABBAR

PREFACE TO THE FIRST EDITION

A number of Indian universities have revised their curricula at the undergraduate level and have included various topics which earlier formed a part of the postgraduate curricula. Ever since the new syllabi were introduced, there had been a dearth of good books which strictly follow the revised syllabi. A number of standard texts are available on Solid State Physics but these are of advanced level. The present book is written specifically to meet the requirements of the undergraduate students and is in accordance with the common prescribed syllabi of most of the Indian universities.

The general approach and aim of this book is to provide a comprehensive introduction to the subject of Solid State Physics to the undergraduate students in a coherent, simple and lucid manner. The coverage of basic topics is concise, brief and self-explanatory. The topics such as Lasers, Magnetic Resonances, and the Mossbauer Effect are excluded as their advanced treatment is generally covered at the postgraduate level. The text is divided into ten chapters and each chapter is followed by a set of solved examples which acquaint the students with the application of the various principles and formulae used in the text and give them a feeling of the magnitude of the physical quantities involved therein. The SI units are followed throughout the book and their conversions to other practical units are appropriately introduced. Some of the conversion factors are also listed in appendix I. A summary of each chapter is given for a quick review of the topics. Each chapter is concluded with a set of questions and unsolved problems to help the students to comprehend these topics. A list of useful references is given for the indepth study of the subject.

We hope that the undergraduate students will find this book useful as well as concise for the subject of Solid State Physics. The comments and feedback from the students as well as teachers about this book will be gratefully appreciated.

We thank our friends and families, particularly our spouses, for their inspiration and encouragement. We also thank the publishers for quality printing and timely publication of this book.

R.K. PURI

V.K. BABBAR

New Delhi
June 2, 1996

CONTENTS

SOLID STATE PHYSICS

CHAPTER - I

CRYSTAL STRUCTURE

1.1 INTRODUCTION

Matter, consisting of one or more elements or their chemical compounds, exists in nature in the solid, liquid and gaseous states. As the atoms or molecules in solids are attached to one another with strong forces of attraction, the solids maintain a definite volume and shape. The solid state physics is the branch of physics dealing with physical properties of solids, particularly crystals, including the behaviour of electrons in these solids. The solids may be broadly classified as *crystalline* and *non-crystalline* depending upon the arrangement of atoms or molecules.

The crystalline state of solids is characterized by regular or periodic arrangement of atoms or molecules. Most of the solids are crystalline in nature. This is due to the reason that the energy released during the formation of an ordered structure is more than that released during the formation of a disordered structure. Thus crystalline state is a low energy state and is, therefore, preferred by most of the solids. The crystalline solids may be subdivided into *single crystals* and *polycrystalline* solids. In single crystals, the periodicity of atoms extends throughout the material as the case of diamond, quartz, mica, etc. A polycrystalline material is an aggregate of a number of small crystallites with random orientations separated by well-defined boundaries. The small crystallites are known as *grains* and the boundaries as *grain boundaries.* It may be noted that although the periodicity of individual crystallites is interrupted at grain boundaries, yet the polycrystalline form of a material may be more stable compared with its single crystal form. Most of the metals and ceramics exhibit polycrystalline structure.

The non-crystalline or *amorphous* solids are characterized by the completely random arrangement of atoms or molecules. The periodicity, if at all present, extends up to a distance of a few atomic diameters only. In other words, these solids exhibit *short range order.* Such type of materials are formed when the atoms do not get sufficient time to undergo a periodic arrangement. Glass is an example of amorphous materials. Most of the plastics and rubbers are also amorphous.

The science which deals with the study of geometrical forms and physical properties of crystalline solids is called *crystallography*. The study of crystallography is necessary to understand the strong correlation between the structure of a material and its physical properties. The present chapter deals with some of the basic concepts of crystallography which are fundamental to the study of solid state physics.

1.2 CRYSTAL LATTICE AND TRANSLATION VECTORS

Before describing the arrangement of atoms in a crystal, it is always convenient to describe the arrangement of imaginary points in space which has a definite relationship with the atoms of the crystal. This set of imaginary points forms a framework on which the actual crystal structure is based. Such an arrangement of infinite number of imaginary points in three-dimensional space with each point having identical surroundings is known as *point lattice* or *space lattice.*

The term 'identical surroundings' means that the lattice has the same appearance when viewed from a point **r** in the lattice as it has when viewed from any other point **r'** with respect to some arbitrary origin. This is possible only if the lattice contains a small group of points, called *pattern unit*, which repeats itself in all directions by means of a *translation operation* **T** given by

$$\mathbf{T} = n_1\mathbf{a} + n_2\mathbf{b} + n_3\mathbf{c} \tag{1.1}$$

where n_1, n_2 and n_3 are arbitrary integers and the vectors **a**, **b** and **c** are called the *fundamental translation vectors.* Thus, we have

$$\mathbf{r'} = \mathbf{r} + \mathbf{T} = \mathbf{r} + n_1\mathbf{a} + n_2\mathbf{b} + n_3\mathbf{c} \tag{1.2}$$

In a perfect lattice, Eq. (1.2) holds good, i.e., point **r'** can be obtained from **r** by the application of the operation (1.1). However, in an imperfect lattice, it is not possible to find **a**, **b** and **c** such that an arbitrary choice of n_1, n_2 and n_3 makes **r'** identical to **r**. The translation vectors **a, b** and **c** are also called the *crystal axes* or *basis vectors* and shall be described later.

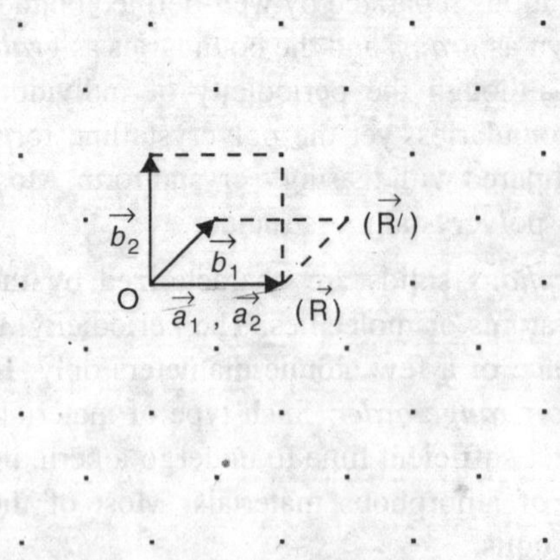

Fig. 1.1. *Primitive* ($\mathbf{a}_1$, $\mathbf{b}_1$) *and non-primitive* ($\mathbf{a}_2$, $\mathbf{b}_2$) *translation vectors in a two-dimensional lattice.*

Consider, for simplicity, a part of a two-dimensional lattice as shown in Fig 1.1. The translation vectors **a** and **b** can be chosen in a number of ways. Two such possibilities are shown in this figure where two sets $\mathbf{a}_1$, $\mathbf{b}_1$ and $\mathbf{a}_2$, $\mathbf{b}_2$ of translation vectors are drawn. Considering first the translation vectors $\mathbf{a}_1$ and $\mathbf{b}_1$, the point $\mathbf{R'}$ can be obtained from $\mathbf{R}$ using the translation operation given by

$$\mathbf{T} = 0\mathbf{a}_1 + 1\mathbf{b}_1$$

which contains integral coefficients. Thus $\mathbf{R'}$ is related to $\mathbf{R}$ by the equation

$$\mathbf{R'} = \mathbf{R} + \mathbf{T} = \mathbf{R} + 0\mathbf{a}_1 + 1\mathbf{b}_1$$

Such translation vectors which produce a translation operation containing integral coefficients are called *primitive translation vectors.* Referring to the second set of translation vectors $\mathbf{a}_2$ and $\mathbf{b}_2$, the point $\mathbf{R'}$ can be obtained from $\mathbf{R}$ by using the equation

$$\mathbf{R'} = \mathbf{R} + \frac{1}{2}\mathbf{a}_2 + \frac{1}{2}\mathbf{b}_2$$

which contains non-integral coefficients of $\mathbf{a}_2$ and $\mathbf{b}_2$. Such translation vectors for which the translation operation contains non-integral coefficients are called *non-primitive translation vectors.* Either type of translation vectors may be used to describe the structure of a crystal inspite of the fact that the non-primitive translation vectors involving non-integral coefficients are not in accordance with the periodicity of the crystal. Usually, a set of orthogonal and the shortest possible translation vectors is preferred for describing a lattice.

1.3 UNIT CELL

The parallelograms formed by the translation vectors (Fig. 1.1) may be regarded as building blocks for constructing the complete lattice and are known as *unit cells* of the lattice. For a three-dimensional lattice, the unit cells are of the form of a parallelopiped. An application of the translation operation (1.1) for some values of n_1, n_2 and n_3 takes the unit cell to another region which is exactly similar to the initial region. On repeatedly applying the same operation with all possible values of n_1, n_2 and n_3, one can reproduce the complete lattice. Thus a unit cell may be defined as the smallest unit of the lattice which, on continuous repitition, generates the complete lattice. Both primitive and non-primitive translation vectors may be used to construct a unit cell. Accordingly, a unit cell is named as a *primitive unit cell* or a *non-primitive unit cell.* In Fig. 1.2, the parallelogram ABCD represents a two-dimensional primitive cell, whereas the parallelograms EFGH and KLMN represent non-primitive cells. Primitive unit cell is the smallest volume cell. All the lattice points belonging to a primitive cell lie at its corners. Therefore, the effective number of lattice points in a primitive unit cell is one. A non-primitive cell may have the lattice points at the corners as well as at other locations both

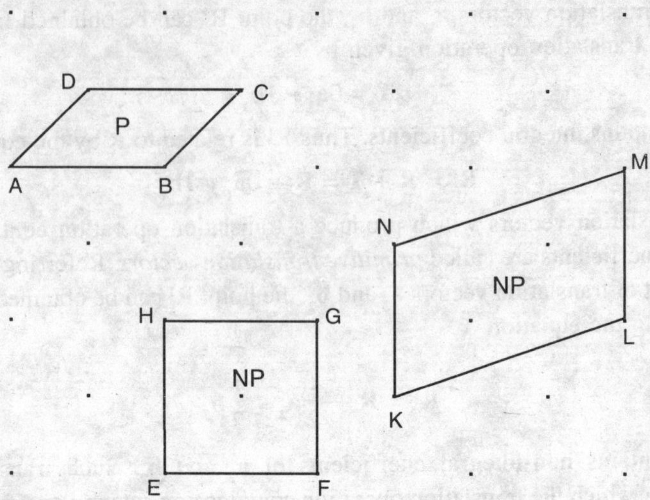

Fig. 1.2. *Primitive (P) and non-primitive (NP) unit cells of a two-dimensional lattice.*

inside and on the surface of the cell and, therefore, the effective number of lattice points in a non-primitive cell is greater than one. A primitive cell can also be constructed using the following procedure :

(i) Connect a given lattice point to all the nearby lattice points.

(ii) Draw normals at the mid-points of lines connecting the lattice points.

The smallest volume enclosed by the normals is the required primitive cell. Such a cell is called *Wigner-Seitz cell* and is shown in Fig. 1.3. The volume of a primitive cell having **a, b** and **c** as the fundamental translation vectors or crystallographic axes is given by

$$V = |\mathbf{a}.\mathbf{b}\times\mathbf{c}|$$

Since there exists a number of ways of choosing a unit cell, the choice of a *conventional unit cell* is a matter of convenience. Ideally, the primitive cell having the smallest volume should be chosen as the conventional unit cell.

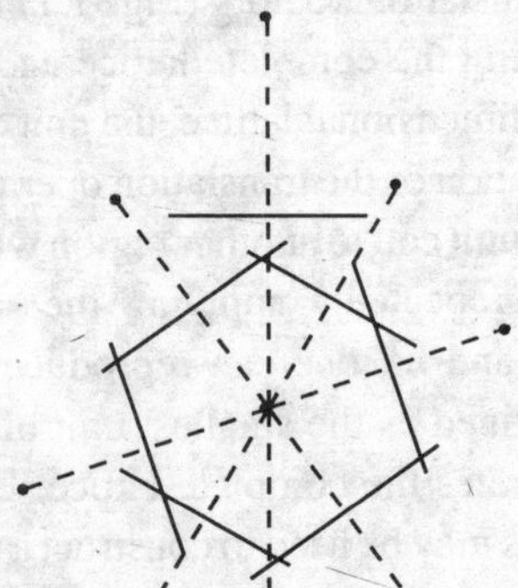

Fig. 1.3. *Construction of Wigner-Seitz primitive cell.*

However, sometimes a non-primitive cell is selected as the conventional unit cell because it possesses higher symmetry than a primitive cell.

1.4 BASIS

The space lattice has been defined as an array of imaginary points which are so arranged in space that each point has identical surroundings. The crystal structure is always described in terms of atoms rather than points. Thus in order to obtain a crystal structure, an atom or a group of atoms must be placed on each lattice point in a regular fashion. Such an atom or a group of atoms is called the *basis* and acts as a building unit or a structural unit for the complete crystal structure. Thus a lattice combined with a basis generates the *crystal structure.* Mathematically, it is expressed as

$$\text{Space lattice + Basis} \rightarrow \text{Crystal structure}$$

Thus, whereas a lattice is a mathematical concept, the crystal structure is a physical concept.

The generation of a crystal structure from a two-dimensional lattice and a basis is illustrated in Fig. 1.4. The basis consists of two atoms, represented by O and ●, having orientation as shown in Fig. 1.4. The crystal structure is obtained by placing the basis on each lattice point such that the centre of the basis coincides with the lattice point.

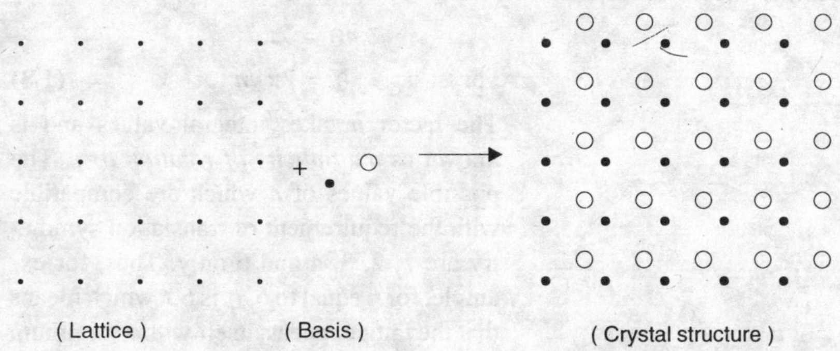

Fig. 1.4. *Generation of crystal structure from lattice and basis.*

The number of atoms in a basis may vary from one to several thousands, whereas the number of space lattices possible is only fourteen as described in a later section. Thus a large number of crystal structures may be obtained from just fourteen space lattices simply because of the different types of basis available. If the basis consists of a single atom only, a monoatomic crystal structure is obtained. Copper is an example of monoatomic face-centred cubic structures. Examples of complex bases are found in biological materials.

1.5 SYMMETRY OPERATIONS

A *symmetry operation* is that which transforms the crystals to itself, i.e., a crystal remains invariant under a symmetry operation. These operations are *translation, rotation, reflection* and *inversion.* The translation operation applies to lattices only while all the remaining operations and their combinations apply to all objects and are collectively known as *point symmetry operations.* The inversion operation is applicable only to three-dimensional crystals. These operations are briefly described below:

(*i*) **Translations**

The translation symmetry follows from the orderly arrangement of a lattice. It means that a lattice point **r**, under lattice translation vector operation **T**, gives another point **r'** which is exactly identical to **r**, i.e.,

$$\mathbf{r'} = \mathbf{r} + \mathbf{T}$$

where **T** is defined by Eq. (1.1).

(*ii*) **Rotations**

A lattice is said to possess the rotation symmetry if its rotation by an angle θ about an axis (or a point in a two-dimensional lattice) transforms the lattice to itself. Also, since the lattice always remains invariant by a rotation of 2π, the angle 2π must be an integral multiple of θ, i.e.,

$$n\theta = 2\pi$$

or

$$\theta = 2\pi/n \tag{1.3}$$

The factor n takes integral values and is known as *multiplicity of rotation axis.* The possible values of n which are compatible with the requirement of translation symmetry are 1, 2, 3, 4 and 6 only. Thus, for example, for n equal to 6, θ is 60° which means that the lattice repeats itself with a minimum rotation of 60°. Such a rotation is illustrated in Fig.1.5. Regular hexagon is an example of such a lattice. A rotation corresponding to the value of n is called n-fold rotation. A two-dimensional square lattice has 4-fold rotation symmetry. It may be noted that a rotation axis may or may not pass through a lattice point. The fact that 5-fold rotation is not compatible with translation symmetry operation and that only 1, 2, 3, 4 and 6 - fold rotations are permissible is proved as follows :

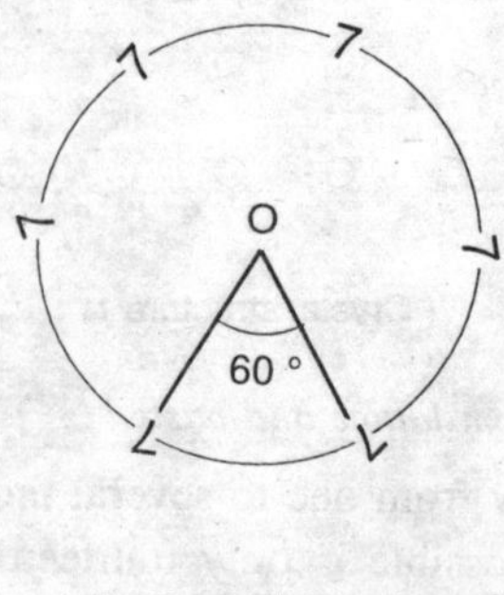

Fig. 1.5. *Six-fold rotation about the point O in two dimensions.*

Consider a row of lattice points A, B, C and D as shown in Fig. 1.6.

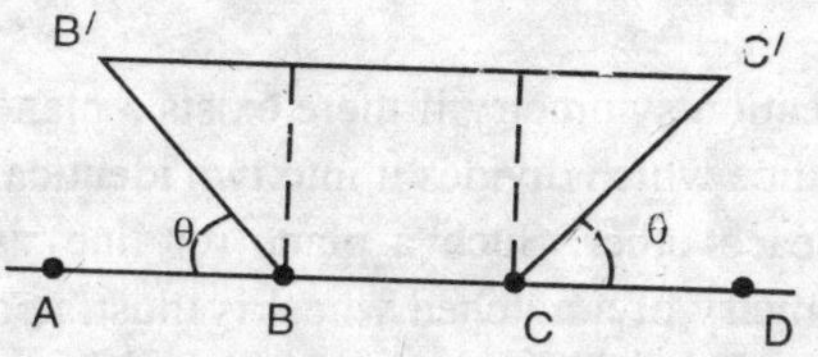

Fig. 1.6. *Geometry used to prove that only 1. 2. 3, 4 and 6 - fold rotation axes are permissible.*

Let **T** be the lattice translation vector and let the lattice have n - fold rotational symmetry with rotation axes passing through the lattice points perpendicular to the plane of paper. Rotations by an angle $\theta = 2\pi/n$ about points B and C in the clockwise and anticlockwise directions respectively yield points B' and C' which must be identical to B and C. Thus the points B' and C' must also be lattice points and should follow lattice translation symmetry. Hence B' C' must be some integral multiple of BC, i.e.,

$$B'C' = m(BC)$$

or

$$2T \cos \theta + T = mT$$

or

$$\cos \theta = (m - 1) / 2 \tag{1.4}$$

where m is an integer. Since $|\cos\theta| \leq 1$, the allowed values of m are 3, 2, 1, 0 and –1. These correspond to the allowed values of θ as 0° or 360°, 60°, 90°, 120° and 180° respectively. Hence from Eq. (1.3), the permissible values of n are 1, 6, 4, 3 and 2. Thus we conclude that 5-fold rotation is not permissible because it is not compatible with lattice translation symmetry. Similarly, other rotations, such as 7-fold rotation, are also not permissible.

Figure 1.7 gives a convincing demonstration of non-existence of 5-fold rotation axis. As shown in the figure, the pentagons placed side by side do not cover the complete space. This is because no sets of vectors exist which satisfy translation symmetry operation throughout and hence this arrangement of pentagons cannot be regarded as a lattice. The array itself, however, has a 5-fold symmetry about the point A.

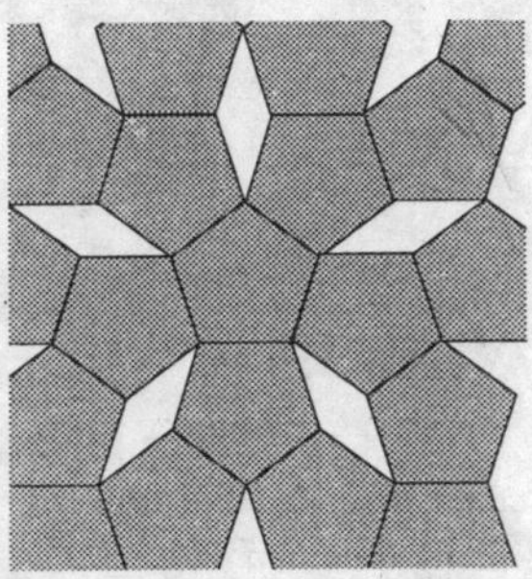

Fig. 1.7. *Demonstration of non-existence of a five-fold rotation axis in a lattice.*

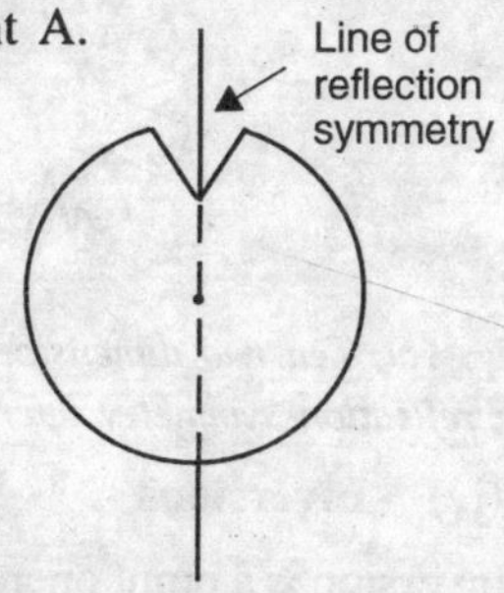

Fig. 1.8. *Reflection symmetry of a notched wheel about a line.*

(iii) **Reflections**

A lattice is said to possess reflection symmetry if there exists a plane (or a line in two dimensions) in the lattice which divides it into two identical halves which are mirror images of each other. Such a plane (or line) is represented by *m*. The reflection symmetry of a notched wheel is illustrated in Fig. 1.8. Considering the combinations of reflections with allowed rotations, we note that each allowed rotation axis can be associated with two possibilities : one is rotation with reflection and the other rotation without reflection. Since there are five allowed rotation axes, the possible number of such combinations is 10. These are designated as

1, 1*m*, 2, 2*mm*, 3, 3*m*, 4, 4*mm*, 6, 6*mm*

where the numerals represent the type of rotation axis, the first *m* represents a plane (or line) parallel to the rotation axis and the second *m* refers to another plane (or line) perpendicular to the rotation axis. These ten groups of symmetry operations are shown in Fig. 1.9.

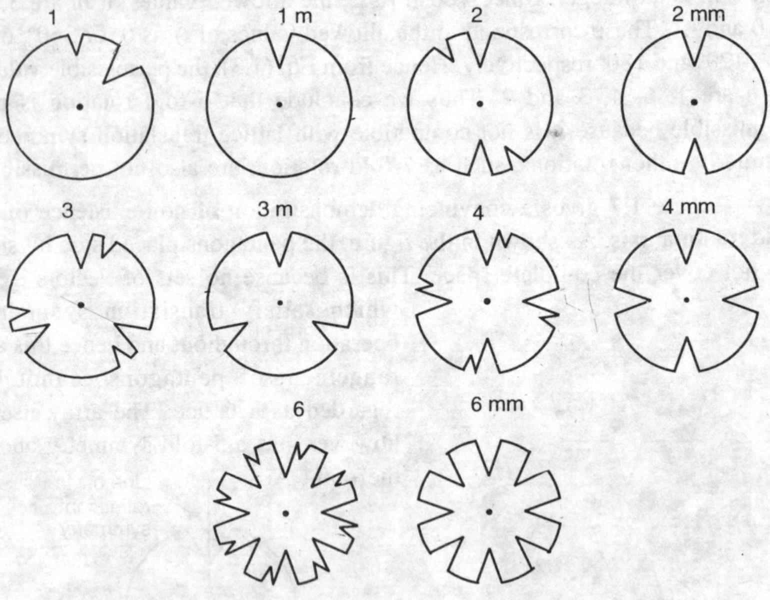

Fig. 1.9. *Ten two-dimensional point groups consisting of rotation and reflection symmetry operations illustrated using notched wheels.*

(iv) **Inversions**

Inversion is a point operation which is applicable to three-dimensional lattices only. This symmetry element implies that each point located at **r** relative to a lattice point has an identical point located at $-\mathbf{r}$ relative to the same lattice point. In other words, it means that the lattice possesses a centre of inversion denoted by $\bar{1}$.

It may be noted that, apart from these symmetry operations, a three-dimensional lattice in particular may have additional symmetry operations formed by the combinations of the above-mentioned operations. One such example is *rotation-inversion operation.* These operations further increase the number of symmetry elements. These symmetry elements are further employed to determine the type of lattices possible in two and three-dimensional spaces.

1.6 POINT GROUPS AND SPACE GROUPS

We have seen that there are mainly four types of symmetry operations, i.e., translation, rotation, reflection and inversion. The last three operations are point operations and their combinations give certain symmetry elements which collectively determine the symmetry of space around a point. The group of such symmetry operations at a point is called a *point group.*

In two-dimensional space, rotation and reflection are the only point operations. As described earlier, their combinations yield 10 different point groups designated as 1, 1*m*, 2, 2*mm*, 3, 3*m*, 4, 4*mm*, 6, and 6*mm* which are shown in Fig. 1.9. In three-dimensional space, however, the situation is complicated due to the presence of additional point operations such as inversion. There are a total of 32 point groups in a three-dimensional lattice.

The crystals are classified on the basis of their symmetry which is compared with the symmetry of different point groups. Also, the lattices consistent with the point group operations are limited. Such lattices are known as *Bravais lattices.* These lattices may further be grouped into distinct crystal systems.

The point symmetry of crystal structure as a whole is determined by the point symmetry of the lattice as well as of the basis. Thus in order to determine the point symmetry of a crystal structure, it should be noted that

(*i*) a unit cell might show point symmetry at more than one locations inside it, and

(*ii*) the symmetry elements comprising combined point and translation operations might be existing at these locations.

The group of all the symmetry elements of a crystal structure is called *space group.* It determines the symmetry of a crystal structure as a whole. There are 17 and 230 distinct space groups possible in two and three dimensions respectively.

1.7 TYPES OF LATTICES

As described earlier, the number of point groups in two and three dimensions are 10 and 32 respectively. These point groups form the basis for construction of different types of lattices. Only those lattices are permissible which are consistent with the point group operations. Such lattices are called *Bravais lattices.* It is beyond the scope of this book to describe the details of formation of various Bravais lattices from the possible point group operations. It can be stated that 10 and 32 point groups in two and three dimensions produce only 5 and 14 distinct Bravais lattices respectively. These Bravais lattices further become parts of 4 and 7 distinct *crystal systems* respectively and are separately described below.

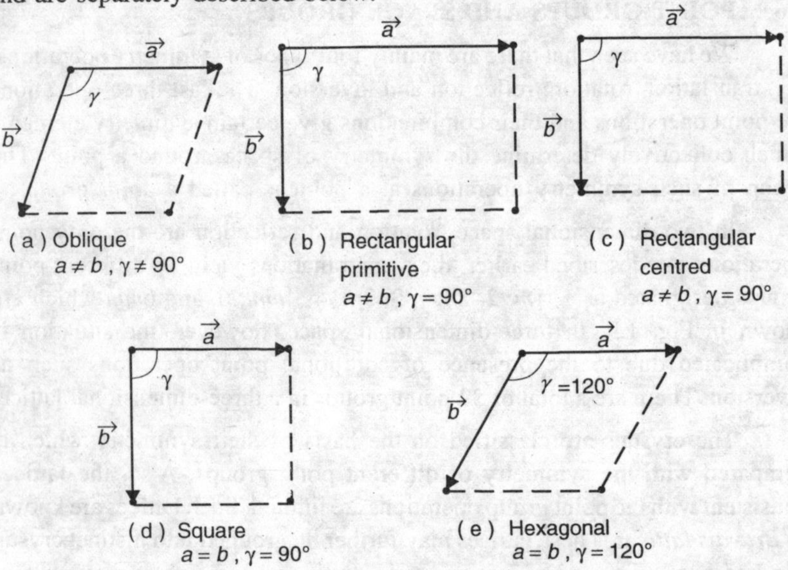

Fig. 1.10. *Bravais lattices in two dimensions.*

(i) Two-Dimensional Lattices

The four crystal systems of two-dimensional space are oblique, rectangular, square and hexagonal. The rectangular crystal system has two Bravais lattices, namely, rectangular primitive and rectangular centred. In all, there are five Bravais lattices which are listed in Table 1.1 along with the corresponding point groups. These lattices are shown in Fig. 1.10.

TABLE 1.1. Crystal systems and Bravais lattices in two dimensions

S. No.	Crystal system	Characteristic point group symmetry	Bravais lattice	Conventional unit cell	Unit cell characteristics
1	Oblique	1, 2	Oblique	Parallelogram	$a \neq b, \gamma \neq 90^0$
2	Rectangular	1*m*, 2*mm*	1. Rectangular primitive 2. Rectangular centred	Rectangle	$a \neq b, \gamma = 90^o$
3	Square	4, 4*mm*	Square	Square	$a = b, \gamma = 90^o$
4	Hexagonal	3, 3*m*, 6, 6*mm*	Hexagonal	60° Rhombus	$a = b, \gamma = 120^o$

(ii) Three-Dimensional Lattices

All the seven crystal systems of three-dimensional space and the corresponding Bravais lattices are listed in Table 1.2 in the decreasing order of symmetry. The crystallographic axes **a, b** and **c** drawn from one of the lattice points determine the size and shape of a unit cell. The angles α, β and γ represent the angles between the vectors **b** and **c, c** and **a,** and **a** and **b** respectively. The lengths *a*, *b* and *c* and angles α, β and γ are collectively known as *lattice parameters* or *lattice constants* of a unit cell. These Bravais lattices are also shown in Fig. 1.11 in the form of their conventional unit cells. The symbols P, F and I represent simple or primitive, face-centred, and body-centred cells respectively. A base or end-centred cell is that which has lattice points at corners and at one of the pairs of opposite faces. It is designated by the letter A, B or C. The designation A refers to the cell in which the faces defined by **b** and **c** axes contain the lattice points, and so on. The symbol R is specifically used for rhombohedral lattice.

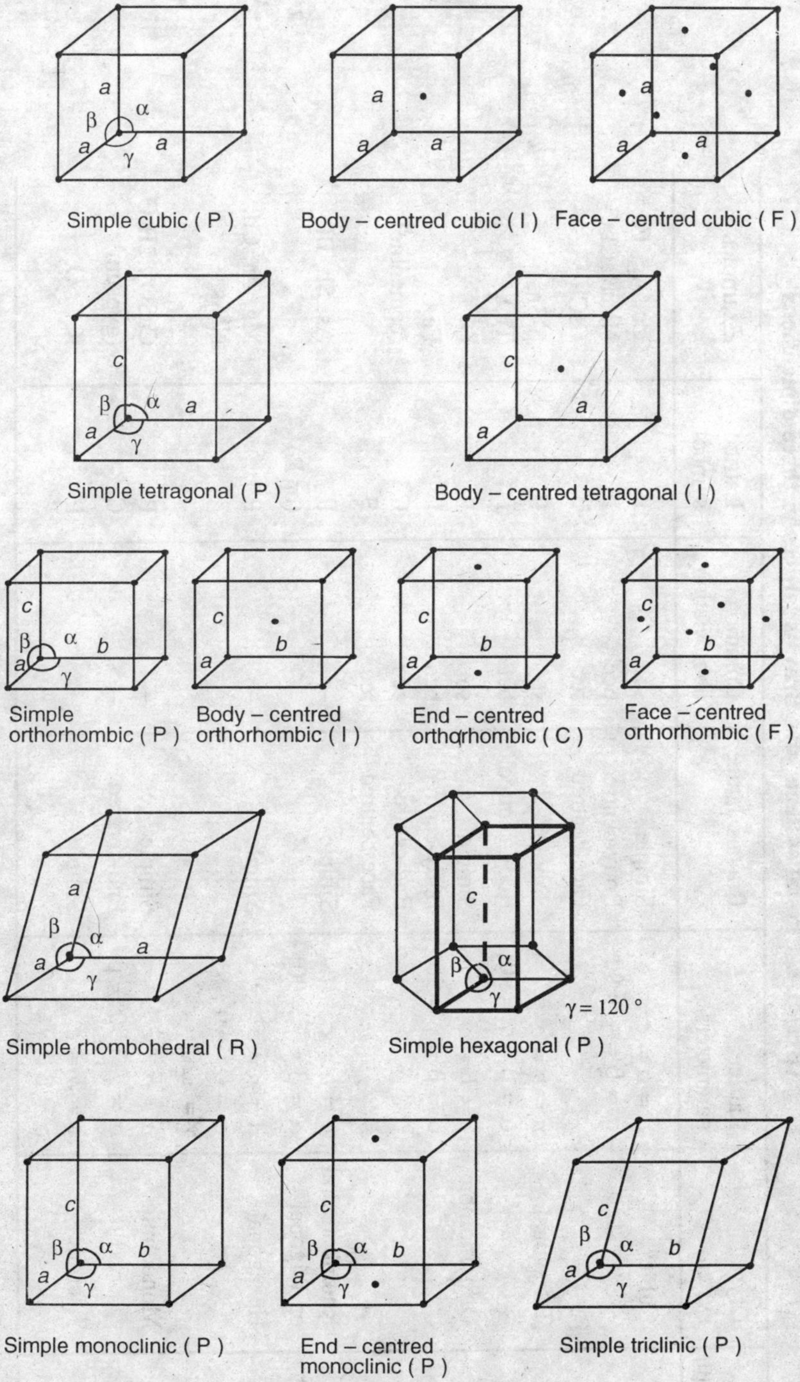

Fig.1.11. *The Bravais lattices in three dimensions*

TABLE 1.2. Crystal systems and Bravais lattices in three dimensions

S. No.	Crystal system	Lattice parameters	Bravais lattice	Common abbreviation	Lattice symbol	Examples
1	Cubic	$a = b = c$ $\alpha = \beta = \gamma = 90^o$	Simple Body-centred Face-centred	sc bcc fcc	P I F	Cu, Ag, Fe, Na, NaCl, CsCl
2	Tetragonal	$a = b \neq c$ $\alpha = \beta = \gamma = 90^o$	Simple Body-centred	st bct	P I	β-Sn, TiO_2
3	Orthorhombic	$a \neq b \neq c$ $\alpha = \beta = \gamma = 90^o$	Simple Body-centred End-centred Face-centred	so bco eco fco	P I C F	Ga, Fe_3C (cementite)
4	Rhombohedral or Trigonal	$a = b = c$ $\alpha = \beta = \gamma \neq 90^o$	Simple	–	P (or R)	As, Sb, Bi
5	Hexagonal	$a = b \neq c$ $\alpha = \beta = 90^o$, $\gamma = 120^o$	Simple	–	P	Mg, Zn, Cd, NiAs
6	Monoclinic	$a \neq b \neq c$ $\alpha = \gamma = 90^o \neq \beta$	Simple End-centred	– –	P C	$CaSO_4 \cdot 2H_2O$ (gypsum)
7	Triclinic	$a \neq b \neq c$ $\alpha \neq \beta \neq \gamma \neq 90^o$	Simple	–	P	$K_2Cr_2O_7$

The hexagonal crystal system has only one Bravais lattice and its unit cell may be either of cubical or of hexagonal type. The cubical type cell (outlined by thick lines) has lattice points only at the corners. The hexagonal cell has lattice points at the corners as well as at the centres of the two hexagonal faces. One hexagonal cell is formed by joining together three cubical type cells.

A lattice point lying at the corner of a cell is shared by eight such cells and the one lying at the face centre position is shared by two cells. Therefore, the contribution of a lattice point lying at the corner towards a particular cell is 1/8 and that of a point lying at the face centre is ½. The following equation is used to calculate the effective number of lattice points, N, belonging to a particular cell :

$$N = N_i + N_f / 2 + N_c/8 \tag{1.5}$$

where N_i represents the number of lattice points present completely inside the cell, and N_f and N_c represent the lattice points occupying face centre and corner positions of the cell respectively. Using this relation, the effective number of lattice points in a simple cubic, body-centred cubic and face-centred cubic lattices comes out to be 1, 2 and 4 respectively.

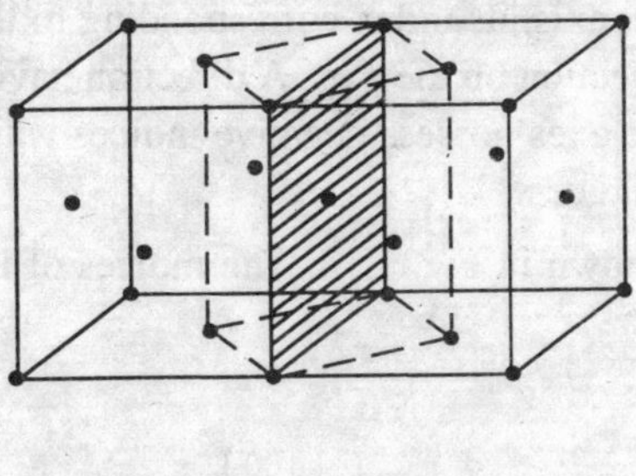

Fig. 1.12. *Two face-centred tetragonal lattices placed side by side result in an end-centred tetragonal lattice shown by dotted lines.*

The list of Bravais lattices given in Table 1.2 appears to be incomplete. The orthorhombic system contains four Bravais lattices whereas the cubic and tetragonal systems contain only three and two lattices respectively. It can be shown that the lattices which are absent in certain crystal systems do not result in new types of arrangements and so need not be considered separately. Figure 1.12 shows two face-centred tetragonal lattices placed side by side. This arrangement of points, shown by dotted lines, produces body-centred tetragonal lattice which already exists in the Bravais list.

1.8 LATTICE DIRECTIONS AND PLANES

The direction of a line in a lattice is defined by assigning certain indices to this line. If the line passes through the origin, its indices are determined by taking a point on this line and finding out the projections of the vector drawn from the origin to that point on the crystallographic axes. Let these projections be u, v and w. Apparently, u, v and w also represent the coordinates of that point. These coordinates are then simplified to get a set of the smallest possible integers which when enclosed in square brackets represent the indices of the line. As an example, to determine the indices of the direction OQ in a cubic crystal (Fig. 1.13), we may take either a point P (½, ½, ½) or Q (1, 1, 1) on this line; either of these points yields the indices of the direction OQ as [111]. The same are the indices of any other direction parallel to OQ because by shifting the origin to an appropriate position, the new direction can be made to pass through the points O, P and Q. The origin is shifted in such a way that the orientation of the axes remains unchanged. If a direction is perpendicular to a certain axis, its index corresponding to that axis is zero as it does not form any projection on the axis. A direction having projections on the negative sides of the axes possess negative indices which are written by putting bars over the indices.

Consider the direction AB as shown in Fig. 1.14. The indices of this direction can be obtained by shifting the origin to the point A. The projections of AB on the axes then become –1, 1, and 0 and hence the indices of the line AB are $[\bar{1}10]$. The indices of a few more directions are illustrated in Fig. 1.14. The cube edges are represented by the indices of the type [100], [010], $[\bar{1}00]$, etc. These constitute a family of cube edges designated as <100> which includes all the directions of this type. Similarly, a

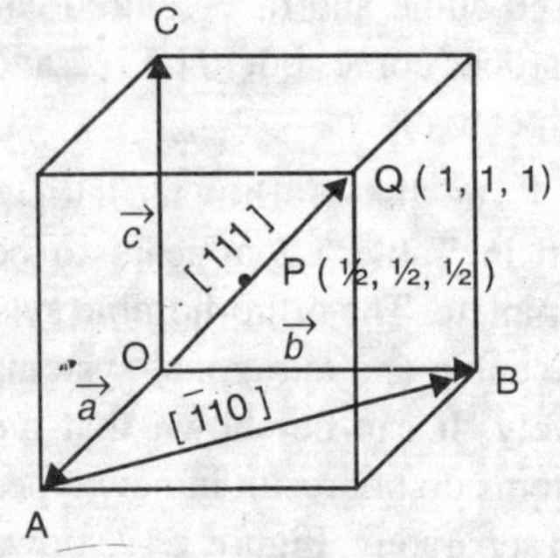

Fig. 1.13. *Determination of indices of a direction.*

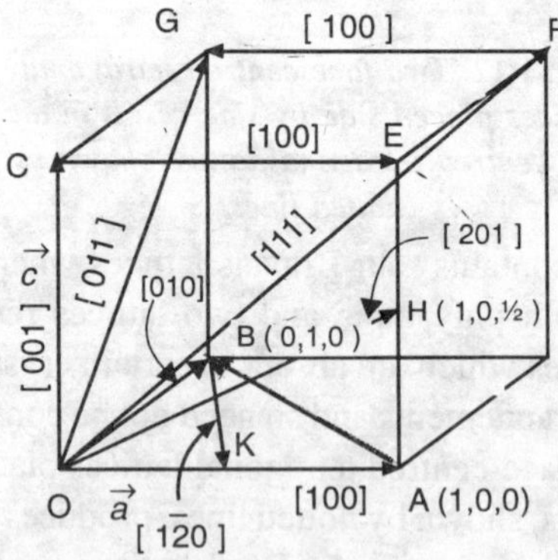

Fig. 1.14. *Indices of some directions in a cubic lattice.*

family of face diagonals is represented by <110> and that of body diagonals by <111>. The number of members in the families of cube edges, face diagonals and body diagonals is 6, 12 and 8 respectively.

The angle θ between the two crystallographic directions $[hkl]$ and $[h'k'l']$ is given by

$$\cos\theta = \frac{hh' + kk' + ll'}{(h^2 + k^2 + l^2)^{1/2}\,(h'^2 + k'^2 + l'^2)^{1/2}} \tag{1.6}$$

The scheme to represent the orientation of planes in a lattice was first introduced by Miller, a British crystallographer. The indices of planes are, therefore, known as the *Miller indices*. The steps involved to determine the Miller indices of a plane are as follows :

(i) Find the intercepts of the plane on the crystallographic axes.

(ii) Take reciprocals of these intercepts.

(iii) Simplify to remove fractions, if any, and enclose the numbers obtained into parentheses.

In step (i), the intercepts are taken in terms of the lengths of fundamental vectors choosing one of the lattice points as the origin. If a plane is parallel to a certain axis, its intercept with that axis is taken as infinity. In step (ii) the reciprocals are taken in order to avoid the occurrence of infinity in the Miller indices.

As an example, consider a plane ABC (Fig. 1.15) having intercepts 1, 2 and 1 with the crystallographic axes **a**, **b** and **c** respectively of a cubic lattice. The Miller indices of this plane are determined as follows :

(i) Intercepts : 1, 2, 1

(ii) Reciprocals : 1, ½, 1

(iii) Simplification : 2, 1, 2

Hence the Miller indices of the plane ABC are (212) ; the numbers within the parentheses are written without commas. The Miller indices of a plane, in general, are written as (hkl).

It may be noted that another plane DEF which is parallel to the plane ABC and lies completely inside the lattice, has intercepts 1/2, 1 and 1/2 with the

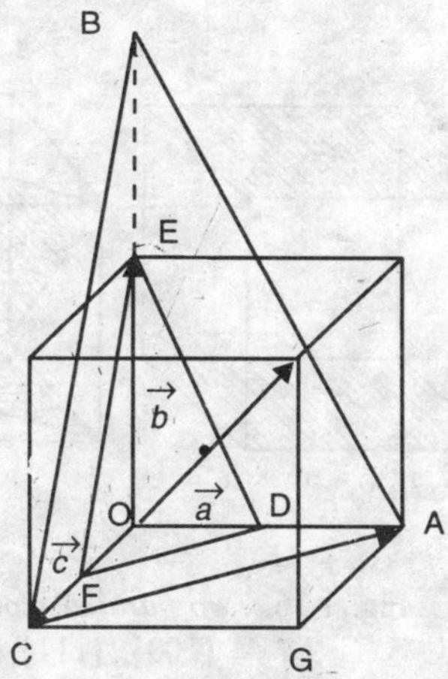

Fig. 1.15. *Miller indices of parallel planes and the planes passing through the origin.*

axes and hence carries the same Miller indices as the plane ABC. Thus we conclude that the parallel planes have the same Miller indices. The plane DEF is rather more convenient to deal with as it lies completely inside the lattice.

If a plane intercepts an axis on the negative side, a bar is put above the corresponding number of the Miller indices. The intercepts of a plane passing through the origin cannot be determined as such. In such a case, we take another plane parallel to this plane and determine its Miller indices. The same are the indices of the given plane. Alternatively, we shift the origin from the plane to some other suitable lattice point without changing the orientation of the axes and then find the Miller indices. For example, the indices of the plane OCGA in Fig. 1.15 become $(0\bar{1}0)$ if the origin is shifted to the point E. The importance of orientation of the axes can be realized with reference to Fig. 1.12. The indices of the shaded plane are of the type (100) when referred to the axes of the face-centred tetragonal cell, whereas these become of the type (110) when referred to the axes of the simple tetragonal cell indicated by dotted lines.

A family of planes of a particular type is represented by enclosing the Miller indices of any one of the planes of that family into braces. Thus {100} represents a family of planes which has the planes (100), (010), (001), $(\bar{1}00)$, $(0\bar{1}0)$ and $(00\bar{1})$ as its members. These six planes represent the faces of the cube. Similarly, the families of diagonal planes and close-packed planes are represented by {110} and {111}, and contain 6 and 8 members respectively. Some of these planes are illustrated in Fig. 1.16.

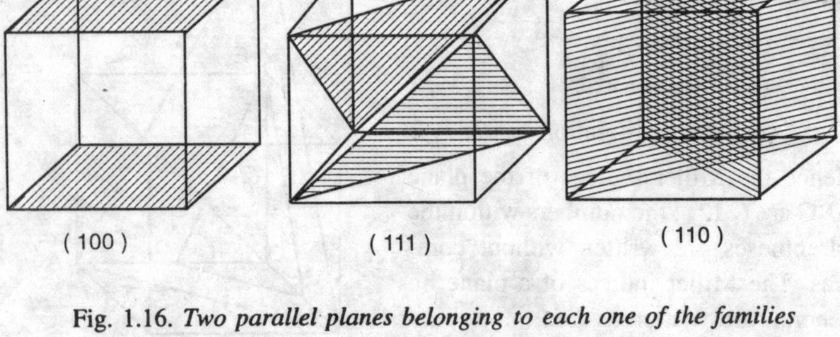

Fig. 1.16. *Two parallel planes belonging to each one of the families* {100}, {111} *and* {110} *in a cubic lattice.*

1.9 INTERPLANAR SPACING

Consider a set of parallel planes with indices (*hkl*). Take origin on a lattice point of one such plane and draw crystallographic axes **a**, **b** and **c**. Now consider another similar plane adjacent to this plane. Since the second plane also has the Miller indices (*hkl*), the lengths of the intercepts on **a**, **b** and **c** are a/h, b/k and c/l respectively. If we draw a normal from the origin to the second plane, the length of the normal represents the interplanar distance *d*. From Fig. 1.17, it follows that

$$d = \text{OP} = \frac{a}{h}\cos\alpha = \frac{b}{k}\cos\beta = \frac{c}{l}\cos\gamma \tag{1.7}$$

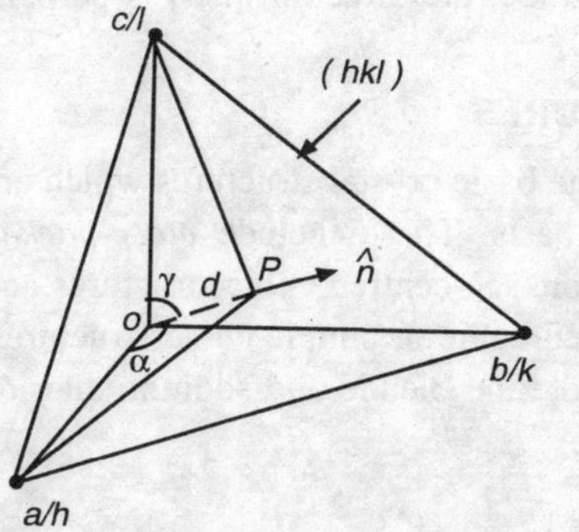

Fig. 1.17. *An (hkl) plane at a distance d from another similar parallel plane passing through the origin.*

where α, β and γ represent the angles between the normal and the axes **a**, **b** and **c** respectively and cos α, cos β and cosγ represent the direction cosines of the normal to the plane (*hkl*). The Eq. (1.7) indicates that the direction cosines of the normal are proportional to h/a, k/b and l/c.

If $\hat{\mathbf{n}}$ be the unit vector of the normal to the plane, then $a\cos\alpha$, $b\cos\beta$ and $c\cos\gamma$ may be written as $\hat{\mathbf{n}}.\mathbf{a}$, $\hat{\mathbf{n}}.\mathbf{b}$ and $\hat{\mathbf{n}}.\mathbf{c}$ respectively and Eqs. (1.7) become

$$d = \hat{\mathbf{n}}.\mathbf{a}/h = \hat{\mathbf{n}}.\mathbf{b}/k = \hat{\mathbf{n}}.\mathbf{c}/l \tag{1.8}$$

Thus the value of *d* can be determined if $\hat{\mathbf{n}}$ is known. In an orthogonal lattice, where **a**, **b** and **c** point along *x*, *y* and *z* directions respectively, the equation of the plane (*hkl*) with intercepts a/h, b/k and c/l on the axes is

$$f(x, y, z) = hx/a + ky/b + lz/c = 1$$

For a surface $f(x, y, z)$ = constant, ∇f represents the vector normal to it.

$$\therefore \qquad \hat{\mathbf{n}} = \frac{\nabla f}{|\nabla f|} = \frac{(h/a)\hat{\mathbf{i}} + (k/b)\hat{\mathbf{j}} + (l/c)\hat{\mathbf{k}}}{(h^2/a^2 + k^2/b^2 + l^2/c^2)^{1/2}}$$

Hence from Eqs. (1.8), we obtain

$$\mathbf{d} = \frac{\hat{\mathbf{n}}.\mathbf{a}}{h} = \frac{\left[(h/a)\hat{\mathbf{i}} + (k/b)\hat{\mathbf{j}} + (l/c)\hat{\mathbf{k}}\right].(a/h)\hat{\mathbf{i}}}{(h^2/a^2 + k^2/b^2 + l^2/c^2)^{1/2}}$$

or
$$d = \frac{1}{(h^2/a^2 + k^2/b^2 + l^2/c^2)^{1/2}} \tag{1.9}$$

This equation is valid for orthogonal lattices only. For non-orthogonal lattices, such an expression may not be obtained easily; one may need to find $\hat{\mathbf{n}}$ by some other method and then use Eq. (1.8) to determine d. For a cubic lattice, a, b and c are equal and we get

$$d = \frac{a}{\left(h^2 + k^2 + l^2\right)^{1/2}} \tag{1.10}$$

It may also be noted that for a cubic lattice, the direction $[hkl]$ is perpendicular to the plane (hkl).

1.10 SIMPLE CRYSTAL STRUCTURES

We shall now describe some of the basic crystal structures which are either monoatomic or contain simple basis. These include *close-packed structures* like hexagonal close-packed or face-centred cubic structures and *loose-packed structures* like body-centred cubic or simple cubic structures. Besides these, the structures of diamond, zinc blende and sodium chloride are also described.

1.10.1. Close-Packed Structures

Close-packed structures are mostly found in monoatomic crystals having non-directional bonding, such as metallic bonding. In these structures, the coordination number of each atom is 12, i.e., each atom is surrounded by twelve similar and equal sized neighbours. Out of these twelve neighbours, six lie in one plane, three in an adjacent parallel plane above this plane and three in a similar plane below it. There are two types of close-packed structures :

(i) Hexagonal close-packed (*hcp*) structure

(ii) Face-centred cubic (*fcc*) structure

These structures are described as follows :

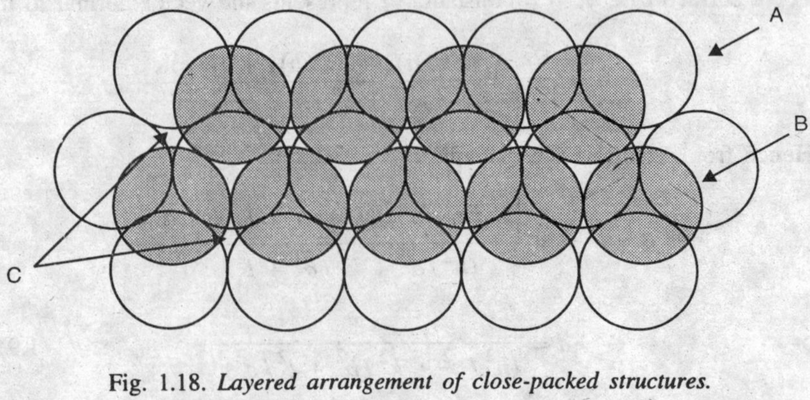

Fig. 1.18. *Layered arrangement of close-packed structures.*

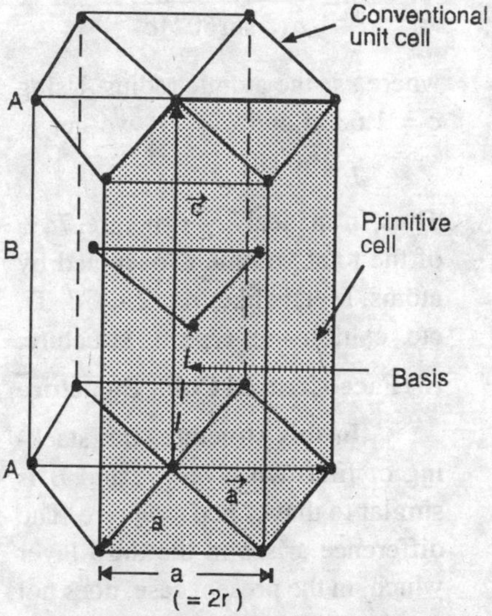

Fig. 1.19. *Conventional and primitive cells of hexagonal close-packed structure.*

(i) Hexagonal Close-Packed Structure

Consider a layer of similar atoms with each atom surrounded by six atoms in one plane as shown in Fig. 1.18. Another similar layer B can be placed on top of layer A such that the atoms of layer B occupy the alternate valleys formed by the atoms of layer A. If a third similar layer is placed on top of the B-layer in such a way that the atoms of B-layer exactly overlap the atoms of A-layer and this type of stacking is repeated successively, the following layered arrangement is obtained :

....ABABAB....

This type of stacking is called *hcp stacking* and the structure is known as *hexagonal close-packed structure*. The name corresponds to the shape of the conventional unit cell which is hexagonal and is shown in Fig. 1.19. There are twelve atoms located at the corners, two at the centres of the basal planes, and three completely inside the hexagon forming a part of the B-layer. The effective number of atoms in a unit cell is

$$12(1/6) + 2\ (1/2) + 3 = 6$$

The interatomic distance for the atoms within a layer is a. The distance between the two adjacent layers is $c/2$, c being the height of the unit cell. For an ideal *hcp* structure, $c = 1.633a$.

It may be noted that although the structure is *hcp*, the space lattice is simple hexagonal with basis consisting of two atoms placed in such a way that if one atom lies at the origin, the other atom lies at the point (2/3, 1/3, 1/2). The shaded portion in Fig. 1.19 represents the primitive cell of this structure. It contains 2 atoms instead of one which is due to the presence of the basis. Also, the volume of the primitive cell is exactly one-third of the volume of the hexagonal cell.

The *packing fraction*, *f*, is defined as the ratio of the volume occupied by the atoms present in a unit cell to the total volume of the unit cell. It is also referred to as the *packing factor* or *packing efficiency* of the unit cell. From the primitive cell, we find

$$f = \frac{2(4/3)\pi r^3}{a(a\sin 60^\circ)c}$$

where r is the atomic radius. Using $c = 1.633a$ and $a = 2r$, we get

$$f = 0.74$$

Thus, in an ideal *hcp* structure, 74% of the total volume is occupied by atoms. Metals like Mg, Zn, Cd, Ti, etc. exhibit this type of structure.

(ii) Face-Centred Cubic Structure

In this structure, the stacking of first two layers A and B is similar to that of hcp structure. The difference arises in the third layer which, in the present case, does not overlap the first layer. The atoms of the third layer occupy the positions of those valleys of the A-layer which are not occupied by the B-layer atoms. The third layer is designated by the letter C. The fourth layer exactly overlaps the first layer and the sequence is repeated. Thus *fcc* structure is represented by the following stacking sequence :

.... ABCABCABC ....

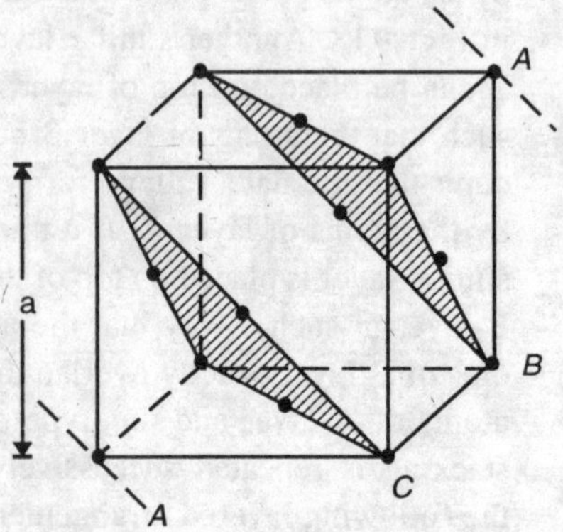

Fig. 1.20. *Conventional unit cell of fcc structure along with the stacking sequence .... ABCABC ....*

The conventional unit cell is *face-centred cubic* and is shown in Fig. 1.20. It is a non-primitive cell having effective number of atoms equal to 8 (1/8) + 6 (1/2) or 4. The atoms touch one another along the face diagonals. The length of the cube edge, *a*, and the atomic radius, *r*, are related to each other as

$$4r = \sqrt{2}\,a$$

The packing fraction, *f*, is given by

$$f = \frac{4(4/3)\pi r^3}{a^3} = 0.74$$

Thus the packing fraction of *fcc* structure is exactly the same as that of hcp structure which is expected because of the close-packed nature of both the structures. Also, the coordination number of each atom is 12. Examples of materials having this type of structure are Cu, Ag, Au, Al, etc.

1.10.2. Loose-Packed Structures

A loose-packed structure is that in which the coordination number of an atom is less than 12 or the packing fraction is less than 0.74. Among the various possible loose-packed structures, the most common and the simplest are the *body-centred cubic (bcc)* and the *simple cubic (sc)* structures. These structures are described as follows :

(i) Body-Centred Cubic Structure (bcc)

The conventional unit cell of *bcc* structure is non-primitive and is shown in Fig. 1.21. It has cubical shape with atoms located at the corners and the body centre. Thus the effective number of atoms per unit cell is 8 (1/8) + 1 = 2. The coordination number of each atom is 8. The atoms touch one another along the body diagonal. Thus a is related to r as

$$4r = \sqrt{3}\, a.$$

The packing fraction is given by

$$f = \frac{2(4/3)\pi r^3}{a^3} = 0.68$$

The examples of materials exhibiting *bcc* structure are Na, K, Mo, W, etc.

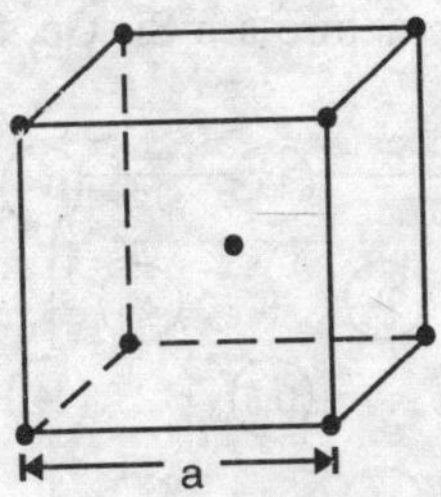

Fig. 1.21. *Conventional unit cell of bcc structure.*

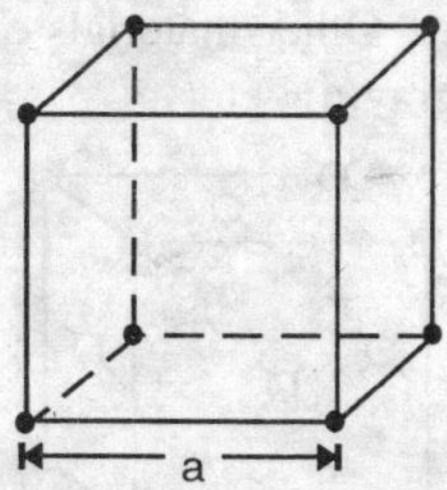

Fig. 1.22. *Unit cell of sc structure.*

(ii) Simple Cubic Structure (sc)

The conventional unit cell of *sc* structure is the same as its primitive cell and is shown in Fig. 1.22. The atoms are located at the corners only and touch one another along the cube edges. Thus in *sc* structures, we have

$$a = 2\, r$$

The coordination number of each atom is 6. The packing fraction is given by

$$f = \frac{1(4/3)\pi r^3}{a^3} = 0.52$$

Only polonium exhibits this type of structure at room temperature.

1.11 STRUCTURE OF DIAMOND

Diamond exhibits both cubic and hexagonal type structures. The diamond cubic (*dc*) structure is more common and is described here. The space lattice of the diamond cubic structure is *fcc* with basis consisting of two carbon atoms, one located at the lattice point and the other at a distance of one quarter of the body diagonal from the lattice point along the body diagonal. The unit cell of the *dc* structure is shown in Fig. 1.23. The carbon atoms placed along the body diagonals, in fact, occupy the alternate tetrahedral void positions in the *fcc* arrangement of carbon atoms. This opens up the otherwise close-packed *fcc* arrangement which decreases the packing efficiency considerably. The packing efficiency of the *dc* structure is only 34% as compared to 74% for the *fcc* structure. The coordination number of each carbon atom is 4 and the nearest neighbour distance is equal to $\sqrt{3}a/4$ where a is the lattice parameter.

The *dc* structure may also be viewed as an interpenetration of two *fcc* sublattices with their origins at (0, 0, 0) and (1/4, 1/4, 1/4). A plan view of the positions of all the carbon atoms in the unit cell is shown in Fig. 1.24. The fractional heights of the carbon atoms relative to the base of the unit cell are given in the circles drawn at the atomic positions. Two numbers in the same circle indicate two carbon atoms at the same position located one above the other. Other materials exhibiting this type of structure are Si, Ge, SiC, GaAs, gray tin, etc.

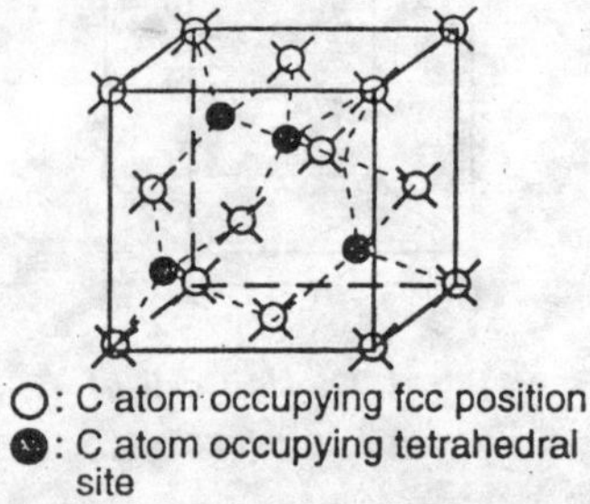

Fig. 1.23. *The unit cell of dc structure. The lattice is fcc with carbon atoms located at fcc positions and at alternate tetrahedral sites.*

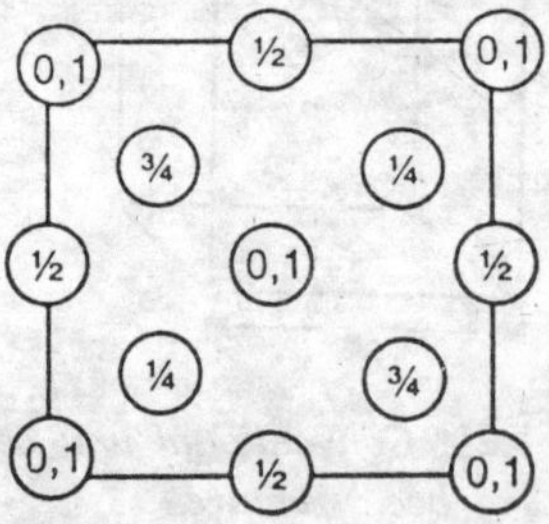

Fig. 1.24. *Plan view of atomic positions in dc unit cell. Numbers in the circles indicate fractional heights of the carbon atoms.*

1.12 ZINC BLENDE (ZnS) STRUCTURE

The zinc blende structure is similar to the *dc* structure except that the two *fcc* lattices in it are occupied by different elements. The structure is similar to the one shown in Fig. 1.23 where the dark circles now represent one type of atoms, say Zn, and the light circles represent the other type of atoms, i.e., S.

1.13 SODIUM CHLORIDE (NaCl) STRUCTURE

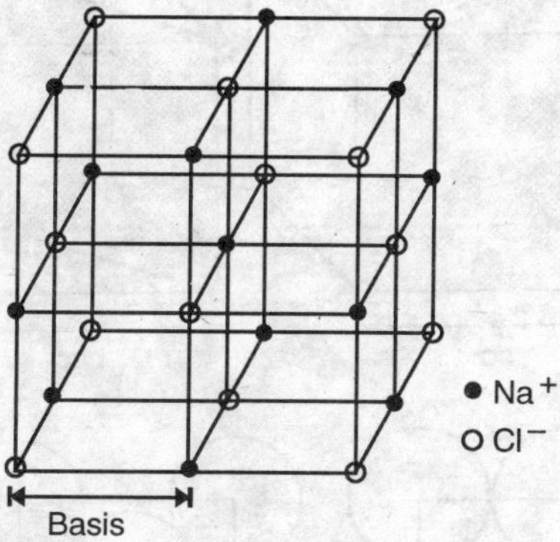

Fig. 1.25. *Unit cell of sodium chloride structure.*

The unit cell of NaCl structure is shown in Fig. 1.25. In NaCl structure, the radii of Na^+ and Cl^- ions are such that each Na^+ ion is octahedrally coordinated to six Cl^- ions. The unit cell is *fcc* with four Cl^- ions occupying all the four *fcc* positions and the four Na^+ ions occupying all the four octahedral voids. The *fcc* positions and the octahedral void positions are, however, interchangeable. The NaCl structure can, therefore, be viewed as two interpenetrating *fcc* sublattices, one belonging to Na^+ ions with its origin at the point (0, 0, 0) and the other belonging to Cl^- ions with its origin at the point (a/2, 0, 0). In the terminology of lattice and basis, the structure can be interpreted as an *fcc* lattice with basis consisting of two ions, one of Na^+ and the other of Cl^-. One of these ions occupies one of the *fcc* positions and the other ion occupies the corresponding octahedral void position. A unit cell of NaCl comprises four molecules. The position of various ions in the unit cell are as follows :

Na^+ : 0, 0, 0; 1/2,1/2,0; 1/2,0,1/2; 0,1/2,1/2

Cl^- : 1/2,1/2,1/2; 0,0,1/2; 0,1/2,0; 1/2,0,0

SOLVED EXAMPLES

Example 1.1. Determine the relationships between the lattice parameter a and the atomic radius r for monoatomic *sc*, *bcc* and *fcc* structures.

Solution. In *sc* structure (Fig. 1.26), the atoms touch one another along the cube edges.

$\therefore$ $$a = 2r$$

In *bcc* structure, the atoms touch along the body diagonals.

$\therefore$ $$\sqrt{3}a = 4r$$

or $$a = 4r/\sqrt{3}$$

In *fcc* structure, the atoms touch along the face diagonals.

$\therefore$ $$\sqrt{2}a = 4r$$

or $$a = 2\sqrt{2}r$$

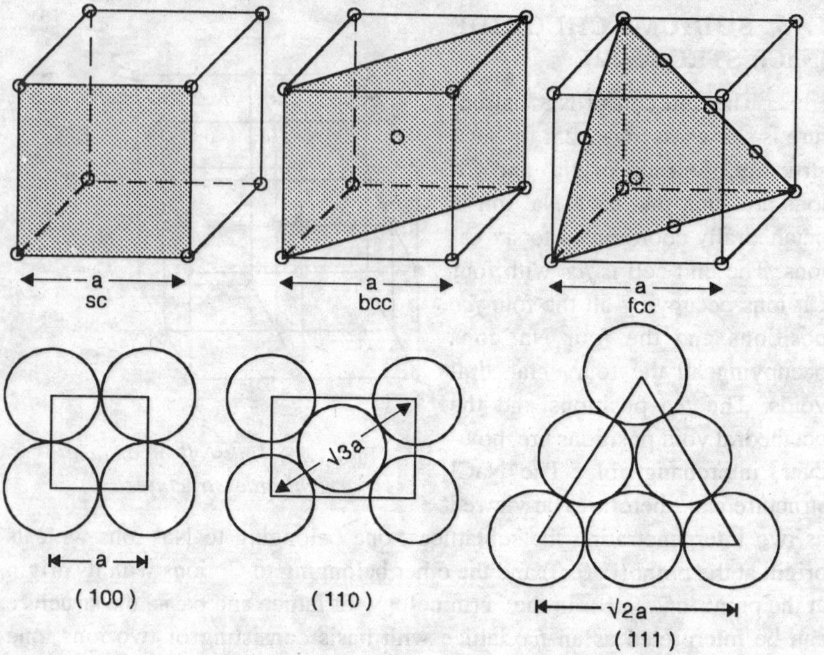

Fig. 1.26. *Monoatomic sc, bcc and fcc structures along with their (100), (110) and (111) type planes respectively.*

Example 1.2. Draw (101) and (111) planes in a cubic unit cell. Determine the Miller indices of the directions which are common to both the planes.

Solution. Intercepts of the plane (101) with the axes

$$= 1/1,\ 1/0 \text{ and } 1/1$$

$$= 1,\ \infty \text{ and } 1$$

Intercepts of the plane (111) with the axes $= 1, 1$ and 1

Taking the point O as origin and the lines OA, OB and OC as the axes **a**, **b** and **c** respectively, the plane with intercepts 1, ∞ and 1 is the plane ADGC and that with intercepts 1, 1 and 1 is plane ABC as shown in Fig. 1.27. Therefore, the line common to both the planes is the line AC. It corresponds to two directions, i.e., AC and CA.

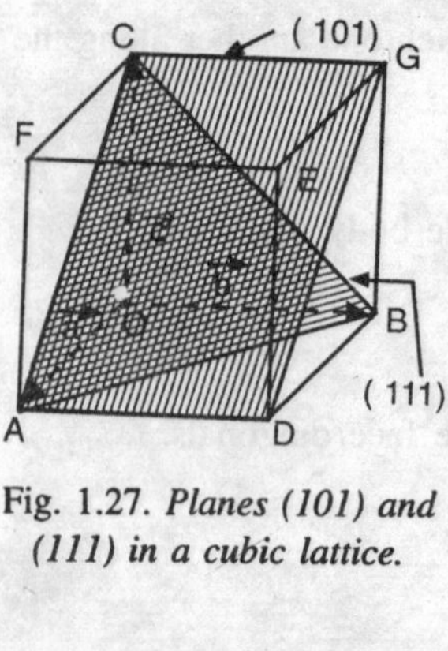

Fig. 1.27. *Planes (101) and (111) in a cubic lattice.*

Projections of the direction AC on the axes $= -1$, 0 and 1

Projections of the direction CA on the axes $= 1$, 0 and -1

Therefore, the required indices are $[\bar{1}01]$ and $[10\bar{1}]$.

Example 1.3. A plane makes intercepts of 1, 2 and 0.5Å on the crystallographic axes of an orthorhombic crystal with $a:b:c = 3:2:1$. Determine the Miller indices of this plane.

Solution. Taking the lengths of the axes OA, OB and OC as 3, 2 and 1Å respectively, the plane with intercepts of 1, 2 and 0.5Å on the axes is the plane DBE as shown in Fig. 1.28. The intercepts of this plane relative to full lengths of the axes are

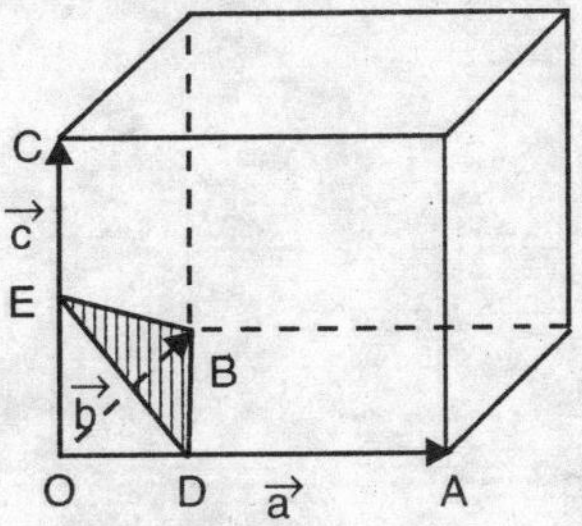

Fig. 1.28. *Plane (312) in an orthorhombic lattice.*

1/3, 2/2 and 0.5/1

or 1/3, 1 and 1/2

Reciprocals : 3, 1 and 2

Therefore, the Miller indices of the plane DBE are (312).

Example 1.4. In a cubic unit cell, find the angle between normals to the planes (111) and (121).

Solution. Since the crystal is cubic, the normals to the planes (111) and (121) are the directions [111] and [121] respectively. Let θ be the angle between the normals.

$$\therefore \quad \cos\theta = \frac{h_1h_2 + k_1k_2 + l_1l_2}{(h_1^2 + k_1^2 + l_1^2)^{1/2}\ (h_2^2 + k_2^2 + l_2^2)^{1/2}}$$

$$= \frac{1\times1+1\times2+1\times1}{(1^2+1^2+1^2)^{1/2}\ (1^2+2^2+1^2)^{1/2}}$$

$$= 0.9428$$

or $\theta = 19.47^{\circ}$ or $19^{\circ}28'$

Example 1.5. Calculate the packing efficiency and density of sodium chloride from the following data :

Radius of sodium ion = 0.98 Å

Radius of chloride ion = 1.81 Å

Atomic mass of sodium = 22.99 amu

Atomic mass of chlorine = 35.45 amu

Solution. The unit cell of NaCl strucutre is shown in Fig. 1.25. The Na^+ and Cl^- ions touch along the cube edges.

$\therefore$ Lattice parameter, a = 2 (Radius of Na^+ + Radius of Cl^-)
= 2 (0.98 + 1.81) = 5.58 Å

$$\text{Packing fraction} = \frac{\text{Volume of ions present in the unit cell}}{\text{Volume of the unit cell}}$$

$$= \frac{4(4/3)\pi r_{N_a^+}^3 + 4(4/3)\pi r_{Cl^-}^3}{a^3}$$

$$= \frac{16\pi}{3}\left[\frac{(0.98)^3 + (1.81)^3}{(5.58)^3}\right]$$

$$= 0.663 \text{ or } 66.3\%$$

$$\text{Density} = \frac{\text{Mass of the unit cell}}{\text{Volume of the unit cell}}$$

$$= \frac{4(22.99+35.45)\times 1.66\times 10^{-27}}{\left(5.58\times 10^{-10}\right)^3} \text{ kg m}^{-3}$$

$$= 2234 \text{ kg m}^{-3} \text{ or } 2.23 \text{ g cm}^{-3}$$

SUMMARY

1. The solids may be broadly classified as crystalline and non-crystalline (or amorphous). The crystalline solids may be further sub-divided into single crystals and polycrystalline materials.

2. Crystallography is the study of formation, structure and properties of crystals.

3. A crystal structure results from the combination of a space lattice and a basis. A space lattice is a regular arrangement of infinite number of imaginary points in three-dimensional space. A basis is a structural unit comprising a single atom or a group of atoms which is placed on each lattice point in a regular fashion to generate the crystal structure.

4. A unit cell is a small group of points which acts as a building block for the entire lattice. It may be primitive or non-primitive. A primitive cell is the smallest volume unit cell and contains only one lattice point per cell. A non-primitive cell contains more than one lattice points per cell. The conventional unit cell has the highest possible symmetry and the lowest possible volume. It may be primitive or non-primitive.

5. The effective number of lattice points belonging to a unit cell is

$$N = N_i + N_f/2 + N_c/8$$

where N_i, N_f and N_c denote the number of lattice points present inside, at the face centres, and at the corners of the cell respectively.

6. A crystal remains invariant under the application of various symmetry operations like translation, rotation, reflection, inversion etc. Some rotational operations, such as 5-fold and 7-fold rotations, are not permissible as these are not compatible with lattice translation symmetry.

7. A point group is the combination of certain symmetry operations like rotation, reflection and inversion. It determines the symmetry of space around a point. The number of point groups in three-dimensional space is 32. These point groups produce only 14 Bravais lattices.

8. The set of all the symmetry elements of a crystal structure is called the space group. The number of distinct space groups possible in three dimensions is 230.

9. The Miller indices of a crystallographic plane and a direction are denoted by (*hkl*) and [*hkl*] respectively where *h*, *k* and *l* are integers. The parallel planes and the parallel directions have the same indices.

10. The angle between two directions [*hkl*] and [*h'k'l'*] is given by

$$\cos\theta = \frac{hh' + kk' + ll'}{(h^2 + k^2 + l^2)^{1/2}\,(h'^2 + k'^2 + l'^2)^{1/2}}$$

11. The interplanar distance for the parallel (*hkl*) planes for an orthorhombic lattice is

$$d = (h^2/a^2 + k^2/b^2 + l^2/c^2)^{-1/2}$$

where *a*, *b* and *c* are the lengths of the axes.

12. A close-packed structure is that in which each atom has twelve identical nearest neighbours. A close-packed structure may be either *fcc* or *hcp* with the following sequence of layers :

hcp : ......ABABABAB......

fcc : ......ABCABCABC......

VERY SHORT QUESTIONS

1. Define a single crystal.
2. How does a crystal differ from a grain?
3. What is short-range order?
4. What are amorphous materials? Give an example of such a material.
5. What is crystallography?
6. What is a unit cell?
7. How does a crystal differ from a lattice?
8. What is the maximum number of possible Bravais lattices?
9. What are Miller indices? What is their importance?
10. Write the indices of all the twelve edges of a cube.
11. Write the indices of all the face diagonals of a cube.
12. What is packing efficiency? What are its values for sc, *bcc*, *fcc* and *hcp* structures?
13. Give at least one example each of materials exhibiting *sc*, *bcc*, *fcc* and *hcp* structures.
14. Calculate the number of carbon atoms per unit cell of diamond.

SHORT QUESTIONS

1. Define primitive and non-primitive translation vectors. Which type of translation vectors are preferred for describing a lattice?
2. Prove that the crystals cannot have five-fold symmetry.
3. Differentiate primitive cell, non-primitive cell and conventional cell from one another. How is a Weigner-Seitz cell constructed?
4. Describe the scheme to determine the Miller indices of a plane. Show that the parallel planes have the same Miller indices.
5. What is the relationship between the Miller indices and reciprocal lattice vectors corresponding to any plane?
6. Find the Miller indices of cube faces and diagonal planes of a unit cube.

7. Explain the concepts of lattice, basis and crystal structure. How are they related ?
8. Draw primitive cells corresponding to *bcc* and *fcc* unit cells.
9. How does *hcp* structure differ from *bcc* structure?
10. What is Bravais lattice? What is the maximum number of Bravais lattices possible? How will you account for the existence of thousands of structures from these lattices?
11. The end-centred orthorhombic is one of the Bravais lattices but the end-centred tetragonal is not. Give reasons.
12. The primitive cell of *fcc* lattice is rhombohedral. Why then is the rhombohedral lattice included separately in the Bravais list?
13. State the points of similarity and difference of the monoatomic, *sc*, monoatomic *bcc*, and CsCl structures?
14. Calculate the volume of the primitive cell and the number of nearest neighbours for an *fcc* lattice.
15. Obtain an expression for the packing fraction for *hcp* structure.
16. Show that the *c/a* ratio for an ideal *hcp* lattice is $\sqrt{8/3}$.
17. Determine the values of packing fraction for *fcc*, *bcc* and *sc* structures.
18. Assuming one of the basis atoms lying at the origin, find the coordinates of the other atoms for an *hcp* structure.
19. Explain, without calculation, why *fcc* and *hcp* structures have the same packing factor.
20. Show that for a cubic lattice, the lattice constant, *a*, is given by

$$a = \sqrt[3]{\frac{n\,M}{N\,\rho}}$$

where the symbols have their usual meanings.

21. What type of lattice and basis do the following structures have :

    (i) Sodium chloride (ii) Diamond cubic ?
22. Diamond is the hardest substance known in spite of the fact that the packing fraction and the coordination number of carbon atom in the *dc* structure are quite low. Explain.
23. There are four vacant tetrahedral sites in a unit cell of the *dc* structure. Can four additional carbon atoms occupy these sites? Give reasons.
24. How many crystal directions constitute the family of body diagonals of a unit cube ? Draw all such directions.

LONG QUESTIONS

1. What are symmetry operations? Describe the principal symmetry operations applicable to a three-dimensional lattice. Show that the five-fold rotational axis is not permissible in case of lattices.
2. What are point group and space group? Give their number for two-and three-dimensional lattices. List all the point groups of a two- dimensional lattice.
3. Determine the interplanar spacing between the two parallel planes with Miller indices (h, k, l) in a cubic crystal of side a.
4. Which is the most densely packed structure amongst the various cubic structures? Determine the packing fraction and porosity of this structure. Can the porosity be reduced by some means? What type of solids generally exhibit this type of structure and why?
5. Draw a plan view of sodium chloride structure. In how many ways can this structure be interpreted?
6. Draw a plan view of *hcp* unit cell and give coordinates of all the atoms. Are all the atoms located at equivalent sites? Discuss implications of your answer.
7. Draw the following:

   (*i*) $\left[1\bar{1}1\right], \left[1\bar{2}1\right]$ and $\left[0\bar{1}2\right]$ directions in cubic and tetragonal lattices.

   (*ii*) $(\bar{1}11)$, $(1\bar{1}2)$ and (210) planes in cubic and orthorhombic lattices.

PROBLEMS

1. Find the Miller indices for planes with each of the following sets of intercepts:

   (*i*) $3a, 3b, 2c$; (*ii*) $a, 2b, \infty$; (*iii*) $5a, -6b, c$;
   (*iv*) $a, b/2, c$; (*vi*) $a, b, -c$; (*vi*) $a/2, b, \infty$

   where a, b and c are lattice parameters.

   ((223), (210), $\left(6\bar{5}\ 30\right)$, (121), (111), (210))

2. Draw a $(1\bar{1}0)$ plane in a cubic unit cell. Show all the <111> directions that lie on this plane and give the Miller indices of each direction.

   $\left([111], [11\bar{1}], [\bar{1}\bar{1}1] \text{ and } [\bar{1}\bar{1}\bar{1}]\right)$

4. A plane makes intercepts of 1, 2 and 3 Å on the crystallographic axes of an orthorhombic crystal with $a : b : c = 3 : 2 : 1$. Determine the Miller indices of this plane. (931)

5. Determine the number of the nearest neighbours and the closest distance of approach in terms of lattice parameter for monoatomic *sc*, *bcc* and *fcc* structures. ($6, a; 8, a\sqrt{3}/2; 12, a/\sqrt{2}$)

6. Calculate the linear density (number of atoms per unit length) along cube edge, face diagonal and body diagonal of an *fcc* unit cell of side length *a*. [$1/a, \sqrt{2}/a, 1/(a\sqrt{3})$]

7. Nickel (*fcc*) has the lattice parameter of 3.52 Å. Calculate the atomic planar density (number of atoms per unit area) on (100), (110) and (111) planes. Is it possible to pack the atoms more closely than in (111) plane ? (1.61×10^{19}, 1.14×10^{19}, 1.86×10^{19} atoms m^{-2}; No)

8. Calculate the angles which [111] direction of a cubic lattice makes with [100] and [110] directions. (54°44', 35°15')

9. Show (111) and (222) planes in a cubic unit cell of side *a*. Compute the distances of these planes from a parallel plane passing through the origin. [$a/\sqrt{3}, a/(2\sqrt{3})$]

10. Calculate the distances between the adjacent parallel planes of the type (100), (110) and (111) in an *fcc* lattice of lattice constant *a*. Check the validity of the statement "The most close-packed planes are the most widely spaced." [$a/2, a/(2\sqrt{2})$ and $a/(\sqrt{3})$]

11. Copper (*fcc*) has density of 8960 kg m^{-3}. Calculate the unit cell dimension and the radius of Cu atom, given the atomic mass of Cu as 63.54 amu. (3.61 Å, 1.28 Å)

12. Prove that c/a ratio for an ideal *hcp* structure is 1.633.

13. Zinc (*hcp*) has lattice parameters *a* and *c* as 2.66 Å and 4.95 Å respectively. Calculate the packing fraction and density of zinc, given the atomic radius and the atomic mass of Zn as 1.31 Å and 65.37 amu respectively. (62%, 7155 kg m^{-3})

14. Calculate the distance between two atoms of a basis of the diamond structure, if the lattice constant of the structure is 5 Å. (2.17 Å)

CHAPTER - II

X-RAY DIFFRACTION AND RECIPROCAL LATTICE

2.1 INTRODUCTION

X-rays, being electromagnetic radiations, also undergo the phenomenon of diffraction as observed for visible light. However, unlike visible light, x-rays cannot be diffracted by ordinary optical grating because of their very short wavelengths. In 1912, a German physicist Max Von Laue suggested the use of a single crystal to produce diffraction of x-rays. Since all the atoms in a single crystal are regularly arranged with interatomic spacing of the order of a few angstroms, a crystal can act as a three-dimensional natural grating for x-rays. Friedrich and Knipping later successfully demonstrated the diffraction of x-rays from a thin single crystal of zinc blende (ZnS). The diffraction pattern obtained on a photographic film consisted of a series of dark spots arranged in a definite order. Such a pattern is called the *Laue's pattern* and reflects the symmetry of the crystal. Apart from this, the phenomenon of x-ray diffraction has become an invaluable tool to determine the structures of single crystals and polycrystalline materials. It is also extensively used to determine the wavelength of x-rays.

2.2 X-RAY DIFFRACTION

When an atomic electron is irradiated by a beam of monochromatic x-rays, it starts vibrating with a frequency equal to that of the incident beam. Since an accelerating charge emits radiations, the vibrating electrons present inside a crystal become sources of secondary radiations having the same frequency as the incident x-rays. These secondary x-rays spread out in all possible directions. The phenomenon may also be regarded as scattering of x-rays by atomic electrons. If the wavelength of incident radiations is quite large compared with the atomic dimensions, all the radiations emitted by electrons shall be in phase with one another. The incident x-rays, however, have the same order of wavelength as that of the atomic dimensions; hence the radiations emitted by electrons are, in general, out of phase with one another. These radiations may, therefore, undergo constructive or destructive interference producing maxima or minima in certain directions.

Consider a one-dimensional row of similar atoms having interatomic spacing equal to *a*. Let a wave front of x-rays of wavelength λ be incident on the row of atoms such that the wave crests are parallel to the row. The atoms emit secondary wavelets which travel in all possible directions. As shown in Fig. 2.1, the reinforcement of secondary wavelets takes place not only in a direction perpendicular to the row of atoms but also in other directions. These directions correspond to different orders of x-ray diffraction. The zeroth, first and second order diffraction directions are shown in Fig. 2.1. It may be noted that reinforcement takes place in some particular directions only, whereas in other directions the wave fronts interfere destructively and the intensity is minimum. Such reinforcements produce Laue's pattern.

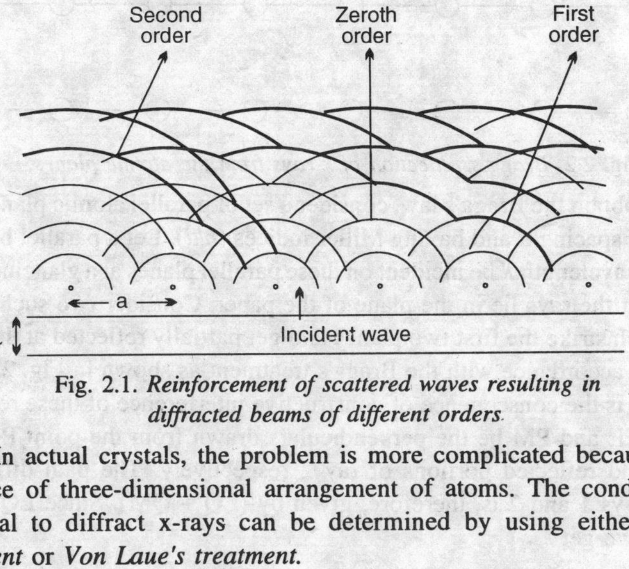

Fig. 2.1. *Reinforcement of scattered waves resulting in diffracted beams of different orders.*

In actual crystals, the problem is more complicated because of the presence of three-dimensional arrangement of atoms. The conditions for a crystal to diffract x-rays can be determined by using either *Bragg's treatment* or *Von Laue's treatment.*

2.2.1 The Bragg's Treatment : Bragg's Law

In 1912, W.H. Bragg and W.L. Bragg put forward a model which generates the conditions for diffraction in a very simple way. They pointed that a crystal may be divided into various sets of parallel planes. The directions of diffraction lines can then be accounted for if x-rays are considered to be reflected by such a set of parallel atomic planes followed by the constructive interference of the resulting reflected rays. Thus the problem of diffraction of x-rays by the atoms was converted into the problem of reflection of x-rays by the parallel atomic planes. Hence the words 'diffraction' and 'reflection' are mutually interchangeable in Bragg's treatment. Based on these considerations, Braggs derived a simple mathematical relationship which serves as a condition for the Bragg reflection to occur. This condition is known as the Bragg's law.

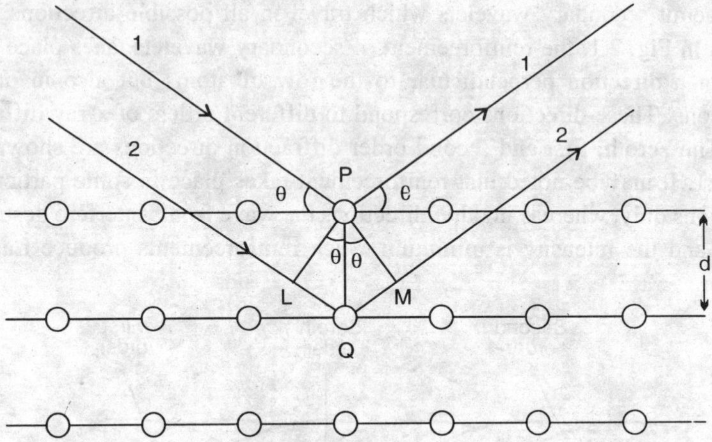

Fig. 2.2. *Bragg's reflection of x-rays from the atomic planes.*

To obtain the Bragg's law, consider a set of parallel atomic planes with interplanar spacing d and having Miller indices (hkl). Let a parallel beam of x-rays of wavelength λ be incident on these parallel planes at a glancing angle θ such that the rays lie in the plane of the paper. Consider two such rays 1 and 2 which strike the first two planes and get partially reflected at the same angle θ in accordance with the Bragg's treatment as shown in Fig. 2.2. The diffraction is the consequence of constructive interference of these reflected rays. Let PL and PM be the perpendiculars drawn from the point P on the incident and reflected portions of ray 2 respectively. The path difference between rays 1 and 2 is, therefore, given by (LQ + QM). Since LQ = QM = $d \sin\theta$, we get

$$\text{Path difference} = 2d \sin\theta$$

For constructive interference of rays 1 and 2, the path difference must be an integral multiple of wavelength λ, i.e.,

$$2d \sin\theta = n\lambda \tag{2.1}$$

where n is an integer. This equation is called the *Bragg's law*. The diffraction takes place for those values of d, θ, λ and n which satisfy the Bragg's condition. In Eq. (2.1), n represents the order of reflection. For $n = 0$, we get the zeroth order reflection which occurs for θ equal to zero, i.e., in the direction of the incident beam and hence it cannot be observed experimentally. For the given values of d and λ, the higher order reflections appear for larger values of θ. The diffraction lines appearing for $n = 1$, 2 and 3 are called first, second and third order diffraction lines respectively and so on. The intensity of the reflected lines decreases with increase in the value of n or θ. The highest possible order is determined by the condition that $\sin\theta$ cannot exceed unity. Also, since $\sin\theta \le 1$, λ must be $\le d$ for Bragg reflection to occur. Taking d $\approx 10^{-10}$ m, we obtain $\lambda \le 10^{-10}$ m or 1Å. X-rays having wavelength in this range are, therefore, preferred for analysis of crystal structures.

2.2.2 The Von Laue Treatment : Laue's Equations

Von Laue treated the phenomenon of diffraction in a more general way by considering the scattering of x-rays from individual atoms in the crystal followed by their recombination to obtain the directions of diffraction maxima. It will be shown below that diffraction maxima appear in some specific directions which obey certain conditions known as the *Laue's equations*. It also proves the validity of Bragg's treatment and the Bragg's law can be derived from the Laue's equations.

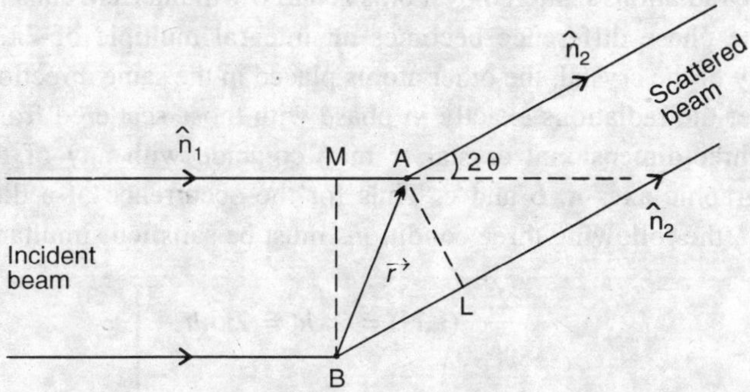

Fig. 2.3. *Scattering of x-rays from two identical scattering centres separated by a distance* **r**.

Consider the scattering of an incident beam from two identical scattering centres A and B placed at a distance r from each other in a crystal as shown in Fig. 2.3. Let $\hat{\mathbf{n}}_1$ and $\hat{\mathbf{n}}_2$ be the unit vectors in the directions of the incident and scattered beams respectively and let the angle between $\hat{\mathbf{n}}_1$ and $\hat{\mathbf{n}}_2$ be 2θ. Draw BM and AL perpendiculars to the directions of the incident and scattered beams respectively. Then the path difference between the rays scattered from A and B is given by

$$\text{Path difference} = AM - BL = \mathbf{r}.\hat{\mathbf{n}}_1 - \mathbf{r}.\hat{\mathbf{n}}_2 = \mathbf{r}.\,(\hat{\mathbf{n}}_1 - \hat{\mathbf{n}}_2) = \mathbf{r}.\mathbf{N}$$

where $\mathbf{N} = \hat{\mathbf{n}}_1 - \hat{\mathbf{n}}_2$. As will be seen later, the vector **N** happens to be a normal to the reflecting plane. It is a plane which may be assumed to be reflecting the incident ray into the direction of the scattered ray following the ordinary laws of reflection. This is one of the planes which forms the basis of Bragg's treatment. From Fig. 2.4, we find

$$|\mathbf{N}| = 2\sin\theta$$

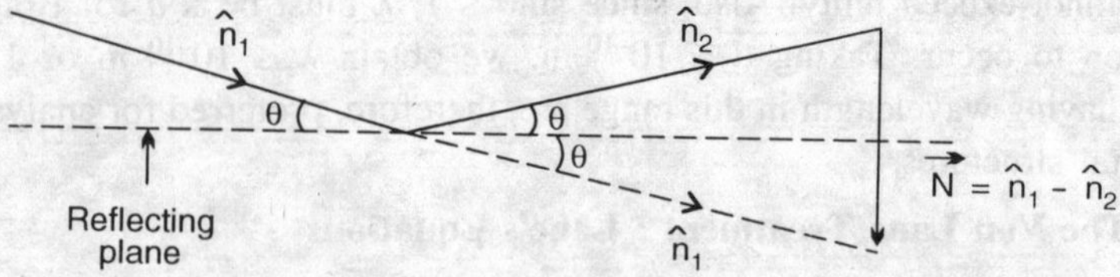

Fig. 2.4. *Geometrical relationship of incident beam, scattered beam, reflecting plane and the normal.*

The phase difference between the rays scattered from A and B is

$$\phi = \frac{2\pi}{\lambda}\,(\mathbf{r.N}) \tag{2.2}$$

The radiations scattered by atoms A and B will interfere constructively only if the phase difference becomes an integral multiple of 2π. Due to periodicity of the crystal, the other atoms placed in the same direction would also scatter the radiations exactly in phase with those scattered from A and B. In a three-dimensional crystal, **r** may coincide with any of the three crystallographic axes **a**, **b** and **c**. Thus for the occurrence of a diffraction maximum, the following three conditions must be satisfied simultaneously :

$$\left.\begin{aligned} \frac{2\pi}{\lambda}(\mathbf{a.N}) &= 2\pi h' = 2\pi nh \\ \frac{2\pi}{\lambda}(\mathbf{b.N}) &= 2\pi k' = 2\pi nk \\ \frac{2\pi}{\lambda}(\mathbf{c.N}) &= 2\pi l' = 2\pi nl \end{aligned}\right] \tag{2.3}$$

where h', k' and l' represent any three integers. While obtaining Eqs. (2.3), it is assumed that atoms A and B are the nearest neighbours and, so, the magnitudes a, b and c represent the interatomic distances along their respective crystallographic directions. The integers h', k' and l' and h, k, l differ only by a common factor n which may be equal to or greater than unity. Thus the integers h, k and l cannot have a common factor other than unity and resemble the Miller indices of a plane which happens to be the reflecting plane. Let α, β and γ be the angles between the scattering normal **N** and the crystallographic axes **a**, **b** and **c** respectively. Then,

$$\mathbf{a.N} = aN\cos\alpha = 2\,a\sin\theta\cos\alpha, \text{ and so on.}$$

Therefore, Eqs. (2.3) become

$$\left.\begin{aligned} \mathbf{a.N} &= 2a \sin\theta \cos\alpha = h'\lambda = nh\lambda \\ \mathbf{b.N} &= 2b \sin\theta \cos\beta = k'\lambda = nk\lambda \\ \mathbf{c.N} &= 2c \sin\theta \cos\gamma = l'\lambda = nl\lambda \end{aligned}\right] \tag{2.4}$$

Equations (2.4) are known as *Laue's equations* and represent the conditions for diffraction to occur. In an orthogonal coordinate system, α, β and γ also satisfy the condition

$$\cos^2\alpha + \cos^2\beta + \cos^2\gamma = 1 \tag{2.5}$$

where $\cos\alpha$, $\cos\beta$ and $\cos\gamma$ represent the direction cosines of the scattering normal. The Eqs. (2.4) and (2.5) yield the values of α, β, γ and θ for which diffraction takes place provided h, k, l and n are known.Thus, for a given reflecting plane, Eqs. (2.4) serve to determine unique values of θ and **N** which define a scattering direction.

From Eqs. (2.4), we also find that, for fixed θ, the direction cosines $\cos\alpha$, $\cos\beta$ and $\cos\gamma$ of the scattering normal are proportional to h/a, k/b and l/c. Also, as described in Sec. 1.9, the direction cosines of the normal to any arbitrary plane (hkl) are proportional to h/a, k/b and l/c. This leads to the conclusion that the scattering normal **N** is the same as the normal to the plane (hkl) and hence the arbitrary plane (hkl) happens to be the reflecting plane.

To obtain the Bragg's law, consider the expressions for interplanar spacing for the (hkl) planes as given by Eq. (1.7), i.e.,

$$d = \frac{a}{h}\cos\alpha = \frac{b}{k}\cos\beta = \frac{c}{l}\cos\gamma$$

In combination with Eqs. (2.4), these yield

$$2d \sin\theta = n\lambda$$

which is the Bragg's law. Here n represents the order of reflection and, as described above, is the greatest common factor among the integers h', k' and l' in Eqs. (2.4). Thus one may have the planes (hkl) and consider different orders of reflection from these; alternatively, one may have the planes (nh nk nl) or ($h'k'l'$) and always consider the first order reflection. The latter practice is normally adopted during the process of structure determination by x-ray diffraction. It is obvious that the nth order reflection from the planes (hkl) would overlap with the first order reflection from the planes (nh nk nl) or ($h'k'l'$). Thus, putting n equal to 1, one can get rid of the factor n in the Bragg's equation provided the reflections from all the planes, real or imaginary, having Miller indices with or without a common factor be considered.

2.3 X-RAY DIFFRACTION METHODS

The phenomenon of x-ray diffraction is employed to determine the structure of solids as well as for the study of x-ray spectroscopy. The underlying principle in both the cases is the Bragg's law as given by Eq. (2.1). Considering only the first order reflections from all the possible atomic planes, real or fictitious, the Bragg's law may be written as

$$2d \sin \theta = \lambda \tag{2.6}$$

The reflections take place for those values of d, θ and λ which satisfy the above equation. For structural analysis, x-rays of known wavelength are employed and the angles for which reflections take place are determined experimentally. The d values corresponding to these reflections are then obtained from Eq. (2.6). Using this information, one can proceed to determine the size of the unit cell and the distribution of atoms within the unit cell. In the x-ray spectroscopy, x-rays are incident on a particular cleavage surface of a single crystal so that the interplanar spacing d is known. The angle for which reflections take place are determined experimentally. The wavelength λ of the incident x-rays is then obtained from Eq. (2.6).

It may be noted that the x-rays used for diffraction purposes should have wavelength which is the most appropriate for producing diffraction effects. Since $\sin\theta$ should be less than unity, Eq. (2.6) yields

$$\lambda < 2d$$

Normally,

$$d \sim 3 \text{ Å}$$

$$\therefore \qquad \lambda < 6 \text{ Å}$$

Longer wavelength x-rays are unable to resolve the details of the structure on the atomic scale whereas shorter wavelength x-rays are diffracted through angles which are too small to be measured experimentally.

In x-ray diffraction studies, the probability that the atomic planes with right orientations are exposed to x-rays is increased by adopting one of the following methods:

(i) A single crystal is held stationary and a beam of white radiations is inclined on it at a fixed glancing angle θ, i.e., θ is fixed while λ varies. Different wavelengths present in the white radiations select the appropriate reflecting planes out of the numerous present in the crystal such that the Bragg's condition is satisfied. This technique is called the *Laue's technique*.

(ii) A single crystal is held in the path of monochromatic radiations and is rotated about an axis, i.e., λ is fixed while θ varies. Different sets of parallel atomic planes are exposed to incident radiations for different values of θ and reflections take place from those atomic planes for which d and θ satisfy the Bragg's law. This method is known as the *rotating crystal method.*

(iii) The sample in the powdered form is placed in the path of monochromatic x-rays, i.e., λ is fixed while both θ and d vary. Thus a number of small crystallites with different orientations are exposed to x-rays. The reflections take place for those values of d, θ and λ which satisfy the Bragg's law. This method is called the *powder method.*

2.3.1 The Laue's Method

An experimental arrangement used to produce *Laue's patterns* is shown in Fig. 2.5. It consists of a flat plate camera which contains a collimator with a fine hole to obtain a very fine beam of x-rays. The sample is placed on a goniometer which can be rotated to change the orientation of the single crystal. Two flat photographic films are used, one for receiving the transmitted diffracted beam and the other for receiving the reflected diffracted beam for back reflection experiments. Such experiments are performed particularly when there is excessive absorption of x-rays in the crystal.

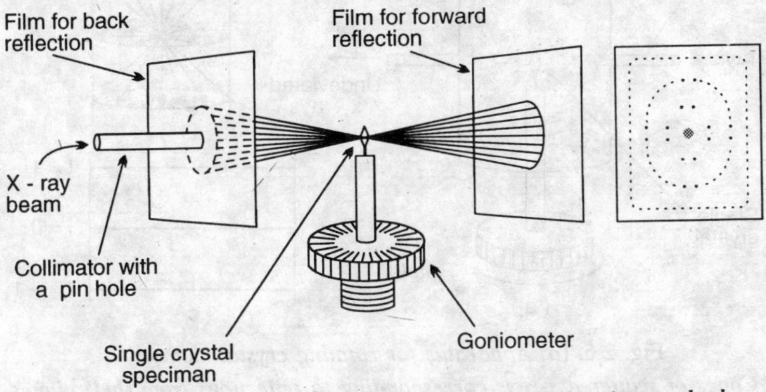

Fig. 2.5. *A flat plate camera used in Laue's diffraction method.*

Initially, a single crystal specimen having dimensions of the order of 1mm × 1mm × 1mm is held stationary in the path of white x-rays having wavelengths ranging from 0.2 to 2 Å. Since the crystal contains a number of sets of parallel atomic planes with different interplanar spacings, diffraction is possible for certain values of λ and d which satisfy the Bragg's condition. Thus diffraction spots are produced on the photographic films as shown in Fig. 2.5. The crystal can be rotated with the help of goniometer to change its orientation with respect to the incident beam. By doing so, the diffraction condition may be satisfied for a new set of atomic planes and hence a different type of pattern may be obtained on the photographic film. The symmetry of the crystal is, however, reflected in each pattern.

The Laue's method is mostly used to determine the crystal symmetry. For example, if a crystal having four-fold axial symmetry is oriented so that its axis is parallel to the beam, the resulting Laue's pattern also exhibits the four-fold symmetry. The symmetry of the pattern helps to determine the shape of the unit cell. It is, however, not practicable to determine the structure of the crystal by this method. It is because a number of wavelengths may be reflected from a single plane in different orders and may superpose at a single point resulting in the loss of a number of reflections. The symmetry of the Laue's pattern also helps to orient the crystals for various solid state experiments. Another application of the Laue's method is the determination of imperfections or strains in the crystal. An imperfect or strained crystal has atomic planes which are not exactly plane but are slightly curved. Thus instead of sharp diffraction spots one gets streaks in the Laue's pattern. This type of streaking on Laue's photographs is called *asterism*.

2.3.2 Rotating Crystal Method

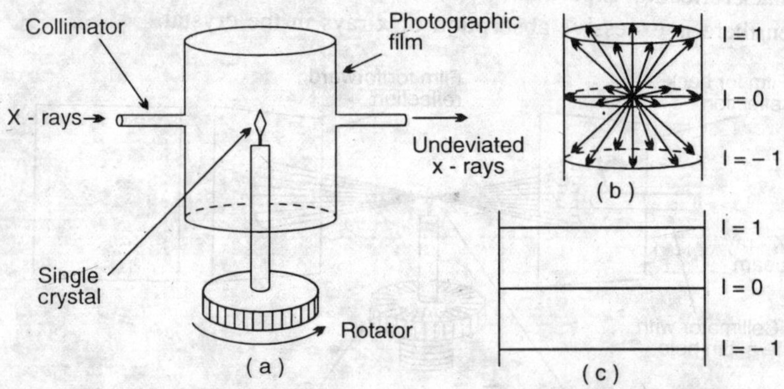

Fig. 2.6. *(a) Apparatus for rotating crystal method*
(b) Cones of scattered x-rays corresponding to reflections from (hkl) planes.
(c) Layer lines produced after flattening the photographic film.

In this method, a monochromatic beam of x-rays is incident on a single crystal mounted on a rotating spindle such that one of its crystallographic axes coincides with the axis of rotation which is kept perpendicular to the direction of the incident beam. The single crystal having dimensions of the order of 1mm is positioned at the centre of a cylindrical holder concentric with the rotating spindle as shown in Fig. 2.6. A photographic film is attached at the inner circular surface of the cylinder.

The diffraction takes place from those planes which satisfy the Bragg's law for a particular angle of rotation. The planes parallel to the axis of rotation diffract the incident rays in a horizontal plane. However, reflections cannot be observed for those planes which always contain the incident beam. The planes inclined to the rotation axis produce reflections above or below the horizontal plane depending upon the angle of inclination. The horizontal lines produced by diffraction spots on the photographic film are called *layer lines*. If the crystal is positioned such that its **c**-axis coincides with the axis of rotation, all the planes with Miller indices of the type $(hk0)$ will produce the central layer line. Likewise, the planes having Miller indices of the type $(hk1)$ and $(hk\bar{1})$ will produce the layer lines above and below the central line respectively, and so on. These layer lines are shown in Fig. 2.6c. The vertical spacing between the layer lines depends on the distance between the lattice points along the **c**-axis. Hence the distance **c** can be measured from the photographic film. Similarly, one can determine the translation vectors **a** and **b** on mounting the crystal along **a** and **b** axes respectively. Thus the dimensions of the unit cell can be easily determined.

2.3.3 Powder Method

This is the most widely used diffraction method to determine the structure of crystalline solids. The sample used is in the form of a fine powder containing a large number of tiny crystallites with random orientations. It is prepared by crushing the commonly available polycrystalline material, thus eliminating the tedious process of growing the single crystals.

The experimental arrangement used to produce diffraction is shown in Fig. 2.7. It consists of a cylindrical camera, called the Debye-Scherrer camera, whose length is small as compared to the diameter. The finely powdered sample is filled in a thin capillary tube or is simply pasted on a wire by means of a binder and mounted at the centre of the camera. The capillary tube or wire and the binder should be of a non-diffracting material. A collimated beam of monochromatic x-rays is produced by passing the x-rays through a filter and a collimator. The x-rays enter the camera through the collimator and strike the powdered sample. Since the specimen contains a large number of small crystallites ($\sim 10^{12}$ in 1mm^3 of powder sample) with random orientations, almost all the possible θ and d values are available. The diffraction takes place for those values of d and θ which satisfy the Bragg's condition, i.e., $2d \sin\theta = n\lambda$, λ being a constant in this case. Also, since for a particular value of the angle of incidence θ, numerous orientations of a particular set of planes are possible, the diffracted rays corresponding to fixed values of θ and d lie on the surface of a cone with its apex at the sample and the semivertical angle equal to 2θ. Different cones are observed for different sets of d and θ for a particular value of n, and also for different combinations

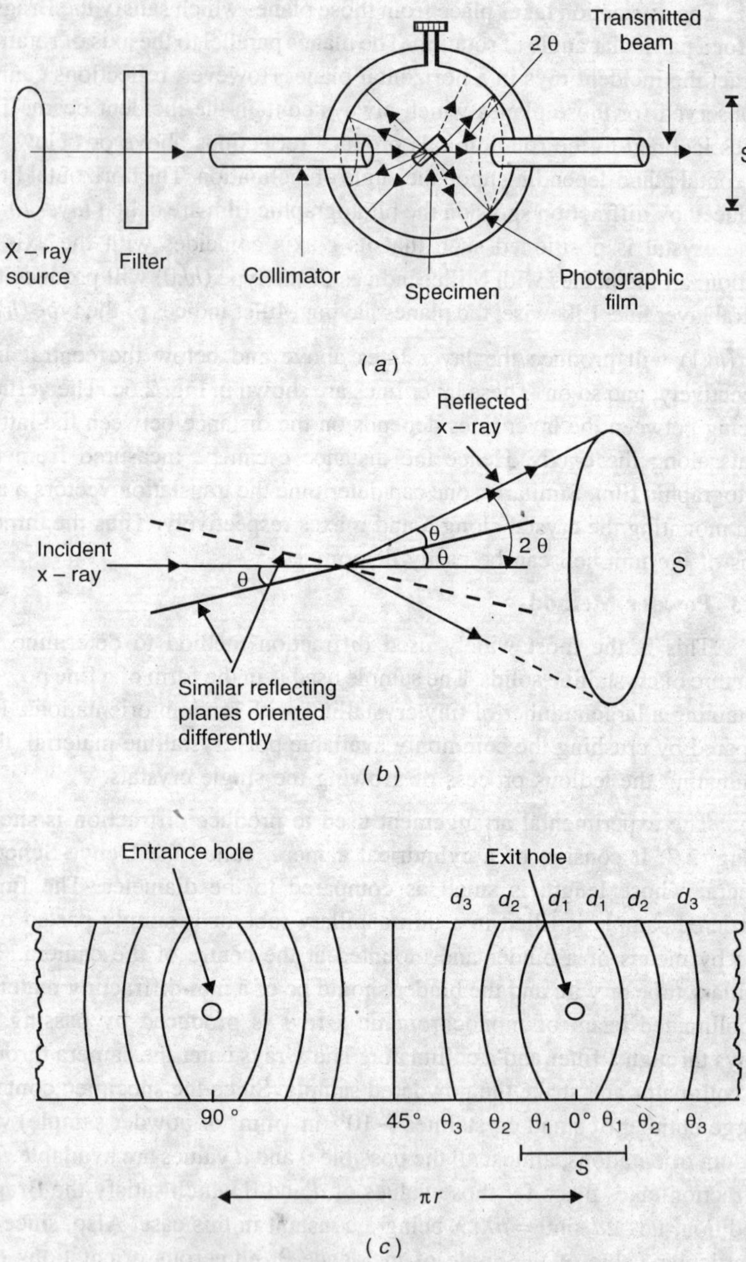

Fig. 2.7. *(a) Front view of the Debye-Scherrer Camera.*
(b) A cone produced by reflection of x-rays from identical planes having different orientations.
(c) Flattened photographic film after developing and indexing of diffraction lines.

of θ and n for a particular value of d. The transmitted x-rays move out of the camera through an exit hole located diametrically opposite to the entrance hole. A photographic film is attached to the inner side of the curved surface of the camera. Each cone of the reflected beam leaves two impressions on the film which are in the form of arcs on either side of the exit hole with their centres coinciding with the hole. Similarly, cones produced by back-reflected x-rays produce arcs on either side of the entrance hole. If the sample consists of coarse grains rather than fine particles, a spotty diffraction pattern may be obtained. This is because a sufficient number of crystallites with all possible orientations may not be available in a coarse-grained sample. In such a case, the sample has to be rotated to obtain almost continuous diffraction arcs. The film is exposed for a long time (~ a few hours) in order to obtain reflected lines of sufficiently high intensity. It is then removed from the camera and developed. The arcs produced by reflected rays appear dark on the developed film. The angle θ corresponding to a particular pair of arcs is related to the distance S between the arcs as

$$4\theta \text{ (radians)} = S/R \tag{2.7}$$

where R is the radius of the camera. If θ is measured in degrees, the above equation is modified as

$$4\theta \text{ (degrees)} = \frac{57.296S}{R} \tag{2.8}$$

The calculations can be made simpler by taking the radius of the camera in multiples of 57.296. For example, taking R = 57.296 mm, we get

$$\theta \text{ (degrees)} = S \text{ (mm)}/4 \tag{2.9}$$

Thus one-fourth of the distance between the corresponding arcs of a particular pair in mm is a measure of the angle θ in degrees. Knowing all the possible θ's and considering only the first order reflections from all the possible planes, Eq. (2.6) is used to calculate the interplanar spacing for various sets of parallel planes which contribute to these reflections. Thus, we have

$$d = \lambda \,/\, (2 \sin\theta)$$

These d values are used to determine the space lattice of the crystal structure.

In modern x-ray diffractometers, the photographic film is replaced by a radiation detector, such as ionization chamber or scintillation detector, which records the positions and relative intensities of the various reflected lines as a function of the angle 2θ. The detector is mounted on a goniometer and is capable of rotation about the sample at different speeds. The whole system is computerised. The availability of a lot of software makes the system versatile.

2.4 RECIPROCAL LATTICE

As described earlier, the diffraction of x-rays occurs from various sets of parallel planes having different orientations (slopes) and interplanar spacings. In certain situations involving the presence of a number of sets of parallel planes with different orientations, it becomes difficult to visualize all such planes because of their two-dimensional nature. The problem was simplified by P.P. Ewald by developing a new type of lattice known as the *reciprocal lattice*. The idea underlying the development was that each set of parallel planes could be represented by a normal to these planes having length equal to the reciprocal of the interplanar spacing. Thus the direction of each normal represents the orientation of the corresponding set of parallel planes and its length is proportional to the reciprocal of the interplanar spacing.

The normals are drawn with reference to an arbitrary origin and points are marked at their ends. These points form a regular arrangement which is called a reciprocal lattice. Obviously, each point in a reciprocal lattice is a representative point of a particular parallel set of planes and it becomes easier to deal with such points than with sets of planes.

A reciprocal lattice to a direct lattice is constructed using the following procedure :

(a) Take origin at some arbitrary point and draw normals to every set of parallel planes of the direct lattice.

(b) Take length of each normal equal to the reciprocal of the interplanar spacing for the corresponding set of planes. The terminal points of these normals form the reciprocal lattice.

Consider, for example, a unit cell of monoclinic crystal in which $a \neq b \neq c$, $\alpha = \gamma = 90^\circ$ and $\beta > 90^\circ$ as shown in Fig. 2.8. For simplicity, we orient the unit cell in such a way that the **b**-axis is perpendicular to the plane of the paper; hence **a** and **c**-axes lie in the plane of the paper as shown in Fig. 2.9.

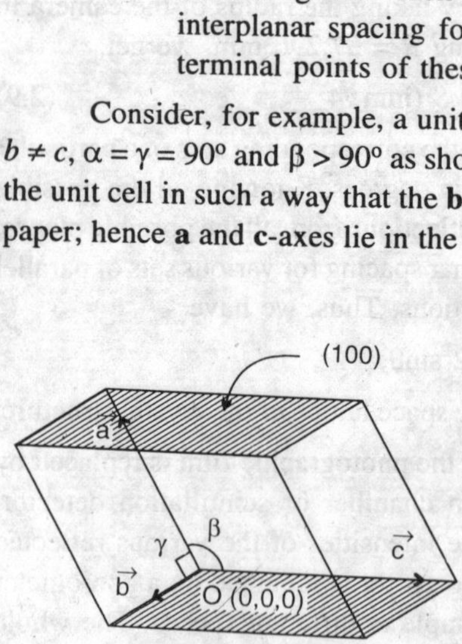

Fig. 2.8. *Unit cell of a monoclinic crystal.*

Consider planes of the type (*h0l*) which are parallel to **b**-axis, i.e., perpendicular to the plane of the paper. Hence normal to these planes lie in the plane of the paper. The planes (*h0l*), being perpendicular to the plane of the paper, are represented by lines. Thus the line (101) in fact means the plane (101), and so on. Taking the point of intersection of the three axes as the origin, normals are drawn to the planes ($h0l$) and their lengths are taken to be $1/d_{h0l}$, where d_{h0l} is the interplanar spacing for the planes ($h0l$). For example, since the planes (200) have half the interplanar spacing as compared to the plane (100), the reciprocal lattice point (200) is twice as far away as point (100) from the origin. If normals to all the (hkl) planes are drawn, a three-dimensional reciprocal lattice is obtained.

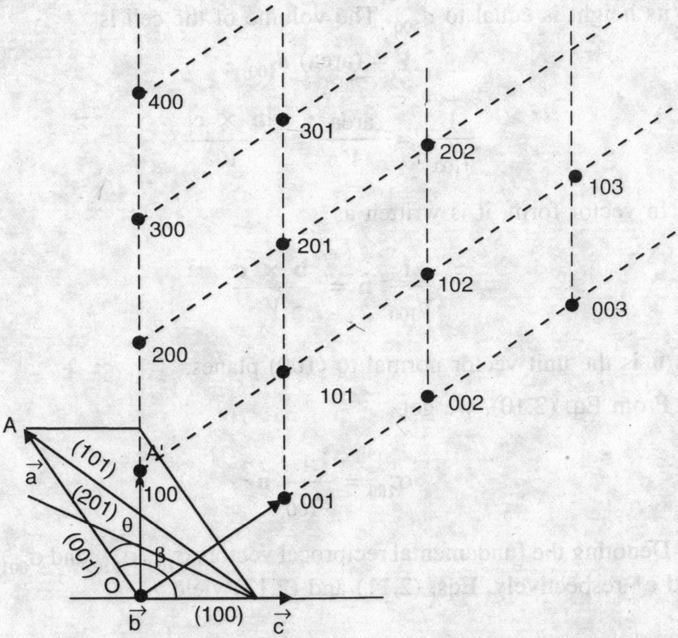

Fig. 2.9. *Two-dimensional reciprocal lattice to a monoclinic lattice. The* **b**-*axis is perpendicular to the plane of the paper.*

2.4.1 Reciprocal Lattice Vectors

A *reciprocal lattice vector*, $\sigma_{\mathbf{hkl}}$, is defined as a vector having magnitude equal to the reciprocal of the interplanar spacing d_{hkl} and direction coinciding with normal to the (hkl) planes. Thus, we have

$$\sigma_{\mathbf{hkl}} = \frac{1}{d_{hkl}}\,\hat{\mathbf{n}} \tag{2.10}$$

where $\hat{\mathbf{n}}$ is the unit vector normal to the (hkl) planes. In fact, a vector drawn from the origin to any point in the reciprocal lattice is a reciprocal lattice vector.

Like a direct lattice, a reciprocal lattice also has a unit cell which is of the form of a parallelopiped. The unit cell is formed by the shortest normals along the three directions, i.e., along the normals to the planes (100), (010) and (001). These normals produce reciprocal lattice vectors designated as σ_{100}, σ_{010} and σ_{001} which represent the *fundamental reciprocal lattice vectors.*

Let **a, b** and **c** be the primitive translation vectors of the direct lattice as shown in Fig. 2.8. The base of the unit cell is formed by the vectors **b** and **c** and its height is equal to d_{100}. The volume of the cell is

$$V = (\text{area})\, d_{100}$$

or

$$\frac{1}{d_{100}} = \frac{\text{area}}{V} = \frac{|\mathbf{b} \times \mathbf{c}|}{V}$$

In vector form, it is written as

$$\frac{1}{d_{100}}\,\hat{\mathbf{n}} = \frac{\mathbf{b} \times \mathbf{c}}{V} \tag{2.11}$$

where $\hat{\mathbf{n}}$ is the unit vector normal to (100) planes.

From Eq. (2.10), we get

$$\sigma_{100} = \frac{1}{d_{100}}\,\hat{\mathbf{n}} \tag{2.12}$$

Denoting the fundamental reciprocal vectors σ_{100}, σ_{010} and σ_{001} by **a***, **b*** and **c*** respectively, Eqs. (2.11) and (2.12) yield

$$\mathbf{a}^* = \sigma_{100} = \frac{\mathbf{b}\times\mathbf{c}}{\mathbf{a}.\mathbf{b}\times\mathbf{c}}$$

Similarly,

$$\mathbf{b}^* = \sigma_{010} = \frac{\mathbf{c}\times\mathbf{a}}{\mathbf{a}.\mathbf{b}\times\mathbf{c}} \tag{2.13}$$

and

$$\mathbf{c}^* = \sigma_{001} = \frac{\mathbf{a}\times\mathbf{b}}{\mathbf{a}.\mathbf{b}\times\mathbf{c}}$$

where **a.b×c** = **b.c×a** = **c.a×b** is the volume of the direct cell. Thus the reciprocal translation vectors bear a simple relationship to the crystal translation vectors as

$$\left.\begin{array}{l} \mathbf{a}^* \text{ is normal to } \mathbf{b} \text{ and } \mathbf{c} \\ \mathbf{b}^* \text{ is normal to } \mathbf{c} \text{ and } \mathbf{a} \\ \mathbf{c}^* \text{ is normal to } \mathbf{a} \text{ and } \mathbf{b} \end{array}\right] \tag{2.14}$$

In vector notation, it means

$$\left.\begin{aligned} \mathbf{a}^*.\mathbf{b} &= 0 & \mathbf{a}^*.\mathbf{c} &= 0 \\ \mathbf{b}^*.\mathbf{c} &= 0 & \mathbf{b}^*.\mathbf{a} &= 0 \\ \mathbf{c}^*.\mathbf{a} &= 0 & \mathbf{c}^*.\mathbf{b} &= 0 \end{aligned}\right] \tag{2.15}$$

Taking scalar product of **a*, b*** and **c*** with **a, b** and **c** respectively and using Eqs. (2.13), we find

$$\mathbf{a}^*.\mathbf{a} = 1, \quad \mathbf{b}^*.\mathbf{b} = 1, \quad \mathbf{c}^*.\mathbf{c} = 1 \tag{2.16}$$

It appears from Eqs. (2.16) that **a*, b*** and **c*** are parallel to **a, b** and **c** respectively. However, this is not always true. In non-cubic crystal systems, such as monoclinic crystal system, as shown in Fig. 2.8, **a*** and **a** point in different directions, i.e., along OA′, and OA respectively. Thus all that is meant by Eqs. (2.16) is that the length of **a*** is the reciprocal of $a \cos\theta$, where θ is the angle between **a*** and **a**.

In some texts on Solid State Physics, the primitive translation vectors **a, b** and **c** of a direct lattice are related to the primitive translation vectors **a***, **b*** and **c*** of the reciprocal lattice as

$$\mathbf{a}^*.\mathbf{a} = \mathbf{b}^*.\mathbf{b} = \mathbf{c}^*.\mathbf{c} = 2\pi \tag{2.17}$$

with Eqs. (2.15) still being valid. These equations can be satisfied by choosing the reciprocal lattice vectors as

$$\left.\begin{aligned} \mathbf{a}^* &= 2\pi \frac{\mathbf{b}\times\mathbf{c}}{\mathbf{a}.\mathbf{b}\times\mathbf{c}} \\ \mathbf{b}^* &= 2\pi \frac{\mathbf{c}\times\mathbf{a}}{\mathbf{a}.\mathbf{b}\times\mathbf{c}} \\ \mathbf{c}^* &= 2\pi \frac{\mathbf{a}\times\mathbf{b}}{\mathbf{a}.\mathbf{b}\times\mathbf{c}} \end{aligned}\right] \tag{2.18}$$

It is now obvious that every crystal structure is associated with two important lattices — the direct lattice and the reciprocal lattice. The two lattices are related to each other by Eqs. (2.13). The fundamental translation vectors of the crystal lattice and the reciprocal lattice have dimensions of [length] and $[\text{length}]^{-1}$ respectively. This is why the latter is called the reciprocal lattice. Also, the volume of the unit cell of a reciprocal lattice is inversely proportional to the volume of the unit cell of its direct lattice.

A crystal lattice is a lattice in real or ordinary space, i.e., the space defined by the coordinates, whereas a reciprocal lattice is a lattice in the *reciprocal space*, *associated* **k**-*space* or *Fourier space*. A wave vector **k** is always drawn in the **k**-space. The points of the crystal lattice are given by

$$\mathbf{T} = m\mathbf{a} + n\mathbf{b} + p\mathbf{c} \tag{2.19}$$

where m, n and p are integers. Similarly, the reciprocal lattice points or reciprocal lattice vectors may be defined as

$$\mathbf{G} = h\mathbf{a}^* + k\mathbf{b}^* + l\mathbf{c}^* \tag{2.20}$$

where h, k and l are integers. Every point in the Fourier space has a meaning, but the reciprocal lattice points defined by Eq. (2.20) carry a special importance. In order to find the significance of **G**'s, we take the dot product of **G** and **T** :

$$\begin{aligned} \mathbf{G.T} &= (h\mathbf{a}^* + k\mathbf{b}^* + l\mathbf{c}^*) \,.\, (m\mathbf{a}+n\mathbf{b}+p\mathbf{c}) \\ &= 2\pi\,(hm+kn+lp) = 2\pi \text{ (an integer)} \end{aligned}$$

or $$\exp(i\mathbf{G.T}) = 1$$

where we have used Eq. (2.17). Thus it is clear from Eq. (2.20) that h, k and l define the coordinates of the points of reciprocal lattice space. In other words, it means that a point (h,k,l) in the reciprocal space corresponds to the set of parallel planes having the Miller indices (hkl). The concept of reciprocal lattice is useful for redefining the Bragg's condition and introducing the concept of Brillouin zones.

2.4.2 Reciprocal Lattice to SC Lattice

The primitive translation vectors of a simple cubic lattice may be written as

$$\mathbf{a} = a\hat{\mathbf{i}},\ \mathbf{b} = a\hat{\mathbf{j}},\ \mathbf{c} = a\hat{\mathbf{k}}$$

$$\begin{aligned} \text{Volume of the simple cubic unit cell} &= \mathbf{a.b\times c} \\ &= a^3\,(\hat{\mathbf{i}}.\hat{\mathbf{j}}\times\hat{\mathbf{k}}) = a^3 \end{aligned}$$

Using Eqs. (2.18), the reciprocal lattice vectors to the *sc* lattice are obtained as

$$\mathbf{a}^* = 2\pi\,\frac{\mathbf{b\times c}}{\mathbf{a.b\times c}} = 2\pi\,\frac{a\hat{\mathbf{j}}\times a\hat{\mathbf{k}}}{a^3} = \frac{2\pi}{a}\hat{\mathbf{i}}$$

Similarly,

$$\mathbf{b}^* = 2\pi\,\frac{\mathbf{c\times a}}{\mathbf{a.b\times c}} = \frac{2\pi}{a}\hat{\mathbf{j}} \tag{2.21}$$

and

$$\mathbf{c}^* = 2\pi\,\frac{\mathbf{a\times b}}{\mathbf{a.b\times c}} = \frac{2\pi}{a}\hat{\mathbf{k}}$$

The Eqs. (2.21) indicate that all the three reciprocal lattice vectors are equal in magnitude which means tha' the reciprocal lattice to *sc* lattice is also simple cubic but with lattice constant equal to $2\pi/a$.

2.4.3 Reciprocal Lattice to BCC Lattice

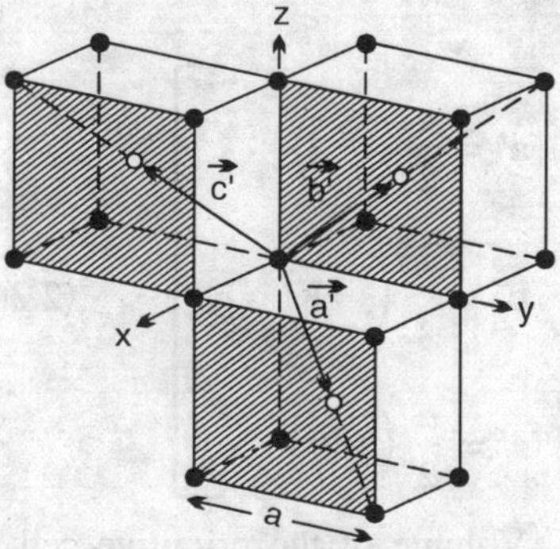

Fig. 2.10. *Primitive translation vectors of a bcc lattice.*

The primitive translation vectors of a body-centred cubic lattice, as shown in Fig. 2.10, are

$$\left.\begin{aligned} \mathbf{a}' &= \frac{a}{2}\left(\hat{\mathbf{i}}+\hat{\mathbf{j}}-\hat{\mathbf{k}}\right) \\ \mathbf{b}' &= \frac{a}{2}\left(-\hat{\mathbf{i}}+\hat{\mathbf{j}}+\hat{\mathbf{k}}\right) \\ \mathbf{c}' &= \frac{a}{2}\left(\hat{\mathbf{i}}-\hat{\mathbf{j}}+\hat{\mathbf{k}}\right) \end{aligned}\right] \tag{2.22}$$

where a is the length of the cube edge and $\hat{\mathbf{i}}$, $\hat{\mathbf{j}}$ and $\hat{\mathbf{k}}$ are the orthogonal unit vectors along the cube edges. The volume of the primitive cell is given by

$$V = \mathbf{a}'.\mathbf{b}'\times\mathbf{c}' = \frac{a}{2}\left(\hat{\mathbf{i}}+\hat{\mathbf{j}}-\hat{\mathbf{k}}\right).\left[\frac{a^2}{4}\left(-\hat{\mathbf{i}}+\hat{\mathbf{j}}+\hat{\mathbf{k}}\right)\times\left(\hat{\mathbf{i}}-\hat{\mathbf{j}}+\hat{\mathbf{k}}\right)\right]$$

$$= \frac{a}{2}\left(\hat{\mathbf{i}}+\hat{\mathbf{j}}-\hat{\mathbf{k}}\right).\frac{a^2}{2}\left(\hat{\mathbf{i}}+\hat{\mathbf{j}}\right)$$

$$= a^3/2$$

Using Eqs. (2.18), the reciprocal lattice vectors are obtained as

$$\mathbf{a}^* = 2\pi\frac{\mathbf{b}'\times\mathbf{c}'}{\mathbf{a}'.\mathbf{b}'\times\mathbf{c}'} = \frac{2\pi\left(a^2/2\right)}{a^3/2}\left(\hat{\mathbf{i}}+\hat{\mathbf{j}}\right) = \frac{2\pi}{a}\left(\hat{\mathbf{i}}+\hat{\mathbf{j}}\right)$$

Similarly,

$$\mathbf{b}^* = 2\pi\frac{\mathbf{c}'\times\mathbf{a}'}{\mathbf{a}'.\mathbf{b}'\times\mathbf{c}'} = \frac{2\pi}{a}\left(\hat{\mathbf{j}}+\hat{\mathbf{k}}\right) \tag{2.23}$$

and

$$\mathbf{c}^* = 2\pi\frac{\mathbf{a}'\times\mathbf{b}'}{\mathbf{a}'.\mathbf{b}'\times\mathbf{c}'} = \frac{2\pi}{a}\left(\hat{\mathbf{k}}+\hat{\mathbf{i}}\right)$$

As will be seen later, these are the primitive translation vectors of an *fcc* lattice. Thus the reciprocal lattice to a *bcc* lattice is *fcc* lattice.

2.4.4 Reciprocal Lattice to FCC Lattice

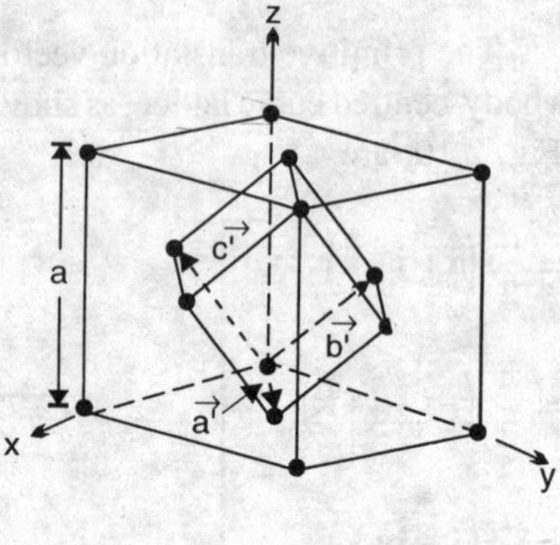

Fig. 2.11. *Primitive translation vectors of an fcc lattice.*

The primitve translation vectors of an *fcc* lattice, as shown in Fig. 2.11, are

$$\left.\begin{aligned}\mathbf{a}' &= \frac{a}{2}\left(\hat{\mathbf{i}}+\hat{\mathbf{j}}\right)\\ \mathbf{b}' &= \frac{a}{2}\left(\hat{\mathbf{j}}+\hat{\mathbf{k}}\right)\\ \mathbf{c}' &= \frac{a}{2}\left(\hat{\mathbf{k}}+\hat{\mathbf{i}}\right)\end{aligned}\right] \tag{2.24}$$

Volume of the primitive cell is given by

$$\begin{aligned}V &= \mathbf{a}'\,.\mathbf{b}' \times \mathbf{c}'\\ &= \frac{a}{2}\left(\hat{\mathbf{i}}+\hat{\mathbf{j}}\right)\,.\;\frac{a^2}{4}\left[\left(\hat{\mathbf{j}}+\hat{\mathbf{k}}\right)\times\left(\hat{\mathbf{k}}+\hat{\mathbf{i}}\right)\right]\\ &= \frac{a}{2}\left(\hat{\mathbf{i}}+\hat{\mathbf{j}}\right)\,.\;\frac{a^2}{4}\left(\hat{\mathbf{i}}+\hat{\mathbf{j}}-\hat{\mathbf{k}}\right)\\ &= a^3/4\end{aligned}$$

Using Eqs. (2.18), the primitive translation vectors of the reciprocal lattice are obtained as

$$\mathbf{a}^* = 2\pi\,\frac{\mathbf{b}'\times\mathbf{c}'}{\mathbf{a}'.\,\mathbf{b}'\times\mathbf{c}'} = 2\pi\frac{\left(a^2/4\right)\left(\hat{\mathbf{i}}+\hat{\mathbf{j}}-\hat{\mathbf{k}}\right)}{a^3/4} = \frac{2\pi}{a}\left(\hat{\mathbf{i}}+\hat{\mathbf{j}}-\hat{\mathbf{k}}\right)$$

Similarly,

$$\mathbf{b}^* = 2\pi\,\frac{\mathbf{c}'\times\mathbf{a}'}{\mathbf{a}'.\,\mathbf{b}'\times\mathbf{c}'} = \frac{2\pi}{a}\left(-\hat{\mathbf{i}}+\hat{\mathbf{j}}+\hat{\mathbf{k}}\right) \tag{2.25}$$

and

$$\mathbf{c} = 2\pi\,\frac{\mathbf{a}'\times\mathbf{b}'}{\mathbf{a}'.\,\mathbf{b}'\times\mathbf{c}'} = \frac{2\pi}{a}\left(\hat{\mathbf{i}}-\hat{\mathbf{j}}+\hat{\mathbf{k}}\right)$$

Comparing Eqs. (2.25) with Eqs. (2.22), we find that these are the primitive translation vectors of a *bcc* lattice having length of the cube edge as $2\pi/a$. Thus the reciprocal lattice to an *fcc* lattice is a *bcc* lattice.

2.5 PROPERTIES OF RECIPROCAL LATTICE

1. Each point in a reciprocal lattice corresponds to particular set of parallel planes of the direct lattice.
2. The distance of a reciprocal lattice point from an arbitrarily fixed origin is inversely proportional to the interplanar spacing of the corresponding parallel planes of the direct lattice.
3. The volume of a unit cell of the reciprocal lattice is inversely proportional to the volume of the corresponding unit cell of the direct lattice.
4. The unit cell of the reciprocal lattice need not be a parallelopiped. It is customary to deal with Wigner-Seitz cell of the reciprocal lattice which constitutes the Brillouin zone.
5. The direct lattice is the reciprocal lattice to its own reciprocal lattice. Simple cubic lattice is self-reciprocal whereas *bcc* and *fcc* lattices are reciprocal to each other.

2.6. BRAGG'S LAW IN RECIPROCAL LATTICE

The Bragg's diffraction condition obtained earlier by considering reflection from parallel lattice planes can be used to express geometrical relationship between the vectors in the reciprocal lattice. Consider a reciprocal lattice as shown in Fig. 2.12. Starting from the point A (not necessarily a reciprocal lattice point), draw a vector $\vec{AO}$ of length $1/\lambda$ in the direction of the incident x-ray beam which terminates at the origin O of the reciprocal lattice. Taking A as the centre, draw a sphere of radius AO which may intersect some point B of the reciprocal lattice.

Let the coordinates of point B be (h', k', l') which may have a highest common factor n, i.e., the coordinates are of the type (nh, nk, nl), where h, k and l do not have a common factor other than unity. Apparently, vector $\vec{OB}$ is the reciprocal vector. It must, therefore, be normal to the plane $(h'k'l')$ or (hkl) and should have length equal to $1/d_{h'k'l'}$ or n/d_{hkl} Thus,

$$|\vec{OB}| = n/d_{hkl} \tag{2.26}$$

It follows from the geometry of Fig. 2.12. that one such plane is the plane AE. If $\angle EAO = \theta$ is the angle between the incident ray and the normal, then from ΔAOB, we have

$$OB = 2\ OE = 2\ OA \sin\theta = (2 \sin\theta)/\lambda \tag{2.27}$$

From Eqs. (2.26) and (2.27), we get

$$(2 \sin\theta)/\lambda = n/d_{hkl}$$

or $$2d_{hkl}\sin\theta = n\lambda$$

which is the Bragg's law, n being the order of reflection. Thus we notice that if the coordinates of a reciprocal point, (nh, nk, nl), contain a common factor n, then it represents nth order reflection from the planes (hkl). It is also evident from the above geometrical construction that the Bragg's condition will be satisfied for a given wavelength λ provided the surface of radius $1/\lambda$ drawn about the point A intersects a point of the reciprocal lattice. Such a construction is called *Ewald construction.*

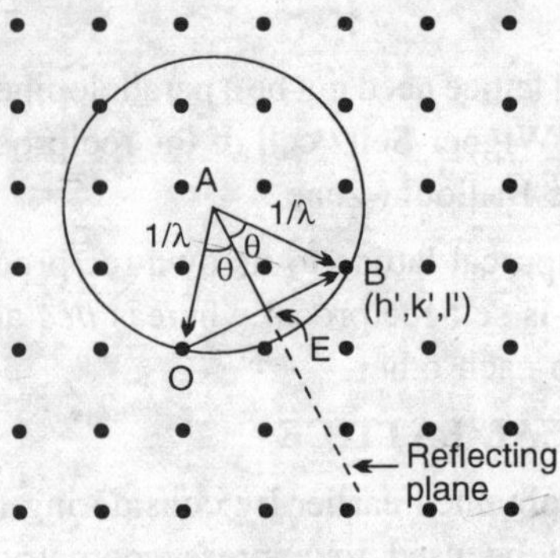

Fig. 2.12. *Ewald construction in the reciprocal lattice.*

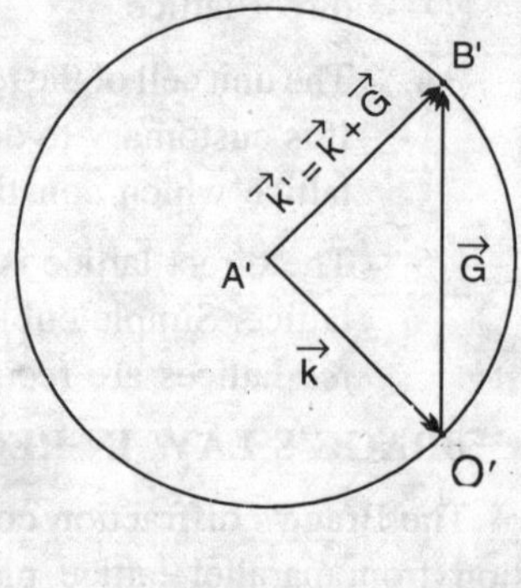

Fig. 2.13. *Magnified Ewald construction relating reciprocal lattice vector to the wave vectors of the incident and reflected radiation.*

The Bragg's law itself takes a different form in the reciprocal lattice. To obtain the modified form of the Bragg's law, we redraw the vectors $\overrightarrow{AO}$, $\overrightarrow{OB}$ and $\overrightarrow{AB}$ such that each is magnified by a constant factor of 2π. Let the new vectors be $\overrightarrow{A'O'}$, $\overrightarrow{O'B'}$ and $\overrightarrow{A'B'}$ respectively as shown in Fig. 2.13. Since

$$A'O' = 2\pi\,(AO) = 2\pi/\lambda,$$

we can represent the wave vector **k** by the vector $\overrightarrow{A'O'}$. The vector $\overrightarrow{O'B'}$ is the reciprocal vector and is written as **G**. Thus according to vector algebra, $\overrightarrow{A'B'}$ must be equal to **(k + G)**. For diffraction to occur, the point B' must be on the sphere, i.e.,

$$|\overrightarrow{A'B'}| = |\overrightarrow{A'O'}|$$

or $$(\mathbf{k} + \mathbf{G})^2 = k^2$$

or $$k^2 + 2\mathbf{k}.\mathbf{G} + G^2 = k^2$$

or $$2\mathbf{k}.\mathbf{G} + G^2 = 0 \qquad (2.28)$$

This is the vector form of Bragg's law and is used in the construction of the Brillouin zones.

The vector $\overrightarrow{A'B'}$ represents the direction of reflected or scattered beam. Denoting it by **k'**, we get

$$\mathbf{k}' = \mathbf{k} + \mathbf{G}$$

which gives

$$k'^2 = k^2 \tag{2.29}$$

and

$$\mathbf{k}' - \mathbf{k} = \Delta\mathbf{k} = \mathbf{G} \tag{2.30}$$

This indicates that the scattering does not change the magnitude of wave vector **k**; only its direction is changed. Also, the scattered wave differs from the incident wave by a reciprocal lattice vector **G**.

2.7 BRILLOUIN ZONES

It has been indicated in the Ewald construction that all the **k**-values for which the reciprocal lattice points intersect the Ewald sphere are Bragg reflected. A *Brillouin zone* is the locus of all those **k**-values in the reciprocal lattice which are Bragg reflected. We construct the Brillouin zones for a simple square lattice of side a. The primitive translation vectors of this lattice are

$$\mathbf{a} = a\hat{\mathbf{i}}; \quad \mathbf{b} = a\hat{\mathbf{j}}$$

The corresponding translation vectors of the reciprocal lattice are

$$\mathbf{a}^* = (2\pi/a)\,\hat{\mathbf{i}}\,; \mathbf{b}^* = (2\pi/a)\,\hat{\mathbf{j}}$$

Therefore, the reciprocal lattice vector is written as

$$\mathbf{G} = (2\pi/a)\,(h\hat{\mathbf{i}} + k\hat{\mathbf{j}})$$

where h and k are integers. The wave vector **k** can be expressed as

$$\mathbf{k} = k_x\hat{\mathbf{i}} + k_y\hat{\mathbf{j}}$$

From the Bragg's condition (2.28), we have

$$2\mathbf{k}.\mathbf{G} + G^2 = 0$$

or $$\frac{4\pi}{a}\,[(k_x\hat{\mathbf{i}} + k_y\hat{\mathbf{j}})\,.\,(h\hat{\mathbf{i}} + k\hat{\mathbf{j}})] + \frac{4\pi^2}{a^2}\,(h^2 + k^2) = 0$$

or $$hk_x + kk_y = -(\pi/a)\,(h^2 + k^2) \tag{2.31}$$

The **k**-values which are Bragg reflected are obtained by considering all possible combinations of h and k.

For $h = \pm 1$ and $k = 0$, $k_x = \pm\pi/a$ and k_y is arbitrary;

Also, for $h = 0$ and $k = \pm 1$, $k_y = \pm\pi/a$ and k_x is arbitrary.

These four lines, i.e., $k_x = \pm\pi/a$ and $k_y = \pm\pi/a$, are plotted in Fig. 2.14. Taking origin as shown, all the **k**-vectors originating from it and terminating on these lines will produce Bragg reflection. The square bounded by these four lines is called the *first Brillouin zone*. Thus the first zone of a square lattice of side a is a square of side $2\pi/a$. In addition to this set of lines, some other sets of lines are also possible which satisfy (2.31). For example, for $h = \pm 1$ and $k = \pm 1$, the condition (2.31) gives the following set of four lines.

$$\pm k_x \pm k_y = 2\pi/a$$

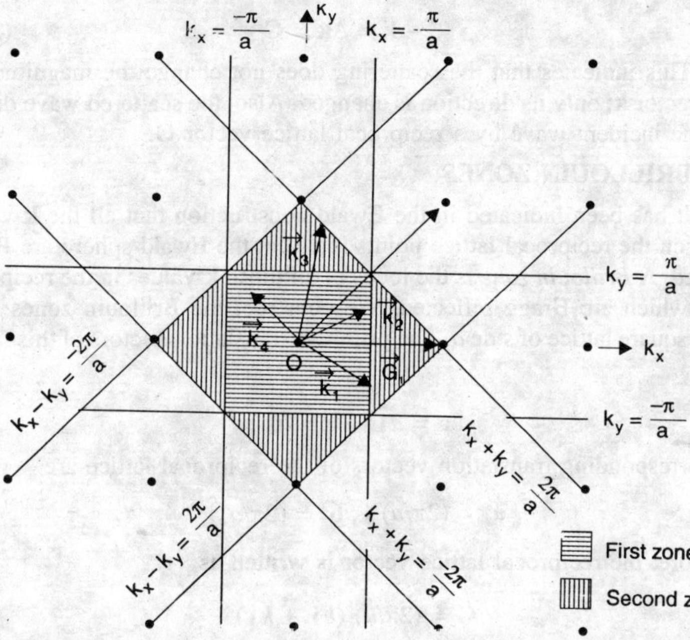

Fig. 2.14. ***Brillouin zones of a square lattice in its reciprocal lattice. The vectors*** $\mathbf{k}_1$, $\mathbf{k}_2$, ***and*** $\mathbf{k}_3$ ***are Bragg reflected whereas*** $\mathbf{k}_4$ ***is not. The vectors*** $\mathbf{k}_1$ ***and*** $\mathbf{k}_2$ ***have the same reciprocal lattice vector*** $\mathbf{G}_1$, ***while*** $\mathbf{G}_2$ ***is the reciprocal vectors of*** $\mathbf{k}_3$.

These lines are also plotted in Fig. 2.14. The additional area bounded by these four lines is the *second Brillouin zone*. Similarly the other zones can be constructed. The boundaries of the Brillouin zones represent the loci of **k**-values that are Bragg reflected and hence may be considered as the reflecting planes. The boundaries of the first zone represent the reflecting planes for the first order reflection, those of the second zone represent the reflecting planes for the second order reflection, and so on. A **k**-vector that does not terminate at a zone boundary cannot produce Bragg reflection. Thus the Brillouin zone pattern can be employed to determine the x-ray diffraction pattern of a crystal and vice versa.

The Brillouin zones for a three-dimensional cubical lattice are constructed using the generalized equation

$$hk_x + kk_y + lk_z = -(\pi/a)\,(h^2 + k^2 + l^2) \tag{2.32}$$

where a is the length of the cube edge. It is clear from Eq. (2.32) that the first zone is a cube having side equal to $2\pi/a$. The second zone is formed by adding pyramids to each face of the cube (first zone) as triangles are added to the square in two dimensions, and so on.

There is another simple method to determine Brillouin zones. We note from Fig. 2.14 that the reciprocal lattice vector **G** which satisfies Eq. (2.28) is a perpendicular bisector of the zone boundary and all the **k**-vectors lying on this boundary have the same **G** for reflection. Thus it is sufficient to consider only the allowed **G**-values and their normal bisectors to construct the Brillouin zones. The first Brillouin zone is the region bounded by the normal bisectors of the shortest possible **G**-vectors, the second zone is the region bounded by the normal bisectors of the next larger **G**-vectors, and so on. This method will be used to determine the Brillouin zones of the *bcc* and *fcc* lattices as given below.

2.7.1 Brillouin Zone of BCC Lattice

The primitive translation vectors of a *bcc* lattice are

$$\mathbf{a} = (a/2)\,(\hat{\mathbf{i}}+\hat{\mathbf{j}}-\hat{\mathbf{k}})$$

$$\mathbf{b} = (a/2)\,(-\hat{\mathbf{i}}+\hat{\mathbf{j}}+\hat{\mathbf{k}})$$

$$\mathbf{c} = (a/2)\,(\hat{\mathbf{i}}-\hat{\mathbf{j}}+\hat{\mathbf{k}})$$

The primitive translation vectors of its reciprocal lattice are (Sec. 2.4.3)

$$\mathbf{a}^* = (2\pi/a)\,(\hat{\mathbf{i}}+\hat{\mathbf{j}})$$

$$\mathbf{b}^* = (2\pi/a)\,(\hat{\mathbf{j}}+\hat{\mathbf{k}})$$

$$\mathbf{c}^* = (2\pi/a)\,(\hat{\mathbf{k}}+\hat{\mathbf{i}})$$

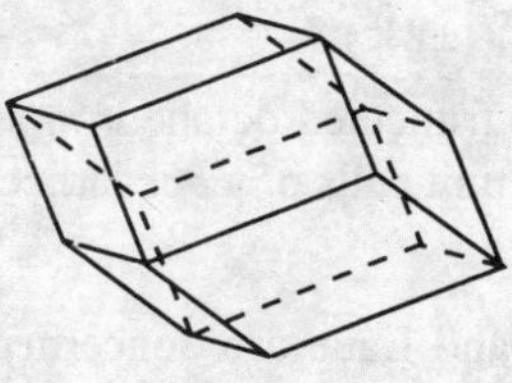

Fig. 2.15. *First Brillouin zone of a bcc lattice.*

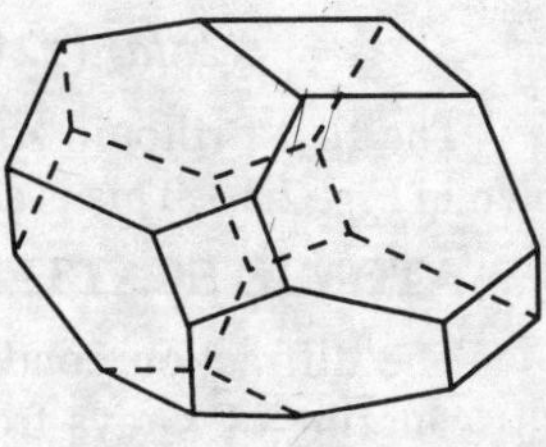

Fig. 2.16. *First Brillouin zone of an fcc lattice.*

The **G**-type reciprocal lattice vector is

$$\mathbf{G} = h\mathbf{a}^* + k\mathbf{b}^* + l\mathbf{c}^*$$

$$= (2\pi/a)\,[(h + l)\,\hat{\mathbf{i}} + (h + k)\,\hat{\mathbf{j}} + (k + l)\,\hat{\mathbf{k}}] \qquad (2.33)$$

The shortest non-zero *G*'s are the following twelve vectors

$$(2\pi/a)\,(\pm\hat{\mathbf{i}} \pm \hat{\mathbf{j}});\ (2\pi/a)\,(\pm\hat{\mathbf{j}} \pm \hat{\mathbf{k}});\ (2\pi/a)(\pm\hat{\mathbf{k}} \pm \hat{\mathbf{i}})$$

The first Brillouin zone is the region enclosed by the normal bisector planes to these twelve vectors. This zone has the shape of a regular twelve-faced solid as shown in Fig. 2.15 and is called rhombic dodecahedron.

2.7.2 Brillouin Zone of FCC Lattice

The primitive translation vectors of an *fcc* lattice are

$$\mathbf{a} = (a/2)\,(\hat{\mathbf{i}} + \hat{\mathbf{j}})$$

$$\mathbf{b} = (a/2)\,(\hat{\mathbf{j}} + \hat{\mathbf{k}})$$

$$\mathbf{c} = (a/2)\,(\hat{\mathbf{k}} + \hat{\mathbf{i}})$$

The primitive translation vectors of its reciprocal lattice are (Sec. 2.4.4)

$$\mathbf{a}^* = (2\pi/a)\,(\hat{\mathbf{i}}+\hat{\mathbf{j}}-\hat{\mathbf{k}})$$

$$\mathbf{b}^* = (2\pi/a)\,(-\hat{\mathbf{i}}+\hat{\mathbf{j}}+\hat{\mathbf{k}})$$

$$\mathbf{c}^* = (2\pi/a)\,(\hat{\mathbf{i}}-\hat{\mathbf{j}}+\hat{\mathbf{k}})$$

The **G**-type vector is

$$\mathbf{G} = h\mathbf{a}^* + k\mathbf{b}^* + l\mathbf{c}^*$$

$$= (2\pi/a)\,[(h - k + l)\,\hat{\mathbf{i}} + (h + k - l)\,\hat{\mathbf{j}} + (-h + k + l)\,\hat{\mathbf{k}}] \quad (2.34)$$

The shortest non-zero *G*'s are the following eight vectors

$$(2\pi/a)\,(\pm\hat{\mathbf{i}} \pm \hat{\mathbf{j}} \pm \hat{\mathbf{k}})$$

The boundaries of the first Brillouin zone are determined mostly by the normal bisector planes to the above eight vectors. However, the corners of the octahedron obtained in this manner are truncated by the planes which are normal bisectors to the following six reciprocal lattice vectors

$$(2\pi/a)\,(\pm2\hat{\mathbf{i}});\ (2\pi/a)\,(\pm2\hat{\mathbf{j}});\ (2\pi/a)\,(\pm2\hat{\mathbf{k}})$$

The first Brillouin zone has the shape of the truncated octahedron as shown in Fig. 2.16. This is also one of the primitive unit cells of a *bcc* lattice.

2.8 ATOMIC SCATTERING FACTOR

The diffraction conditions given by Bragg and Laue are concerned with scattering of x-rays from point scattering centres arranged on a space lattice. Since an electron is the smallest scattering centre, the diffraction conditions would ideally be applicable to a lattice in which every lattice point is occupied by an electron. This is, however, not a realistic situation. Lattice points are always occupied by atoms which may contain a number of electrons. Also, since the wavelength of x-rays used for diffraction purposes is of the order of atomic dimensions, the x-rays scattered from different portions of an atom are, in general, not in phase. Thus the amplitude of radiation scattered by a single atom is not necessarily equal to the product of the amplitude of radiation scattered by a single electron and the number of electrons (atomic number, Z) present in the atom. It is generally less than this value. The *atomic scattering factor* or *form factor*, f, describes the scattering power of a single atom in relation to the scattering power of a single electron and is given by

$$f = \frac{\text{amplitude of radiation scattered from an atom}}{\text{amplitude of radiation scattered from an electron}}$$

In general, $f < Z$. It approaches Z in the limiting case.

Another type of scattering centres in the atoms may be nuclei, but due to their weak interaction with x-rays, the scattering due to nuclei is neglected compared with that due to electrons.

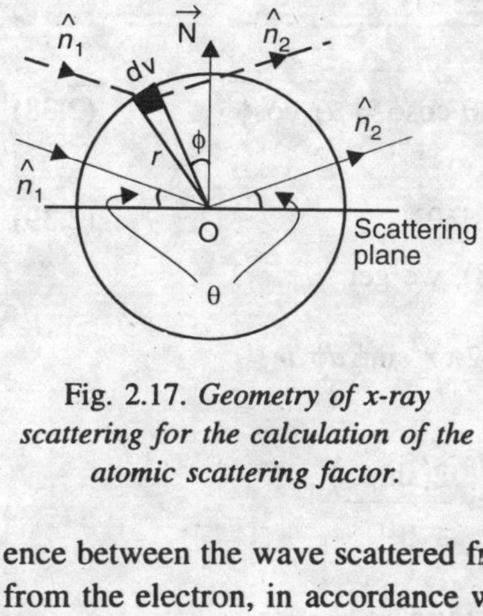

Fig. 2.17. *Geometry of x-ray scattering for the calculation of the atomic scattering factor.*

To calculate f, consider an atom containing electrons arranged in a spherically symmetric configuration around its centre which is taken as origin. Let r be the radius of the atom and $\rho(\mathbf{r})$ the charge density at a point $\mathbf{r}$. Considering a small volume element dv at $\mathbf{r}$, the charge located at $\mathbf{r}$ is $\rho(\mathbf{r})dv$. We first consider the scattering from the charge $\rho(\mathbf{r})\, dv$ and an electron located at the origin. If $\hat{\mathbf{n}}_1$ and $\hat{\mathbf{n}}_2$ represent the directions of the incident and scattered beam respectively as shown in Fig. 2.17, the phase difference between the wave scattered from the charge $\rho(\mathbf{r})\, dv$ and that scattered from the electron, in accordance with Eq. (2.2), is given by

$$\phi_r = (2\pi/\lambda)\ \mathbf{r}.\mathbf{N} \tag{2.35}$$

where $\mathbf{N}$ is the scattering normal. Let the scattering amplitude from the point electron in the direction $\hat{\mathbf{n}}_2$ be written as $Ae^{i(kx-\omega t)}$ where x is the distance covered along $\hat{\mathbf{n}}_2$ and k is the wave number. Then the scattering amplitude from the charge $\rho(\mathbf{r})dv$ in the same direction will be proportional to the magnitude of the charge and will contain the phase factor $e^{i\phi_r}$, i.e., it is of the form

$$Ae^{i(kx-\omega t)+i\phi_r}\rho(\mathbf{r})dv$$

Therefore, the ratio of the amplitude of the radiation scattered by the charge element to that scattered by a point electron at the origin is

$$df = \frac{Ae^{i(kx-\omega t)+i\phi_r}\,\rho(\mathbf{r})\,dv}{Ae^{i(kx-\omega t)}}$$

$$= e^{i\phi_r}\,\rho(\mathbf{r})dv$$

Thus the ratio of the amplitude from the whole atom to that from an electron is

$$f = \int_V \rho(\mathbf{r})\; e^{i\phi_r}\; dv \tag{2.36}$$

where V is the volume of the atom. Since electrons in the atom have spherically symmetric charge distribution, $\rho(\mathbf{r})$ is a function of r only. Using spherical polar coordinates, we obtain

$$dv = 2\pi\, r^2 \sin\phi\; d\phi\; dr \tag{2.37}$$

Also, from Sec. 2.2.2, we have

$$|\mathbf{N}| = 2\sin\theta$$

Therefore, the Eq. (2.35) becomes

$$\phi_r = (2\pi/\lambda)\, rN\cos\phi = (4\pi/\lambda)\, r\sin\theta\cos\phi = \mu r\cos\phi \tag{2.38}$$

where

$$\mu = (4\pi/\lambda)\sin\theta \tag{2.39}$$

From Eqs. (2.36), (2.37) and (2.38), we get

$$f = \int_{r=0}^{\infty}\int_{\phi=0}^{\pi} \rho(r)\, e^{i\mu r\cos\phi}\; 2\pi\, r^2 \sin\phi\; d\phi\; dr$$

$$\int_0^{\pi} e^{i\mu r\cos\phi}\sin\phi\; d\phi = \frac{2(\sin\mu r)}{\mu r}$$

$$\therefore \qquad f = \int_0^{\infty} 4\pi r^2 \rho(\mathbf{r})\frac{\sin\mu r}{\mu r}\, dr \tag{2.40}$$

This is the general expression for atomic scattering factor. A further evaluation needs information about the charge distribution. This type of information may be obtained from a Hartee approximation, or from the statistical model of Thomas and Fermi if atoms contain a large number of electrons (beyond rubidium). These models, however, are strictly applicable to free atoms only. Nevertheless, the results obtained by the application of these models are fairly accurate and match with the experimental values of x-ray intensities. For example, the variation of f with $(\sin\theta)/\lambda$ for magnesium (Z = 12) is shown in Fig.2.18. It is clear that $f \to Z$ as $\theta \to 0$. The same result can also be obtained from Eq. (2.40), since for $\theta \to 0$, $\mu \to 0$ and $(\sin \mu r)/\mu r \to 1$, and we get

$$f = \int_0^\infty 4\pi r^2 \rho(\mathbf{r})\, dr$$

The integrand represents the charge inside a spherical shell of radius r and thickness dr. Hence the integral gives the total electronic charge stored inside the atom, i.e., Z.

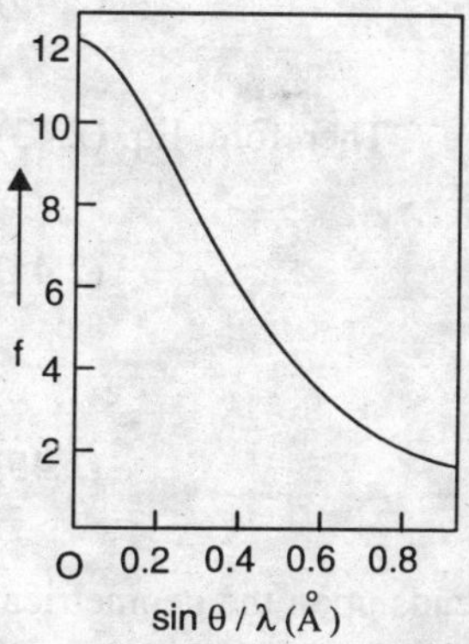

Fig. 2.18. *Atomic scattering factor versus* $\frac{\sin\theta}{\lambda(\text{Å})}$ *for magnesium.*

2.9 GEOMETRICAL STRUCTURE FACTOR

The intensity of an x-ray beam diffracted from a crystal not only depends upon the atomic scattering factors of the various atoms involved but also on the contents of the unit cell, i.e., on the number, type and distribution of atoms within the cell. The x-rays scattered from different atoms of the unit cell may or may not be in phase with each other. It is, therefore, important to know the effect of various atoms present in the unit cell on the total scattering amplitude in a given direction. The total scattering amplitude $F(h'k'l')$ for the reflection $(h'k'l')$ is defined as the ratio of the amplitude of radiation scattered by the entire unit cell to the amplitude of radiation scattered by a single point electron placed at origin for the same wavelength. It is given by

$$F(h'k'l') = \sum_j f_j e^{i\phi_j}$$

$$= \sum_j f_j e^{i(2\pi/\lambda)\left(\mathbf{r}_j . \mathbf{N}\right)} \qquad (2.41)$$

where f_j is the atomic scattering factor for the jth atom, ϕ_j is the phase difference between the radiation scattered from the jth atom of the unit cell and that scattered from the electron placed at the origin. The expression for ϕ_j follows from Eq. (2.2), $\mathbf{r}_j$ is the position of jth atom and $\mathbf{N}$ is the scattering normal. Also, the summation in Fig. (2.41) extends over all the atoms present in the unit cell. If (u_j, v_j, w_j) represent the coordinates of jth atom, we can write

$$\mathbf{r}_j = u_j\,\mathbf{a} + v_j\mathbf{b} + w_j\mathbf{c} \tag{2.42}$$

From Eqs. (2.3), $\mathbf{a}.\mathbf{N} = h'\lambda$, etc.

$$\therefore \qquad \mathbf{r_j}.\mathbf{N} = \lambda\,(u_j h' + v_j k' + w_j l')$$

and Eq. (2.41) becomes

$$F\,(h'k'l') = \sum_j f_j e^{i2\pi\left(u_j h' + v_j k' + w_j l'\right)} \tag{2.43}$$

For identical atoms, all the f_j's have the same value f. Therefore, Eq. (2.43) takes a simple form

$$F\,(h'k'l') = f\,\mathbf{S} \tag{2.44}$$

where

$$\mathbf{S} = \sum_{\mathrm{j}} e^{2\pi i\left(u_j h' + v_j k' + w_j l'\right)} \tag{2.45}$$

is called the *geometrical structure factor* as it depends upon the geometrical arrangement of atoms within the unit cell. Equation (2.44) defines the structure factor as the ratio of the total scattering amplitude to the atomic scattering factor. For dissimilar atoms, Eq. (2.43) should be used instead of Eq. (2.44).

Now since the intensity of a radiation is proportional to the square of its amplitude, the intensity of diffracted beam may be written, by using Eq. (2.43), as

$$I = |F|^2 = F^*F$$

$$= \left[\sum_j f_j \cos 2\pi\left(u_j h' + v_j k' + w_j l'\right)\right]^2 + \left[\sum_j f_j \sin 2\pi\left(u_j h' + v_j k' + w_j l'\right)\right]^2 \tag{2.46}$$

where F^* is the complex conjugate of F. It is to be noted that, whereas the space lattice of a crystal structure may be determined from the position of diffraction lines in the x-ray diffraction pattern, the determination of basis is much more complicated and requires the knowledge of intensity of various diffraction lines. Thus the concept of structure factor carries a special importance in the context of structure determination. The structure factors for some simple crystals are calculated below and the intensity of various orders of reflections associated with these structures is discussed.

(i) Simple Cubic Crystals

The effective number of atoms in a unit cell of simple cubic structure is one. Assuming that it lies at the origin, the structure factor given by Eq. (2.45) comes out to be unity. The diffraction amplitude, from Eq. (2.44), becomes

$$F\,(h'k'l') = f$$

Thus all the diffraction lines predicted by the Bragg's law would appear in the diffraction pattern provided the value of *f* is large enough to produce peaks of observable intensity.

(ii) Body-Centred Cubic Crystals

The effective number of atoms in a *bcc* unit cell is two; one occupies a corner position and the other occupies the centre of the cube. If the coordinates of corner atom be arbitrarily taken as (0,0,0), then the coordinates of the other atom become (½,½,½). Since both the atoms are identical, Eq. (2.44) gives the value of F as

$$F(h'k'l') = f\sum_j e^{2\pi i\left(u_j h' + v_j k' + w_j l'\right)}$$

$$= f\,[1 + e^{i\pi\,(h' + k' + l')}] \tag{2.47}$$

The expression within the square brackets represents the structure tactor for *bcc* crystal. Here it has been assumed that only one of the eight corners of the cube is occupied and has the coordinates (0,0,0). The validity of this assumption can be verified by considering all the corner positions and using the fact that the contribution of each corner atom is 1/8. This yields the same structure factor as included in Eq. (2.47).

We also find from Eq. (2.47) that the structure factor becomes zero for odd values of $(h' + k' + l')$, since $e^{i\pi n}$ equals -1 if n is odd. For even values of $(h' + k' + l')$, $F(h'k'l')$ equals $2f$ and, from Eq. (2.46), the intensity becomes proportional to $4f^2$. Thus in a *bcc* structure, reflections like (100), (111), (210), etc. are missing, whereas the diffraction lines corresponding to (110), (200), (222), etc. reflections are present. It is to be noted that the presence or absence of a reflection is considered only in terms of the first order reflection. This is because the Miller indices of the planes $(h'k'l')$ used in Eq. (2.43) may have a common factor n; thus we determine reflections from the planes $(nh\;nk\;nl)$. As described earlier, the appropriate Bragg's law applicable to such a case is

$$2d_{h'k'l'}\sin\theta = \lambda \tag{2.48}$$

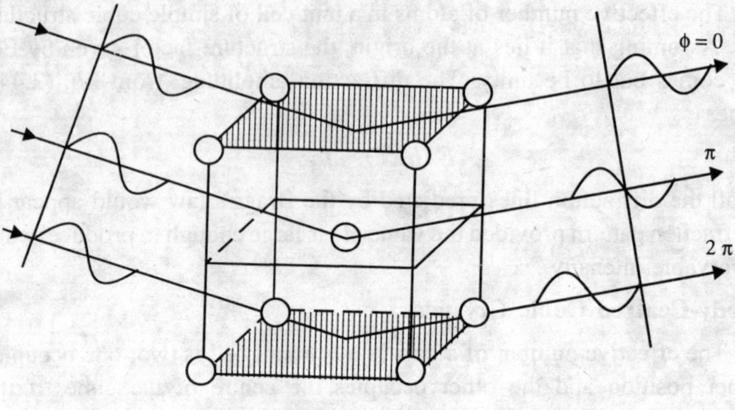

(*a*)

Fig. 2.19. *(a) First order reflections from (100) planes (shaded) in a bcc crystal. The reflections from any two neighbouring planes exactly cancel each other.*

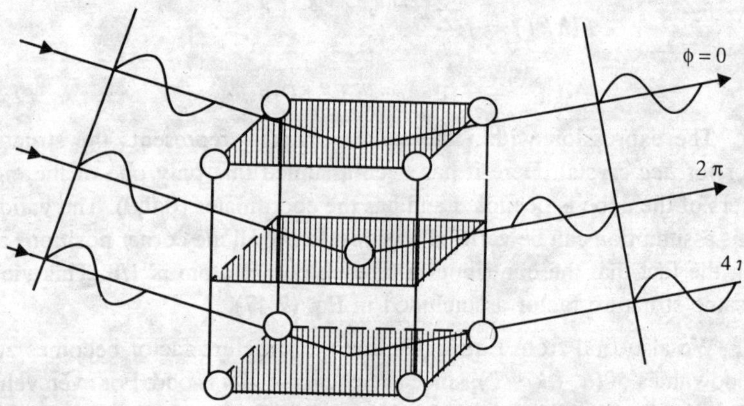

(*b*)

Fig. 2.19. *(b) Second order reflections from (100) planes or first order reflections from (200) planes (shaded). The path difference between the rays reflected from any two neighbouring planes is* λ *which means constructive superposition of waves.*

We have noted that the first order reflection from (100) planes in a *bcc* crystal is absent, whereas a similar reflection from (200) planes is present. It is also easy to find that the second order reflection from (100) planes is present and it is this reflection which appears at the position of the first order reflection from (200) planes. This can be understood from Fig. 2.19a. The Eq. (2.48) indicates that the reflections from two neighbouring

$(h'k'l')$ planes have a path difference of λ or the phase difference of 2π. Thus considering reflections from (100) planes as shown in Fig. 2.19a, we find that the waves reflected from the top and bottom surfaces of the cube differ in phase by 2π. Since in *bcc* crystals, there exists a central plane which is exactly identical to the (100) planes, a wave reflected from this plane must have a phase difference of π relative to its neighbouring (100) planes. Thus the diffracted beams from a regular (100) plane and a body centre plane interfere destructively in pairs causing absence of (100) reflection. Similarly, by using Eq. (2.48), it can be shown that the first order reflections from any two neighbouring (200) planes must differ in phase by 2π radians and hence undergo constructive interference causing the occurrence of these reflections in *bcc* crystals. The presence of second order reflections from (100) planes can be shown by using Eq. (2.1) The second order reflection from two neighbouring (100) planes has a phase difference of 4π radians, which means that the reflection from the middle plane would differ from the reflections from top and bottom planes by a phase difference of 2π. The situation is exactly identical to that shown in Fig. 2.19b, thus indicating that the second order reflection from (100) planes is present. As described earlier, it overlaps with the first order reflection from (200) planes.

(iii) Face-Centred Cubic Crystal

An *fcc* unit cell has four indentical atoms. One of these atoms is contributed by corners and may arbitrarily be assigned coordinates (0,0,0), whereas the other three are contributed by face centres and have the coordinates (½,0,½), (½,½,0) and (0,½,½). From Eq. (2.39), the diffraction amplitude becomes

$$F\,(h'k'l') = f\,[1 + e^{\pi i(h'+l')} + e^{\pi i(h'+k')} + e^{\pi i(k'+l')}]$$

where the expression within the square brackets is the structure factor for *fcc* crystals. It is obvious that the structure factor is non-zero only if h, k and l are all even or all odd and has a value equal to 4. Thus the diffraction amplitude becomes 4f and the intensity becomes proportional to $16f^2$. The structure factor vanishes for all other odd-even combinations of h, k and l. Hence reflections of the type (111), (200), (220), etc. are present, whereas those of the type (100), (110), (211), etc. are absent for an *fcc* crystal.

The conclusions drawn above regarding allowed reflections for *sc*, *bcc* and *fcc* crystals are summarized in Table 2.1 and are called extinction rules. The extinction rules for *dc* structure are also included.

TABLE 2.1 Extinction rules for cubic crystals.

Crystal	Reflections allowed for
SC	all possible values of h, k and l
BCC	even values of $(h + k + l)$
FCC	all odd or all even values of h, k and l
DC	all odd h, k and l, or all even h, k and l with $(h + k + l)$ divisible by 4

The ratios of $(h^2 + k^2 + l^2)$ values for allowed reflections from cubic crystals as obtained from the extinction rules are given as follows :

SC : 1 : 2 : 3 : 4 : 5 : 6 : 8.......

BCC : 2 : 4 : 6 : 8 : 10 : 12 : 14......

or 1 : 2 : 3 : 4 : 5 : 6 : 7 : ....

FCC : 3 : 4 : 8 : 11 : 12 : 16 : 19 .....

DC : 3 : 8 : 11 : 16 : 19 ......

A comparison of these ratios with the observed ratios of $\sin^2\theta$ values is made to identify the cubic crystal structures.

SOLVED EXAMPLES

Example 2.1. An x-ray beam of wavelength 0.71Å is diffracted by a cubic KCl crystal of density 1.99×10^3 kgm^{-3}. Calculate the interplanar spacing for (200) planes and the glancing angle for the second order reflection from these planes. The molecular weight of KCl is 74.6 amu and the Avogadro's number is 6.023×10^{26} kg^{-1} mole^{-1}.

Solution. For cubic crystals, we have

$$a^3 = \frac{n' M}{N\rho}$$

where a is the lattice constant, n' is the number of molecules in a unit cell, M is the molecular weight, N is the Avogadro's number and ρ is the density.

KCl has the same structure as NaCl

$$\therefore \qquad n' = 4$$

$$\therefore \qquad a^3 = \frac{4 \times 74.6}{6.023 \times 10^{26} \times 1.99 \times 10^3} = 0.249 \times 10^{-27} \text{ m}^3$$

or $$a = 6.29 \times 10^{-10} \text{ m}$$

The interplanar spacing for (200) planes is

$$d_{200} = \frac{a}{(4+0+0)^{1/2}} = \frac{6.29 \times 10^{-10}}{2} = 3.145\text{Å}$$

From Bragg's law, we have

$$2d \sin\theta = n\lambda$$

For second order reflection, $n = 2$

$$\therefore \qquad \sin\theta = \lambda/d = 0.71/3.145 = 0.4610$$

or $$\theta = 27.5^\circ$$

Example 2.2. A powder camera of radius 57.3 mm is used to obtain diffraction pattern of gold (*fcc*) having a lattice parameter of 4.08Å. The monochromatic Mo-K_α radiation of wavelength 0.71Å is used. Determine the first four *S*-values.

Solution. We have,

$$R = 57.3 \text{ mm}$$
$$\lambda = 0.71 \text{ Å}$$
$$a = 4.08 \text{ Å}$$

The $(h^2 + k^2 + l^2)$ values for the first four reflections from an *fcc* crystal are 3, 4, 8 and 11.

The Bragg's law for first order reflection is

$$2d \sin\theta = \lambda \tag{2.49}$$

Also, for cubic crystals, we have

$$d_{hkl} = \frac{a}{\left(h^2 + k^2 + l^2\right)^{1/2}} \tag{2.50}$$

From Eqs. (2.49) and (2.50), we obtain

$$\sin^2\theta = \frac{\lambda^2}{4a^2} (h^2 + k^2 + l^2) \tag{2.51}$$

For $h^2 + k^2 + l^2 = 3$, it gives

$$\sin^2\theta_1 = \frac{3(0.71)^2}{4(4.08)^2} = 0.0227$$

or $$\theta_1 = 8.67^\circ$$

Similarly, for $h^2 + k^2 + l^2 = 4$, 8 and 11, the angles θ_2, θ_3 and θ_4 are obtained as 10.02°, 14.25° and 16.78° respectively.

Since radius of the camera is 57.3 mm, Eq. (2.9) gives

$$S \text{ (mm)} = 4\theta \text{ (degrees)}$$

$$\therefore \quad S_1 = 4\theta_1 = 34.68^\circ$$

Likewise, $S_2 = 40.08^\circ$, $S_3 = 57.0^\circ$ and $S_4 = 67.12^\circ$

Example 2.3. In a powder diffraction experiment using Cu-K_α radiation of wavelength 1.54Å, the first five lines are observed from a monoatomic cubic crystal when the angle 2θ is 38.0, 44.2, 64.4, 77.2 and 81.4 degrees. Determine the crystal structure and the lattice parameter.

Solution. From Eq. (2.51), we have

$$\sin^2\theta = \frac{\lambda^2}{4a^2}(h^2 + k^2 + l^2)$$

or $$\sin^2\theta \propto (h^2 + k^2 + l^2)$$

The $\sin^2\theta$ values calculated from the given data are tabulated below :

Line	2θ (degrees)	θ (degrees)	$\sin^2\theta$	Ratio (approx.)
1	38.0	19.0	0.1060	3
2	44.2	22.1	0.1415	4
3	64.4	32.2	0.2840	8
4	77.2	38.6	0.3892	11
5	81.4	40.7	0.4252	12

Since, within experimental errors, $\sin^2\theta$ values for the first five lines are in the ratio 3 : 4 : 8 : 11 : 12, the structure is *fcc*.

From Eq. (2.51), we have

$$a^2 = \frac{\lambda^2}{4\sin^2\theta}(h^2 + k^2 + l^2)$$

For $\sin^2\theta = 0.1060$, $h^2+k^2+l^2 = 3$

$$\therefore \quad a = \left[\frac{3(1.54)^2}{4(0.1060)^2}\right]^{1/2} = 4.10 \text{ Å}$$

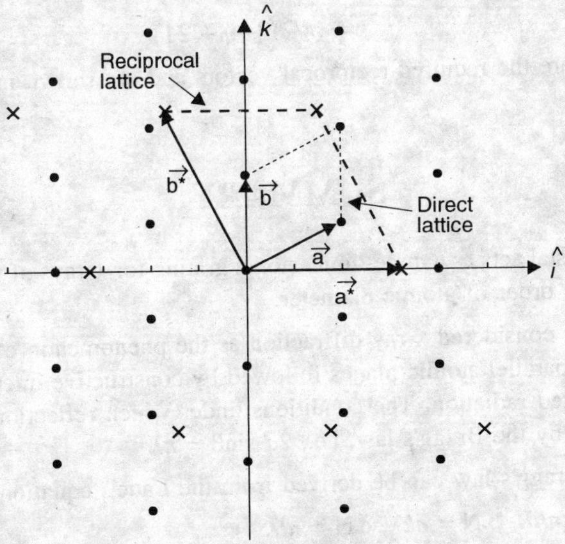

Fig. 2.20. *Two-dimensional direct and reciprocal lattices.*

Example 2.4. The primitive translation vectors of a two-dimensional lattice are

$$\mathbf{a} = 2\hat{\mathbf{i}} + \hat{\mathbf{j}},\ \mathbf{b} = 2\hat{\mathbf{j}}$$

Determine the primitive translation vectors of its reciprocal lattice.

Solution. We assume that the third translation vector **c** of the given lattice lies along the *z*-axis and is of unit magnitude, i.e.,

$$\mathbf{c} = \hat{\mathbf{k}}$$

From Eqs. (2.18), the reciprocal lattice vectors are given by

$$\mathbf{a^*} = 2\pi\ \frac{\mathbf{b \times c}}{\mathbf{a.b \times c}},\ \mathbf{b^*} = 2\pi\ \frac{\mathbf{c \times a}}{\mathbf{a.b \times c}}$$

It is obvious that vectors **a*** and **b*** lie in the plane of **a** and **b**.

$$\begin{aligned} \mathbf{a.(b \times c)} &= (2\hat{\mathbf{i}} + \hat{\mathbf{j}}) \, . \, (2\hat{\mathbf{j}} \times \hat{\mathbf{k}}) \\ &= 2(2\hat{\mathbf{i}} + \hat{\mathbf{j}}) \, . \, \hat{\mathbf{k}} \\ &= 2(2 + 0) = 4 \end{aligned}$$

$$\therefore \quad \mathbf{a^*} = (2\pi/4)\,(2\hat{\mathbf{j}} \times \hat{\mathbf{k}}) = \pi\hat{\mathbf{i}}$$

$$\mathbf{b}^* = (2\pi/4)\,[\hat{\mathbf{k}} \times (2\hat{\mathbf{i}} + \hat{\mathbf{j}})]$$

$$= (\pi/2)\,[-\hat{\mathbf{i}} + 2\hat{\mathbf{j}}]$$

a* and **b*** are the required reciprocal vectors and are shown in Fig. 2.20.

SUMMARY

1. A crystal acts as a three-dimensional grating for x-rays of wavelength of the order of atomic diameter.
2. Bragg considered x-ray diffraction as the phenomenon of reflection from parallel atomic planes followed by constructive interference of reflected radiation. The conditions under which reflection occurs is given by the Bragg's law, i.e., $2d \sin\theta = n\lambda$
3. The Bragg's law can be derived from the Laue's equations given as $\mathbf{a.N} = nh\lambda$, $\mathbf{b.N} = nk\lambda$, $\mathbf{c.N} = nl\lambda$
4. X-ray diffraction is utilized to determine the structure of solids and for the study of x-ray spectroscopy. The position of diffraction lines determines the space lattice and their intensity determines the basis.
5. Every direct lattice in real space is associated with a reciprocal lattice in **k**-space or Fourier space. A reciprocal lattice point corresponds to a particular set of parallel planes of the direct lattice. The distance of a reciprocal lattice point from an arbitrary origin is inversely proportional to the interplanar spacing of the corresponding parallel planes of the normal lattice.
6. The fundamental translation vectors **a**, **b**, and **c** of direct lattice and **a***, **b*** and **c*** of reciprocal lattice are mutually related as

$$\mathbf{a}^* = \frac{\mathbf{b}\times\mathbf{c}}{\mathbf{a.b}\times\mathbf{c}}, \quad \mathbf{b}^* = \frac{\mathbf{c}\times\mathbf{a}}{\mathbf{a.b}\times\mathbf{c}}, \quad \mathbf{c}^* = \frac{\mathbf{a}\times\mathbf{b}}{\mathbf{a.b}\times\mathbf{c}}$$

or $$\mathbf{a}^* = 2\pi\frac{\mathbf{b}\times\mathbf{c}}{\mathbf{a.b}\times\mathbf{c}}, \quad \mathbf{b}^* = 2\pi\frac{\mathbf{c}\times\mathbf{a}}{\mathbf{a.b}\times\mathbf{c}}, \quad \mathbf{c}^* = 2\pi\frac{\mathbf{a}\times\mathbf{b}}{\mathbf{a.b}\times\mathbf{c}}$$

7. A reciprocal lattice vector is expressed as

$$\mathbf{G} = h\mathbf{a}^* + k\mathbf{b}^* + l\mathbf{c}^*$$

where h, k and l are integers or zero. A reciprocal lattice point (nh, nk, nl) for which Bragg reflection occurs corresponds to nth order reflection from (hkl) planes.

8. The Bragg's law is also expressed as

$$2\mathbf{k}.\ \mathbf{G} + G^2 = 0;\ \Delta\mathbf{k} = \mathbf{G}$$

9. A Brillouin zone is the locus of all those **k**-values in the reciprocal lattice which are Bragg reflected.
10. The atomic scattering factor gives the scattering power of an atom relative to a single electron and is given by

$$f = \int_0^\infty 4\pi r^2 \rho(\mathbf{r}) \frac{\sin \mu r}{\mu r} dr$$

where $\mu = (4\pi/\lambda) \sin\theta$

11. The total scattering amplitude for a particular direction is the ratio of the amplitude of radiation scattered by the entire unit cell to that scattered by a single electron. It is expressed as

$$F(hkl) = \sum_j f_j e^{2\pi i (u_j h + v_j k + w_j l)}$$

12. The geometrical structure factor is given by

$$\mathbf{S} = \sum_j e^{2\pi i (u_j h + v_j k + w_j l)}$$

It indicates the presence or absence of a particular reflection in the diffraction pattern.

VERY SHORT QUESTIONS

1. What are x-rays ?
2. What is Bragg's law ?
3. Why zeorth order diffraction is not considered in x-ray diffraction ?
4. Write the Laue's equations for x-ray diffraction.
5. Define a reciprocal lattice.
6. Give the dimensions of translation vectors of a direct lattice and its reciprocal lattice.
7. Write the Bragg's law in vector form and give the meaning of each term.
8. What is Brillouin zone?
9. Define atomic form factor.
10. Define the geometrical structure factor.
11. What types of diffraction patterns are obtained for crystalline and amorphous solids?
12. Give extinction rules for allowed reflections for *bcc* crystals.
13. For some crystals the Bragg's diffraction condition is satisfied but x-ray diffraction line is not observed. Explain.
14. Give the first four values of $(h^2 + k^2 + l^2)$ for which reflections are allowed for *fcc* crystals.

15. What is the structure factor for simple cubic crystal?
16. For what values of *h*, *k* and *l* reflections are allowed in simple cubic crystals?

SHORT QUESTIONS

1. Why cannot ordinary optical grating diffract x-rays?
2. What is the optimum order of x-ray wavelength used to observe the diffraction effects? What happens if the wavelength deviates too much from this value?
3. Why are imaginary planes also taken into account while using Bragg's law for determination of crystal structure?
4. Prove that the *m*th order reflection from (*hkl*) planes overlap with the first order reflection from (*nh nk nl*) planes.
5. Obtain the Bragg's law from the Laue's equations.
6. Why only first order diffraction is usually considered while applying the Bragg's law for the crystal structure determination.
7. What are the basic principles of the Laue's method, the rotating crystal method and the powder method of x-ray diffraction?
8. How can the Laue's method be employed to determine the symmetry of a crystal?
9. Describe the rotating crystal method to observe x-ray diffraction of any material. What additional information do you get as compared to the Laue method?
10. What are layer lines? How are they produced in the rotating crystal method?
11. Describe the procedure for finding the *d*-values of reflecting planes in a powder diffraction method.
12. State the properties of a reciprocal lattice. How is a reciprocal lattice constructed from a direct lattice?
13. Distinguish between reciprocal lattice and direct lattice. How can you observe reciprocal lattice of a crystal experimentally?
14. Prove that *fcc* lattice is reciprocal to *bcc* lattice.
15. Show that a simple cubic lattice is self-reciprocal but with different cell dimensions.
16. Find the reciprocal lattice to a *fcc* lattice.
17. Obtain the vector form of Bragg's law using the concept of reciprocal lattice.
18. Discuss how the concept of reciprocal lattice helps in the Ewald construction and determination of crystal structure.
19. What is Ewald construction? How does it help to interpret x-ray diffraction photographs?

20. What are Brillouin zones? Discuss the construction of the first three Brillouin zones for a square lattice.
21. Obtain the structure factor for *fcc* crystal.
22. Derive an expression for the scattering amplitude in terms of geometrical structure factor for *fcc* crystals. Find the values of *h*, *k* and *l* for allowed reflections.
23. Calculate the geometrical structure factor for the *bcc* structure and explain the fact that the (100) reflection line vanishes for metallic sodium but not for CsCl, both having the *bcc* structure.
24. Calculate the geometrical structure factor for NaCl structure. Will you get the same diffraction pattern for KCl? Explain.
25. The first order reflections from (100) planes for a *bcc* crystal are absent while reflections of the same order from (200) planes are present. Explain.

LONG QUESTIONS

1. Obtain Laue's equations for x-ray diffraction by crystals. Show that these are consistent with the Bragg's law.
2. Describe the principle of Laue's diffraction method. Explain the origin of Laue's spots. What is the utility of Laue's diffraction pattern?
3. Describe the rotating crystal method for diffraction of x-rays. How do layer lines form?
4. Describe the powder method for x-ray diffraction. Discuss the formation of diffraction pattern on the photographic film.
5. What is the reciprocal lattice and why is it named so? Derive the relationships for the primitive translation vectors of the reciprocal lattice in terms of those of the direct lattice.
6. What are Brillouin zones? Determine the reciprocal lattice vectors, which define the Brillouin zones of *bcc* and *fcc* lattices.
7. What is atomic scattering factor? Derive the general expression for the atomic scattering factor using spherical polar coordinates.
8. Define the geometrical structure factor. How is it related to the atomic scattering factor? Write the structure factor for *bcc* crystal and account for the missing reflections for this crystal.
9. Define a Brillouin zone. If the ratio of the length and the width of a two dimensional rectangular lattice is 3, what kind of first Brillouin zone will you expect? Explain your answer with the help of diagrams.

PROBLEMS

1. The Bragg's angle for (220) reflection from nickel (*fcc*) is 38.2° when x-rays of wavelength 1.54Å are employed in a diffraction experiment. Determine the lattice parameter of nickel. (3.52 Å)

2. In an x-ray diffraction experiment using Cu-K_α radiation of wavelength 1.54 Å, the first reflection from an *fcc* crystal is observed when 2θ is 84°. Determine the indices of this reflection and the corresponding interplanar spacing. Show that only one more reflection is possible. Determine the indices of that reflection and the corresponding interplanar spacing. ({111}, 1.15Å; {200}, 0.996 Å)
3. A crystal reflects monochromatic x-rays strongly when the Bragg's glancing angle for a first order reflection is 15°. What are the glancing angles for the second and third order reflections of the same type ? (31.2°, 50.9°)
4. Copper (*fcc*) has a lattice parameter of 3.61 Å. The first order Bragg reflection from (111) planes appears at an angle of 21.7°. Determine the wavelength of x-rays used. (1.54 Å)
5. Show that the error in determining lattice parameters decreases with increase in diffraction angle.
6. A powder pattern is obtained from an *fcc* crystal having lattice parameter of 3.52 Å using x-rays of wavelength 1.79 Å. Determine the lowest and the highest reflections possible. ({111} and {222})
7. The powder pattern of a cubic crystal gives the first three *S*-values as 56.8, 94.4 and 112.0 mm. The radius of the camera is 57.3 mm and the wavelength of monochromatic radiation used is 1.54 Å. Determine the crystal structure and the lattice parameter of the material. (DC, 5.44Å)
8. The primitive translation vectors of a hexagonal space lattice may be taken as

$$\mathbf{a} = (a/2)\hat{\mathbf{i}} + (\sqrt{3}a/2)\hat{\mathbf{j}},\ \mathbf{b} = (-a/2)\hat{\mathbf{i}} + (\sqrt{3}a/2)\hat{\mathbf{j}} \text{ and } \mathbf{c} = c\hat{\mathbf{k}}$$

Determine the primitive translation vectors of the reciprocal lattice. Show that the lattice is its own reciprocal but with a rotation of axes.

$$\left(\frac{2\pi}{a}\hat{\mathbf{i}} + \frac{2\pi}{\sqrt{3}a}\hat{\mathbf{j}}, -\frac{2\pi}{a}\hat{\mathbf{i}} + \frac{2\pi}{\sqrt{3}a}\hat{\mathbf{j}}, \frac{2\pi}{c}\hat{\mathbf{k}}\right)$$

9. Find out reciprocal lattice vectors for a space lattice defined by the following primitive translation vectors :

$$\mathbf{a} = 5\hat{\mathbf{i}} + 5\hat{\mathbf{j}} - 5\hat{\mathbf{k}}, \mathbf{b} = -5\hat{\mathbf{i}} + 5\hat{\mathbf{j}} + 5\hat{\mathbf{k}}, \mathbf{c} = 5\hat{\mathbf{i}} - 5\hat{\mathbf{j}} + 5\hat{\mathbf{k}}$$

where $\hat{\mathbf{i}}$, $\hat{\mathbf{j}}$ and $\hat{\mathbf{k}}$ are the unit vectors along x, y and z axes. Also find out the volume of the primitive cell,

$$\left((\pi/5)\left(\hat{\mathbf{i}}+\hat{\mathbf{j}}\right), (\pi/5), (\pi/5)\left(\hat{\mathbf{j}}+\hat{\mathbf{k}}\right), (\pi/5)\left(\hat{\mathbf{k}}+\hat{\mathbf{i}}\right)\right)$$

10. Two dimensional lattice has the basis vectors

$$\mathbf{a} = 2\hat{\mathbf{x}}, \mathbf{b} = \hat{\mathbf{x}} + 2\hat{\mathbf{y}}$$

Find the reciprocal lattice vectors. $\left(\pi(\hat{\mathbf{x}} - \hat{\mathbf{y}}/2), \pi\hat{\mathbf{y}}\right)$

CHAPTER - III

BONDING IN SOLIDS

3.1 INTRODUCTION

This chapter deals with the general nature of the forces which bind the atoms together in a crystal. These forces are classified on the basis of the nature of electrostatic interaction between the neighbouring atoms.The crystals are categorised depending on the types of forces existing amongst the atoms. The description of the classical theory of cohesive energy of ionic crystals and inert gases is also given.

3.2 INTERATOMIC FORCES AND TYPES OF BONDING

The interatomic forces exist amongst the atoms of a crystal and are responsible for holding the atoms together to form solid strucutres. The very existence of solids makes us to draw the following conclusions :

(a) Some attractive forces must be present between the atoms and molecules of a solid which hold them together.

(b) Some repulsive forces must also be present between the atoms or molecules since a large external pressure is needed to compress a solid to any appreciable extent.

In order to understand the nature of these forces, we consider a pair of atoms which is capable of forming a stable chemical bond in the solid state. When separated by a large distance from each other, each atom of the pair may be considered to be free from the influence of the other atoms and hence the potential energy, U, of the system may be arbitrarily taken as zero. As the distance between the atoms is decreased, they start interacting with each other resulting in a change of potential energy of the system. The atoms exert the following two types of forces on one another :

(i) The attractive forces arising from the interaction of the negative electron cloud of one atom with the positive nuclear charge on the other. Its magnitude is proportional to some power of the interatomic distance r.

(ii) The repulsive forces which come into existence when the distance between the atoms is decreased to such an extent that their electronic clouds start overlapping, thus violating the Pauli's exclusion principle. The repulsion between the positively charged nuclei also contributes to the repulsive forces. The magnitude of the total repulsive force is also proportional to some power of r.

Since the attractive forces decrease the potential energy of the system and the repulsive forces increase it, the net energy of the system is equal to the algebraic sum of these two energies and is written as

$$U = U_{att} + U_{rep}$$
$$= -A/r^m + B/r^n \tag{3.1}$$

where A, B, m and n are constants which depend upon the nature of the participating atoms; A and B are known as attraction and repulsion constants respectively. Equation (3.1) indicates that the magnitudes of both attractive and repulsive energies increase with decrease in interatomic distance. Generally, $n > m$ which indicates that the increase in repulsive energy is faster than the increase in attractive energy particularly for very small values of interatomic distance. The repulsive forces are, therefore, known as *short range forces*. This means that the repulsive interaction between the nuclei becomes appreciable only for very small distances.

The variations of the attractive energy, repulsive energy and total energy versus interatomic distance are shown in Fig. 3.1a.The total energy first decreases gradually as the atoms constituting the pair approach each other, attains a maximum value for the interatomic distance r equal to r_o, and then increases rapidly as the value of r is decreased further. The interatomic distance r_o at which the energy of the system becomes minimum is known as the *equilibrium distance* and signifies the formation of a stable chemical bond. At this distance, the system is in the most stable state and energy is required to displace the atoms in either direction.

Differentiating Eq. (3.1) with respect to r, we get,

$$F = -dU/dr = -mA/r^{m+1} + nB/r^{n+1} \tag{3.2}$$

This gives the total force between the two atoms placed at a distance r from each other. The first term on the right hand side represents the attractive force and the second one represents the repulsive force. The variations of the attractive force, repulsive force and total force with interatomic distance are shown in Fig. 3.1b. At equilibrium distance r_o, the attractive force must be equal and opposite to the repulsive force and hence the total force F is zero. The potential energy of the system corresponding to this distance is, therefore, minimum. Thus from Eq. (3.2), for $r = r_o$, we get

$$mA/r_o^{\ m+1} = nB/r_o^{\ n+1}$$

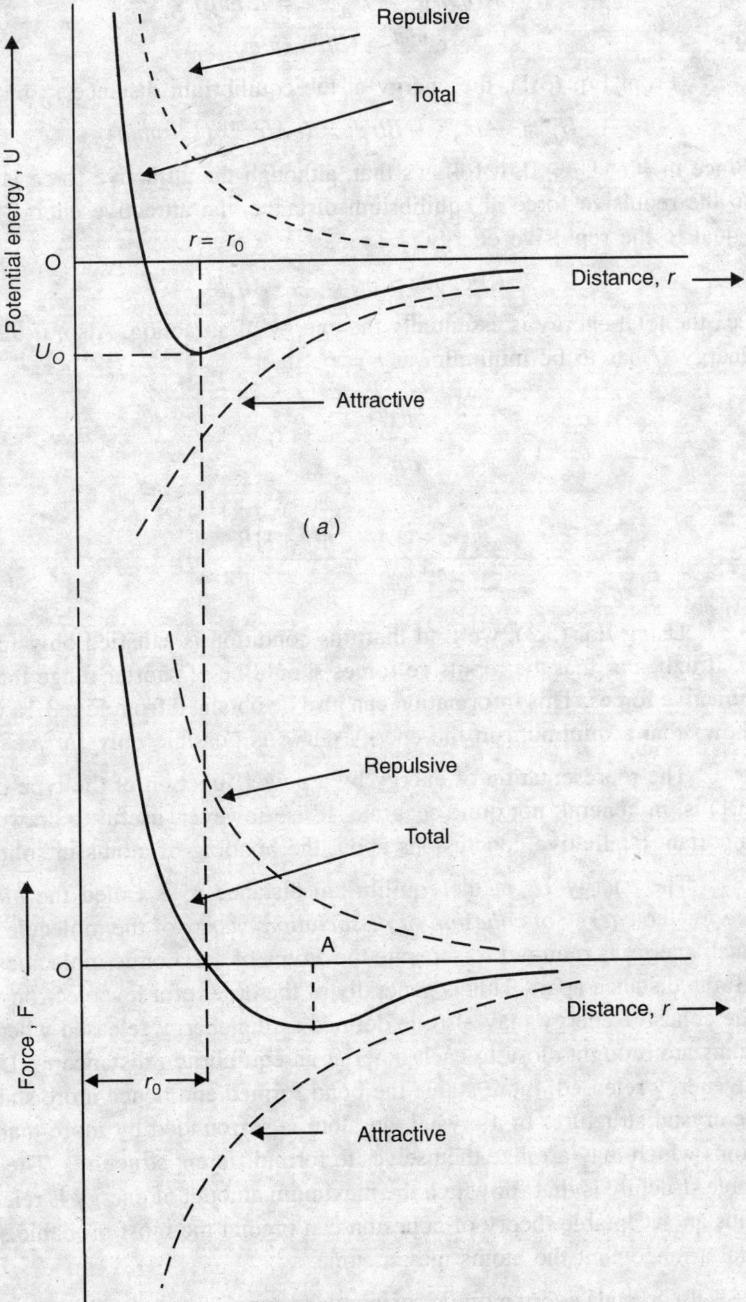

Fig. 3.1. *Potential energies (a) and interatomic forces (b) versus interatomic distance in a system of two atoms.*

or $$B = A(m/n)r_o^{n+1}/r_o^{m+1} = A(m/n)\, r_o^{n-m} \tag{3.3}$$

or $$r_o^{n-m} = (B/A)\,(n/m)$$

From Eq. (3.1), the energy at the equilibrium distance r_o becomes

$$U_o = -A/r_o^{m} + B/r_o^{n} = (-A/r_o^{m})\,(1 - m/n) \tag{3.4}$$

Since m ≠ n, $U_0 \neq 0$, it follows that, although the attractive force is equal to the repulsive force at equilibrium distance, the attractive energy is not equal to the repulsive energy.

$$\text{For } n >> m,\ U_o \simeq -A/r_0^{m},$$

i.e., the total energy is essentially the energy of attraction. Also, if the total energy U has to be minimum at $r = r_o$, then

$$\left.\frac{d^2U}{dr^2}\right|_{r=r_o} > 0$$

i.e., $$-\frac{m(m+1)A}{r_o^{m+2}} + \frac{n(n+1)B}{r_o^{n+2}} > 0$$

Using Eq. (3.3), we find that this condition is satisfied only for $n > m$. It indicates that the repulsive forces should be of shorter range than the attractive forces. This information can also be obtained from Fig. 3.1a which shows that a minimum in the energy curve is possible only for $n > m$.

The representation of energy by a power function of the type of Eq. (3.1) is, in general, not quite accurate. It is, however, useful to draw some important qualitative conclusions about the bonding of atoms in solids.

The energy U_o at the equilibrium distance r_o is called the *binding energy,* the *energy of cohesion* or *dissociation energy* of the molecule. This much energy is required to separate the atoms of a diatomic molecule to an infinite distance apart. This is generally of the order of a few electron volts. The cohesive energy may also be defined as the energy released when two atoms are brought close to each other at an equilibrium distance r_o. Larger the energy released, more stable the bond formed and hence more stable is the crystal structure. In a crystal, an atom is surrounded by more than one atoms which may arrange themselves to form different structures. The most stable structure is that for which the maximum amount of energy is released. Thus an acceptable theory of cohesion can predict the most probable structural arrangement the atoms may assume.

In crystals every atom is surrounded by a number of other atoms and the simple expression for attractive and repulsive energies given by Eq.(3.1) is not applicable. To know the exact form of these energy terms, one must investigate their origins in detail. It further requires the knowledge of charge distributions particularly of the valence electrons of the atoms. In certain crystals, the valence electrons are transferred from one atom to the other during bond formation. In some crystals, the sharing of electrons takes place among the neighbouring atoms while in some others the valence electrons behave as free electrons and move from one part of the crystal to another. There may still be other types of electronic interactions present in crystals. The nature of crystals formed depends upon the nature of these interactions. These interactions or bonds may be broadly classified into the following five main categories :

(i) Ionic bonds, as in NaCl (transfer of valence electrons)

(ii) Covalent bonds, as in diamond (sharing of valence electrons)

(iii) Metallic bonds, as in Ag, Cu (free nature of valence electrons)

(iv) Van der Waals forces, as in solid nitrogen (electrons remain associated with original molecules)

(v) Hydrogen bonds, as in ice.

The first three types of bonds are called *primary bonds* and the last two types of bonds are called *secondary bonds.* The classification of crystals based on these bonds is rather arbitrary. Many of the crystals exhibit mixed bonding. For example, ZnS crystal is believed to be partly ionic and partly covalent. Likewise, graphite has intrachain covalent bonding and interchain weak van der Waals type of bonding. The following sections describe the different types of bonding and the solids resulting from these bonds.

3.2.1 Ionic Bonds

An ionic bond is formed by the actual transfer of electrons from one atom to the other so that each atom acquires a stable electronic configuration similar to the nearest inert gas atoms. The atom which loses an electron becomes electropositive and the one which gains an electron becomes electronegative. The ions arrange themselves in such a way that the Coulomb's attractive forces among the oppositely charged ions dominate over the Coulomb's repulsive forces among the ions of the same sign. Thus the source of cohesive energy which binds the ions together is mainly the Coulomb's electrostatic interaction. The crystals resulting from this type of bonding are called *ionic crystals.* Since after the transfer of electrons, the ions attain electronic configurations similar to inert gas atoms, the charge distribution on the ions is spherically symmetric. Hence an ion of one type tries to have as many neighbours of the opposite type as possible. The coordination number of a cation is limited by the radius ratio of cation to anion while that of an anion is limited by the condition that the charge neutrality of the crystal must be maintained. The cohesive energy of the ionic crystals is, therefore, quite large; it is of the order of 5 to 10 eV.

A good example of ionic crystals is the crystal of NaCl. The electronic configurations of Na and Cl atoms are as follows :

Na : $1s^2\ 2s^2\ 2p^6\ 3s^1$

Cl : $1s^2\ 2s^2\ 2p^6\ 3s^2\ 3p^5$

After the transfer of an electron from 3*s* orbital of Na to 3*p* orbital of Cl, the configurations become

Na^+ : $1s^2\ 2s^2\ 2p^6$ (same as Ne)

Cl^- : $1s^2\ 2s^2\ 2p^6\ 3s^2\ 3p^6$ (same as Ar)

As permitted by radius ratio rule, each Na^+ ion is surrounded by six Cl^- ions and, in turn, each Cl^- ion is surrounded by six Na^+ ions to maintain the charge neutrality. Thus the coordination number of each ion is six. The binding energy per molecule of NaCl is 7.8 eV. The structure of NaCl has been described earlier in Sec. 1.13 and the unit cell is shown in Figs. 1.25 and 3.7. The position of ions on any cube face is as shown in Fig. 3.2. Some other examples of ionic crystals are LiF, KCl, CsCl, Al_2O_3, etc.

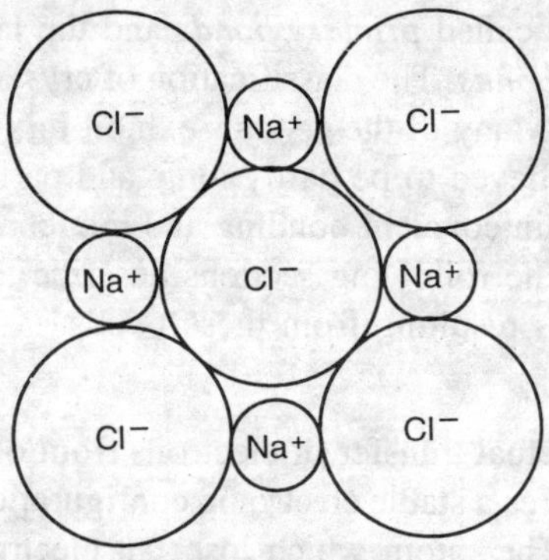

Fig. 3.2. *Ionic arrangement on a face of fcc unit cell of NaCl. The complete unit cell is shown in Fig. 1.25.*

Since ionic crystals have large binding energy, these are, in general, hard and exhibit high melting and boiling points. At normal temperatures, these are poor conductors of electricity but the conductivity increases with increase in temperature owing to the increased mobility of ions. These crystals are generally transparent to visible light but exhibit characteristic absorption peaks in infrared region. These are also soluble in polar solvents such as water.

3.2.2. Covalent Bonds

A covalent bond is formed by an equal sharing of electrons between two neighbouring atoms each having incomplete outermost shell. The atoms do so in order to acquire a stable electronic configuration in accordance with the octet rule. Unlike ionic bond, the atoms participating in the covalent bond have such electronic configurations that they cannot complete their octets by the actual transfer of electrons from one atom to the other. Hence there is no charge associated with any atom of the crystal.

A covalent bond is formed between similar or dissimilar atoms each having a deficiency of an equal number of electrons. When two atoms, each having a deficiency of one electron, come so close that their electronic shells start overlapping, the original atomic charge distributions of the atoms are distorted and each atom transfers its unpaired electron to the common space between the atoms. Thus the common space contains a pair of electrons which belongs equally to both the atoms and serves to complete the outermost shell of each atom. This is what is meant by 'sharing of electrons'. The sharing is effective if the shared electrons have opposite spins. In such a case the atoms attract each other and a covalent bond is formed. Since the participating atoms have the same valence state, this bond is also called the *valence bond.* The formation of a covalent bond between two chlorine atoms to produce Cl_2 molecule is illustrated as follows :

: Cl . + . Cl : → : Cl : Cl :

unpaired electrons — shared pair of electrons

Similarly, a water molecule is produced because of the formation of covalent bonds between an oxygen atom and two hydrogen atoms.

$$2\text{H}\cdot + \cdot\ddot{\underset{\cdot\cdot}{\text{O}}}\cdot \rightarrow \text{H}:\ddot{\underset{\cdot\cdot}{\text{O}}}:\text{H}$$

If each participating atom has a deficiency of two electrons, the atoms may combine to form a double covalent bond. The formation of an oxygen molecule is an example of double covalent bond. In this case, each atom contributes two electrons to the common space. Similarly, a triple covalent bond may be formed as in case of a nitrogen molecule. The number of covalent bonds an atom can form is determined by 8–*N* rule, where *N* is the number of the column of the periodic table to which the atom belongs. Since oxygen belongs to VI group, it can form (8–6) = 2 covalent bonds.

For the formation of stable covalent bond, there should be a net decrease in the potential energy of the system as a result of mutual sharing of electrons. This happens when the participating orbitals overlap effectively and also when the vacant electronic states are available in the outermost orbital of each atom. The overlap is more effective when the participating orbitals are directionally oriented rather than having spherical symmetry. Hence a covalent bond is always directional in character. Since *p*-orbitals are directional in nature, they readily participate in the formation of covalent bonds.

The structure of diamond is a good manifestation of the directional properties of covalent bonds. The carbon atom in the ground state has the configuration $1s^2\, 2s^2\, 2p^2$, i.e., it has two unpaired electrons in the outermost orbital. Since the energy difference between $2s$ and $2p$ states is small and the carbon atom is known to form four covalent bonds, it is proposed that one of the electrons of carbon is transferred from $2s$ state to $2p$ state. This modifies the electronic configuration of carbon as

$$1s^2\ 2s^1\ 2p_x^{\ 1}\ 2p_y^{\ 1}\ 2p_z^{\ 1}$$

Thus in the excited state, the C atom has four unpaired electrons and hence is able to make four covalent bonds with four neighbouring C atoms as in case of diamond. Also, all the four covalent bonds formed by each C atom are known to have equal strength with C-C-C bond angle equal to 109° 28'. To explain this, it is proposed that each C atom in diamond is sp^3 hybridized and has four identical sp^3 hybrid orbitals which are oriented in space along the corners of a regular tetrahedron as shown in Fig. 3.3. Each tetrahedron is covalently bonded to four similar tetrahedrons to produce a three-dimensional regular network type structure. The structure of diamond has been described in detail in Sec. 1.11 and its unit cell as well as plan view have been shown in Figs. 1.23 and 1.24 respectively. The other examples of covalent crystals are Si, Ge, α-Sn, etc.

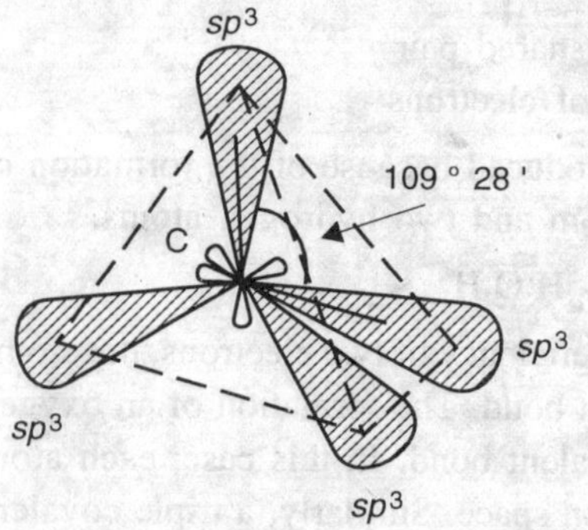

Fig. 3.3. *Four sp^3 hybrid orbitals of a carbon atom oriented along the corners of a regular tetrahedron in diamond.*

The covalent bond is a strong bond. This is apparent from the binding energy of carbon in diamond which is 7.4 eV. Thus the crystals having purely covalent bonding are generally very hard and brittle. These have high melting and boiling points. These exhibit low conductivity at ordinary temperatures which increases slightly with increase in temperature. Certain covalent crystals with moderate binding forces behave as semiconductors, e.g., Si and Ge.

3.2.3 Metallic Bonds

Metallic bond is formed by the partial sharing of valence electrons by the neighbouring atoms. Unlike the case of covalent bond, the sharing in metallic bond is not localized. Hence metallic bond may also be considered as delocalized or unsaturated covalent bond. As proposed by Drude, the atoms in metals contribute their valence electrons to form a common pool of electrons which has freedom to move anywhere in the frame-work of positive ion cores. This common pool of electrons is also known as the *free electron cloud or gas* and acts like a mobile glue to bind all the ion cores together through electrostatic attraction. The bond so formed is called the *metallic bond.* In this type of bonding, the atoms can share the required number of electrons with their neighbours through the common pool to complete their octets. As already mentioned, this sharing is not localized. The free nature of electron gas makes the bonding electrons resonate between different atoms and consequently the metallic sharing changes with time.

Consider the case of sodium which is an alkali metal. Each sodium atom has an electronic configuration of $1s^2\ 2s^2\ 2p^6\ 3s^1$, i.e., it contains one upaired electron in the $3s$ orbital. This electron is loosely held in sodium and may be called valence or conduction electron. When two sodium atoms approach each other, their $3s$ orbitals begin to overlap and one electron-pair bond is formed provided the electrons have opposite spins. Now the $3s$ orbital being full, if a third sodium atom approaches the pair, the valence electron of this atom may occupy the $3p$ state without violating the Pauli's exclusion principle. This is favoured by the fact that the energy difference between $3s$ and $3p$ orbitals is very small. Thus the third sodium atom can also form an electron-pair bond with either of the two sodium atoms. But since each sodium atom has only one unpaired electron, it cannot form two electron-pair bonds with the other two atoms simultaneously. It may, therefore, be assumed that, on the average, one sodium atom forms one half of an electron-pair bond with each of the other two sodium atoms. The actual structure of sodium metal is *bcc*, i.e., each sodium atom forms one-eighth of an electron-pair bond with each of its neighbours. It is possible only if the electron-pair resonates amongst the eight pairs of sodium atoms. Hence such bonds are incomplete or unsaturated covalent bonds. The valence electrons are thus delocalized and behave as if they are free to move about in a crystal.

Due to delocalized nature of valence electrons, the metallic bond is much less directional than covalent bond. Hence metals prefer to form close-packed structures, i.e., either *fcc* or *hcp* structures. However, there exist certain exceptional cases too. For example, alkali metals like Na and K form *bcc* structure which is attributed to their low melting points. The atoms have large vibrational amplitudes near the melting point and hence prefer to have loose-packed structures. The *bcc* structure of some transition metals like iron near room temperature is because of their partial covalent character. In these metals, the electrons present in *d*-orbitals participate in the formation of metallic bonds. Since *d*-orbitals form directional bonds, the tendency of these metals to form close-packed structures is inhibited.

The metallic bond is weaker than ionic or covalent bond. The binding energy ranges from 1 to 5 eV per bond. The melting and boiling points of metallic solids are lower than ionic or covalent solids. Also, these solids have high ductility and malleability, high electrical and thermal conductivities, and high optical reflection and absorption coefficients.

3.2.4 Van der Waals' Bonds

This type of bonding exists in atoms or molecules which have their outermost shells completely filled and hence have no tendency to gain, lose or share valence electrons with other atoms or molecules in the solid. The crystals resulting from this type of bonding are called *molecular crystals.* The examples of such solids are crystalline states of inert gases, such as He, Ne, Ar, etc., and other gases like O_2, Cl_2, CO_2, CH_4, etc. This type of bonding arises due to dipolar interaction between the atoms or molecules of the crystal. The electrons of the adjacent atoms in a molecule tend to repel each other and thus produce a momentary polarization in the molecule by disturbing the symmetry of the electronic distribution. A similar type of disturbance in electronic clouds also takes place in inert gas atoms which otherwise have spherically symmetric electronic distribution. Due to disturbance in the electron cloud of an atom, the centres of the positive and negative charge distributions no more coincide and an electronic dipole with a non-zero dipole moment is generated as shown in Fig. 3.4. This dipole is, however, not a permanent dipole and keeps oscillating with the movement of electron cloud around the nucleus. The electric field originating from this imbalance of charge induces a dipole moment in a neighbouring atom in such a way that the atom gets attracted to it. Similarly, the electric field of the second dipole produces another dipole by inducing dipole moment in a neighbouring atom, and so on. This type of dipole induced dipole interaction is called *Van der Waals' bond* or *dispersion bond.* It is non-directional in character. The presence of this interaction is manifested by the fact that the inert gases can be liquified at very low temperatures.

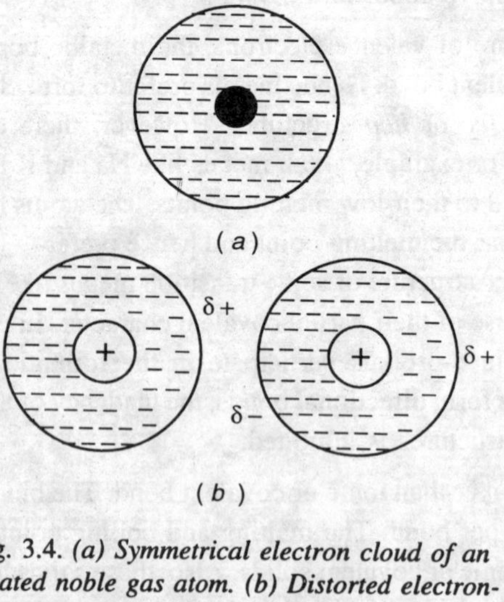

Fig. 3.4. *(a) Symmetrical electron cloud of an isolated noble gas atom. (b) Distorted electronic distribution around two neighbouring noble gas atoms generates two fluctuating dipoles.*

The van der Waals' forces are very weak forces. At room temperature, the thermal energy acquired by atoms or molecules is sufficient to make these forces ineffective. Thus molecular solids exist in gaseous state at room temperature. At low temperatures, however, these forces dominate over the thermal forces and play a significant role in the transformation of gaseous state to liquid or solid state. The energy released during the formation of van der Waals' bonds is of the order of 0.1 eV per bond only. Hence molecular solids are characterized by very low melting and boiling points, low mechanical strength and easy deformability. These solids are also poor conductors of heat and electricity.

3.2.5 Hydrogen Bonds

When a covalent bond is formed between a hydrogen atom and a highly electronegative atom, such as an atom of oxygen, fluorine, chlorine, etc., the shared electron pair gets attracted more towards the electronegative atom than the hydrogen atom. Thus the electronegative atom acquires a slight negative charge and the hydrogen atom acquires an equal amount of positive charge. The molecule so formed is said to be polarized and behaves like a permanent dipole. A number of such dipoles get attracted to one another due to the Coulombic force of attraction. This type of interaction between the oppositely charged ends of permanently polarized molecules each containing a hydrogen atom is called the *hydrogen bond.* In case the hydrogen atom does not participate in bond formation, the bond is called the *dipole bond.* A special significance is attached to hydrogen atom because it can be regarded simply as a proton fixed to one end of a covalent bond, the positive charge of which is not shielded by the surrounding electrons. This is not the case with other atoms participating in a covalent bond ; their positive charges are shielded by the outer electrons which weaken the attraction of the positive nuclear charge with the negative ends of other polarized molecules. Therefore, the positively charged hydrogen atom can interact more strongly with the negative ends of other molecules as compared to any other positively charged atom. Thus hydrogen bonds are stronger than dipole bonds. Materials exhibiting hydrogen bonding possess high melting and boiling points compared with molecular solids.

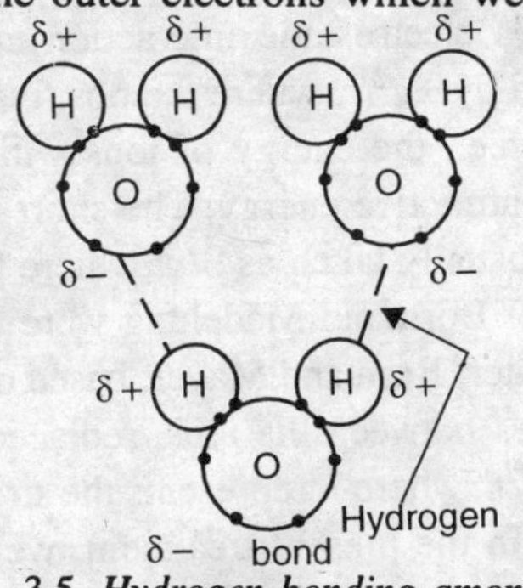

Fig. 3.5. *Hydrogen bonding among water molecules.*

The hydrogen bond plays an important role in the formation of ice and water. The existence of hydrogen bonding among water molecules is illustrated in Fig. 3.5. The binding energy of water or ice is about 0.5 eV. Hydrogen bonding is responsible for the striking physical properties of water and ice. In the absence of hydrogen bonding, the boiling point of water at atmospheric pressure would have been –80°C instead of 100°C and its viscosity much lower than 0.01 Poise at room temperature. Since a hydrogen bond is formed between oppositely charged ends of two permanent dipoles, it is directional in character.

3.3 BINDING ENERGY OF IONIC CRYSTALS

The calculation of binding energy of crystals is one of the major problems of the theory of solids and requires the knowledge of constituting particles. These calculations are relatively simple for ionic crystals and are described below for the case of NaCl crystal. The binding energy of ionic crystals was first calculated by Born and Madelung in 1910 and was later modified by Born and Mayer.

The theory developed by Born and Madelung is based on the assumption that the ionic crystals are built up of positive and negative ions each having a spherically symmetric charge distribution as in rare gas atoms. Thus the force between any two ions depends only on the distance between the ions and is independent of the direction of approach. The force between the ions is assumed to be mainly electrostatic. Thus the main contribution to the binding energy arises from electrostatic interaction and is called the *Madelung energy*. The ions tend to acquire such an arrangement in a crystal structure which results in the maximum attractive interaction amongst themselves.

There exist two types of interactions in ionic crystals, one is long range Coulomb's electrostatic interaction which may be attractive or repulsive in nature, and the other is short range repulsive interaction which comes into play when the interionic distance becomes so small that the electronic clouds of ions start overlapping. The Coulomb's electrostatic interaction energy between two ions with charges $\pm q$ is given by $\pm q^2/r$, where the positive sign stands for electrostatic repulsive energy, i.e., the energy of ions with like charges, and the negative sign stands for attractive energy. The short range repulsive energy as given by Eq. (3.1) is usually taken as β/r^n, where β and n are constants. The earlier calculations of Born and Madelung were based on this expression for repulsive energy. Later, Born and Mayer, based on the quantum mechanical calculations of forces between the ions, replaced this expression by another one of the form $\lambda e^{-r/\rho}$, where λ represents the strength and ρ the range of repulsive interaction. In the present treatment, we shall use the latter expression for the repulsive energy which is also known as the *central field repulsive potential.*

Since each ion in an ionic crystal is surrounded by a large number of ions of the similar or opposite type, the total interaction energy or *cohesive energy* of an ith ion is given by

$$U_i = \sum_j{}' U_{ij} \tag{3.5}$$

where U_{ij} is the interaction energy between the ith and jth ions and may be written as

$$U_{ij} = \lambda \exp(-r_{ij}/\rho) \pm q^2/r_{ij} \tag{3.6}$$

The primed summation in Eq. (3.5) indicates that the summation includes all the ions except those with $j = i$. This is because the interaction energy of an ith ion with itself carries no meaning. The constants ρ and λ can be empirically determined from the observed values of lattice constant and compressibility. As already mentioned, the positive sign in Eq. (3.6) is used for interaction between like charges and the negative sign for interaction between opposite charges. However, in an ionic crystal like NaCl, the value of U_i does not depend on the type of the reference ion i particularly when it is not present near the surface of the crystal. The surface effects are ignored here.

If the crystal contains N molecules, i.e., N positive ions and N negative ions, the total lattice energy or the total binding energy of the lattice becomes

$$U_{tot} = NU_i \tag{3.7}$$

Here we have used N rather than $2N$ because each pair of ij interactions must be considered only once while determinig the total binding energy. For convenience, we introduce a dimensionless quantity p_{ij} such that

$$r_{ij} = p_{ij}R \tag{3.8}$$

where R is the nearest neighbour separation in the crystal. The factor p_{ij} thus defines the distance between any two ions in terms of the nearest neighbour distance. If ions i and j are the nearest neighbours, then

$$r_{ij} = R$$

or

$$p_{ij} = 1$$

From Eqs. (3.6) and (3.8), we get

$$U_{ij} = \lambda \exp(-p_{ij}R/\rho) \pm q^2/(p_{ij}R)$$

The Eq. (3.5) becomes

$$U_i = \sum_j{}' [\lambda \exp(-p_{ij}R/\rho) \pm q^2/(p_{ij}R)] \tag{3.9}$$

Assuming that the repulsive interaction (first term) is effective for the nearest neighbours only and there are z nearest neighbours of the ith ion, the expression (3.9) for the cohesive energy of the ith ion takes the form

$$\begin{aligned} U_i &= z\lambda \exp(-R/\rho) \pm \sum_j q^2/(p_{ij}R) \\ &= z\lambda \exp(-R/\rho) - \alpha q^2/R \end{aligned} \tag{3.10}$$

From Eq. (3.7), the lattice energy becomes

$$U_{tot} = NU_i = N(z\lambda e^{-R/\rho} - \alpha q^2/R) \tag{3.11}$$

Here α is a constant called the *Madelung constant* and is given by

$$\alpha = \sum_j{}' \mp 1/p_{ij} \tag{3.12}$$

The choice of sign depends upon the type of the reference ion. If the reference ion is negative, the positive sign is used for a positive ion and the negative sign for a negative ion. Thus the value of the Madelung constant depends on the lattice structure. It is basically a correction factor that determines the magnitude of the error introduced by considering only the nearest neighbour interaction.

At the nearest neighbour equilibrium distance, R_o, we have

$$dU_{tot}/dR = 0$$

Therefore, from Eq. (3.11), we get

$$N\frac{dU_i}{dR}\bigg|_{R=R_o} = -\frac{Nz\lambda}{\rho}\, e^{-R_o/\rho} + \frac{N\alpha q^2}{R_o^2} = 0$$

or

$$e^{-R_o/\rho} = \frac{\rho\alpha q^2}{z\lambda R_o^2} \tag{3.13}$$

The Eq. (3.11) can now be written as

$$U_{tot} = N\left(z\lambda.\frac{\rho\alpha q^2}{z\lambda R_o^2} - \frac{\alpha q^2}{R_o}\right)$$

$$= -(N\alpha q^2/R_o)\,[1 - \rho/R_o] \tag{3.14}$$

The first term within parentheses resembles the second term of Eq. (3.11). It, therefore, represents the electrostatic interaction energy or the *Madelung energy*. The term within square brackets then represents the contribution of the short range repulsive interaction.

The similar expression for cohesive energy of the ith ion is

$$U_i = U_{tot}/N = -(\alpha q^2/R_o)\,[1 - \rho/R_o] \tag{3.15}$$

For the range $\rho \sim 0.1R_o$, U_i is dominated by the Madelung contribution. It increases rapidly for low values of ρ/R_o which indicates that the repulsive interaction has a very short range.

From Eqs. (3.14) and (3.15), it follows that the lattice energy of a crystal and the cohesive energy of an ion in the crystal can be determined if the values of R_o, α and ρ are known. The equilibrium distance R_o is, in general, determined empirically, the range ρ is determined from the knowledge of the bulk modulus of the crystal, and the Madelung constant is determined theoretically from the geometry of the crystal structure.

3.3.1 Evaluation of the Madelung Constant

Considering the simplest case of a one-dimensional crystal consisting of alternate positive and negative ions with interionic distance R as shown in Fig. 3.6. The Madelung constant, α, given by Eq. (3.12) can be expressed as

$$\alpha = \sum_j{}' \mp 1/p_{ij} = 2[1 - 1/2 + 1/3 - 1/4 + ...] \qquad (3.16)$$

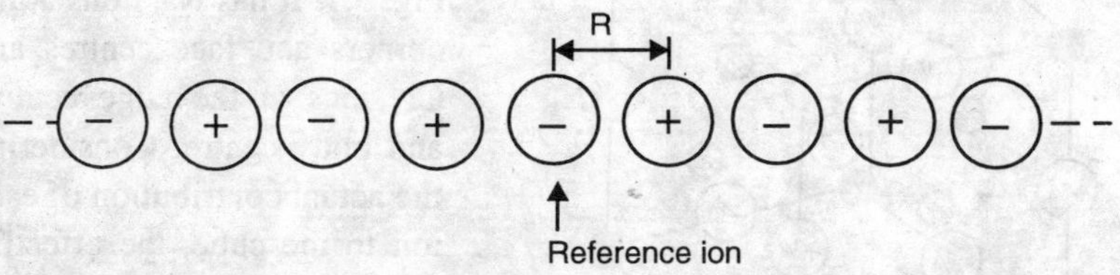

Fig. 3.6. *One-dimensional ionic crystal consisting of alternate positive and negative ions.*

This expression is written taking a negative ion as the reference ion. The factor of 2 appears because similar ions are present on both sides of the reference ion. Using the series expansion

$$\ln(1+x) = x - x^2/2 + x^3/3 - x^4/4 + ...$$

and putting $x = 1$ in it, we get

$$\ln 2 = 1 - 1/2 + 1/3 - 1/4 + ...$$

Therefore, Eq. (3.16) gives

$$\alpha = 2 \ln 2 = 1.38 \qquad (3.17)$$

For the actual three-dimensional crystals, the evaluation of the Madelung constant is not so simple. It is very difficult to write the successive terms by a casual inspection and the series converges quite slowly. Consider, for example, the unit cell of NaCl structure as shown in Fig. 3.7. Each Cl^- ion has 6 nearest neighbour Na^+ ions, 12 second-nearest neighbour Cl^- ions, 8 third-nearest neighbour Na^+ ions, and so on. Since the distance of the first, second, third, fourth, etc. nearest neighbours from the central Cl^- ion is 1, $\sqrt{2}$, $\sqrt{3}$, $\sqrt{4}$ etc. respectively, the Madelung constant becomes

$$\alpha = 6/\sqrt{1} - 12/\sqrt{2} + 8/\sqrt{3} - 6/\sqrt{4} + 24/\sqrt{5} - ......$$

$$= 6.000 - 8.485 + 4.619 - 3.000 + 10.733 - .....$$

Clearly, the convergence of the series is poor. In fact, the series may never converge quickly unless the terms are so arranged that the contributions from the successive positive and negative terms nearly cancel.

The convergence of the series can be improved further if we work with neutral groups, i.e., groups of ions which are more or less neutral. The physical reason of working with neutral groups is that the potential due to a neutral group decreases faster with distance than the potential due to charged ions. One such cubical neutral group is shown in Fig. 3.7. It has Na^+ ions at the corners and face centres and Cl^- ions at the edge centres and body centre. Considering the actual contribution of each ion to the cube, the effective charge carried by this group of ions is

$$-1+(1/2)6-(1/4)12+(1/8)8 = 0$$

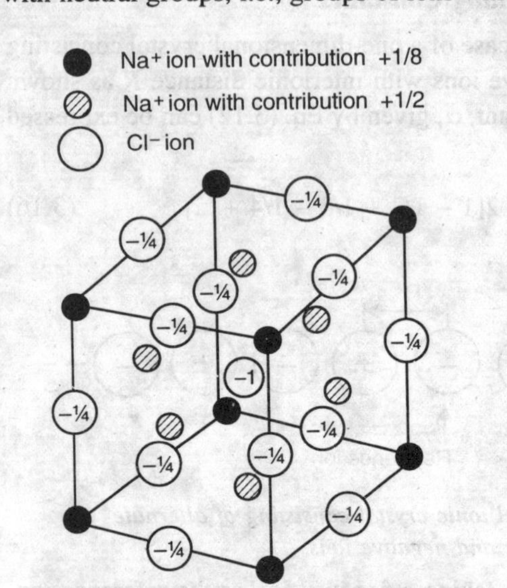

Fig. 3.7. *The cubic arrangement of Na^+ and Cl^- ions around the central Cl^- ion constituting the first neutral group developed by Evjen for calculating the Madelang constant.*

Taking Cl^- ion present at the body-centred position as the reference ion, the contribution of this neutral group to α is

$$\frac{6/2}{\sqrt{1}} - \frac{12/4}{\sqrt{2}} + \frac{8/8}{\sqrt{3}} = 1.46$$

The first term corresponds to the contribution of 6 nearest neighbour Na^+ ions occupying the face centres, the second term corresponds to the contribution of 12 next nearest neighbour Cl^- ions, and so on. Similarly, taking into account the ions of the next larger cube surrounding the first cube, the value of α calculated from the first two cubes comes out to be 1.75 which is close to the accurate value as given below. The values of the Madelung constant for CsCl and ZnS structures are also given.

NaCl	: 1.747565
CsCl	: 1.762675
ZnS (Zinc blende)	: 1.6381

In general, the higher values of the Madelung constant indicate the stronger Madelung contribution to the cohesive energy and hence greater stability of the structure. This is particularly applicable when the stability of the structure like NaCl or CsCl is compared with that of ZnS. This becomes inapplicable when NaCl structure is compared with CsCl structure where we know that, besides its low value of α, the former is more stable than the latter.

The discrepancy is explained on the basis of the magnitude of repulsive energy which is more in case of CsCl than in case of NaCl because of larger number of nearest neighbours in the former. Thus, although the Madelung energy of CsCl is more than that of NaCl, the larger value of repulsive energy for CsCl outweighs the Madelung energy difference and acts to decrease the stability of CsCl structure slightly.

3.3.2 Determination of Range

The range of repulsive interaction, ρ, can be expressed in terms of the bulk modulus, B, of the crystal which is defined as

$$B = -\frac{dp}{dV/V} \tag{3.18}$$

where dp is the small change in pressure required to produce volumetric strain dV/V. Considering adiabatic compression, i.e., assuming that no heat is exchanged with the surroundings, and applying the first law of thermodynamics, the change in energy associated with the decrease in volume dV is given by

$$dU = -p\,dV$$

or

$$dU/dV = -p$$

or

$$\frac{dp}{dV} = -\frac{d^2U}{dV^2}$$

Therefore, from Eq. (3.18), we get

$$B = V\frac{d^2U}{dV^2} \tag{3.19}$$

Now

$$\frac{dU}{dV} = \frac{dU}{dR}\cdot\frac{dR}{dV}$$

and

$$\frac{d^2U}{dV^2} = \frac{d^2U}{dR^2}(dR/dV)^2 + \frac{dU}{dR}\cdot\frac{d^2R}{dV^2} \tag{3.20}$$

Since at equilibrium distance R_o,

$$\frac{dU}{dR} = 0$$

the second term in Eq. (3.20) vanishes and we get

$$\frac{d^2U}{dV^2} = \frac{d^2U}{dR^2}(dR/dV)^2 \quad \text{for } R = R_o$$

Hence Eq. (3.19) becomes

$$B = V\left(\frac{d^2U}{dR^2}\right)(dR/dV)^2 \quad \text{for } R = R_o \tag{3.21}$$

In NaCl structure with lattice parameter $a = 2R$, the volume of the unit cell containing 4 molecules of NaCl is $8R^3$. Therefore, the volume occupied by N molecules of NaCl is

$$V = 2NR^3 \tag{3.22}$$

which gives

$$\frac{dV}{dR} = 6NR^2$$

Substituting these in Eq. (3.21) , we get

$$B = 2NR_o^{\ 3}\left(\frac{1}{36N^2R_o^4}\right)\frac{d^2U}{dR^2}\bigg|_{R=R_o}$$

or

$$B = \left(\frac{1}{18NR_o}\right)\frac{d^2U}{dR^2}\bigg|_{R=R_o} \tag{3.23}$$

From Eq. (3.11), we obtain

$$\frac{dU_{tot}}{dR} = -\frac{Nz\lambda}{\rho}e^{-R/\rho} + \frac{N\alpha q^2}{R^2}$$

which, on differentiation, gives

$$\frac{d^2U_{tot}}{dR^2} = \frac{Nz\lambda}{\rho^2}e^{-R/\rho} - \frac{2N\alpha q^2}{R^3}$$

Substituting it in Eq. (3.23), we get

$$B = \frac{1}{18NR_o}\left[\left(\frac{Nz\lambda}{\rho^2}\right)e^{-R_o/\rho} - \frac{2N\alpha q^2}{R_o^{\ 3}}\right]$$

Using Eq. (3.13), it becomes

$$B = \frac{1}{18NR_o}\left[\left(\frac{Nz\lambda}{\rho^2}\right)\left(\frac{\rho\alpha q^2}{z\lambda R_o^2}\right) - \frac{2N\alpha q^2}{R_o^3}\right]$$

or

$$\mathrm{B} = \frac{\alpha q^2}{18R_o^4}(R_o/\rho - 2) \tag{3.24}$$

The equation can be employed to determine ρ using the observed values of R_0 and B. The cohesive as well as lattice energies can then be determined using Eqs. (3.7) and (3.14).

3.4 BINDING ENERGY OF CRYSTALS OF INERT GASES

As described in Sec. 3.2.4, the molecular crystals of the inert gases have low melting points and are transparent insulators. The atoms of these crystals have their outermost shells completely filled and thus have very little tendency to gain, lose or even share the valence electrons with neighbouring atoms. Due to completely filled shells, these atoms have spherically symmetric electronic charge distribution. Hence these gases, except He^3 and He^4, usually form *fcc* crystals in the solid state.

Consider a molecular crystal consisting of N atoms. If U_{ij} represents the interaction energy between the atoms i and j of the crystal, then the cohesive energy of the ith atom is given by

$$U_i = \sum_j{}' U_{ij} \tag{3.25}$$

and the lattice energy of the crystal is

$$U_{tot} = \tfrac{1}{2} NU_i = \tfrac{1}{2} N \sum_j{}' U_{ij} \tag{3.26}$$

Here the factor $N/2$ is used instead of N to represent the number of distinct ij atomic pairs which contribute to the interaction energy. The primed summation in Eqs. (3.25) and (3.26) indicates that the summation includes all the atoms except those with $j = i$. As in case of ionic crystals, the interaction energy, U_{ij}, of molecular crystals is due to the following two types of interactions:

(a) Van der Waals or London interaction, and

(b) Repulsive interaction

(a) Van der Waals Interaction

The origin of van der Waals interactions has already been discussed in Sec. 3.2.4. It is a dipole induced dipole interaction which exists in two neighbouring atoms so as to produce weak attractive force between the atoms. These interactions are also known as *dispersion bonds.*

Let p_i be the instantaneous dipole moment of the ith atom caused by some fluctuations in the charge distribution on the jth atom. This dipole moment produces an electrostatic dipole field **E** at the centre of the jth atom whose magnitude in CGS units is

$$E = 2p_i/r_{ij}^3 \tag{3.27}$$

This field further produces an instantaneous dipole moment p_j at the jth atom of magnitude

$$p_j = \alpha_e E = 2\alpha_e p_i/r_{ij}^3 \tag{3.28}$$

where α_e is the electronic polarizability. These two diploes interact with each other in such a way as to produce a net attractive force as shown in Fig. 3.8. The potential energy associated with the attraction of two dipoles is given as

$$U_{att} \approx -2p_i p_j/r_{ij}^3 = -4\alpha_e p_i^2/r_{ij}^6 \tag{3.29}$$

From Eq. (3.28), it follows that α_e has dimensions of $[\text{length}]^3$. Therefore, $\alpha_e p_i^2$ has dimensions of $[\text{length}]^5 \times [\text{charge}]^2$. Hence, in Eq. (3.29), $\alpha_e\, p_i^2$ may be written as $e^2 r_o^5$ where r_o is the atomic radius. The Eq. (3.29) then becomes

$$U_{att} \approx -4e^2 r_o^5/r_{ij}^6 \approx -10^{-58}/r_{ij}^6 \text{ ergs}$$
$$= -B/r_{ij}^6 \tag{3.30}$$

where we have used $r_o \approx 10^{-8}$ cm and $B \approx 10^{-58}$ erg cm^6. For example, for Krypton, $r_{ij} = 4$Å and $U_{att} \approx 2\times10^{-14}$ erg which corresponds to a temperature of about 100 K. The same is the order of magnitude of the melting point of inert gases.

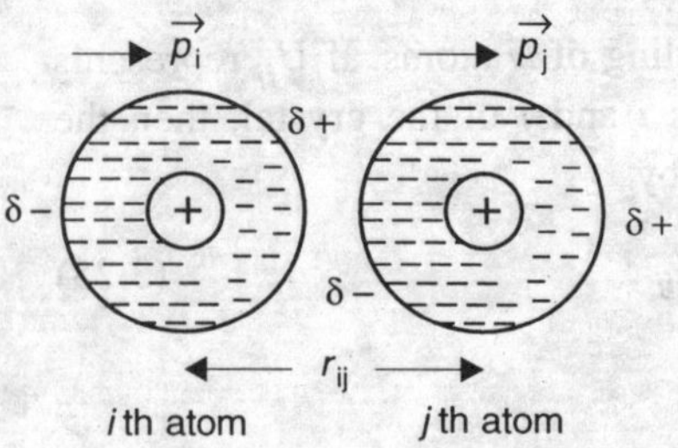

Fig. 3.8. *An instantaneous dipole moment* $\mathbf{p}_i$ *of the ith atom induces a dipole moment* $\mathbf{p}_j$ *on the jth atom.*

Since U_{att} varies as r_{ij}^{-6}, it follows that the van der Waals interaction is a short range interaction which increases rapidly with decrease in r_{ij}.

(b) Repulsive Interaction

As described in case of ionic crystals, the repulsive interaction comes into play for very short distances between the atoms when the electron clouds of the atoms begin to overlap and the Pauli's exclusion principle is disobeyed. The energy due to this interaction may be expressed as the power law of the type U_{rep} = constant / r_{ij}^n or by the expression $U_{rep} = \lambda e^{-r_{ij}/\rho}$, where λ and ρ represent the strength and range of repulsive interaction respectively. In the present case, we shall express U_{rep} by an empirical relation of the type.

$$U_{rep} = C/r_{ij}^{12} \tag{3.31}$$

where C is a positive constant.

From Eqs. (3.30) and (3.31), the total interaction energy, U_{ij}, between the atoms i and j of the crystal can be written as

$$U_{ij} = -B/r_{ij}^{\;6} + C/r_{ij}^{\;12}$$

or, in another form, it is expressed as

$$U_{ij} = 4\varepsilon[(\sigma/r_{ij})^{12} - (\sigma/r_{ij})^6] \qquad (3.32)$$

Here ε and σ are the parameters related to B and C as

$$4\varepsilon\sigma^6 = B, \text{ and } 4\varepsilon\sigma^{12} = C.$$

The values of ε and σ are empirically determined. The Eq. (3.32) represents a potential known as *Lennard- Jones Potential* and is plotted in Fig. 3.9.

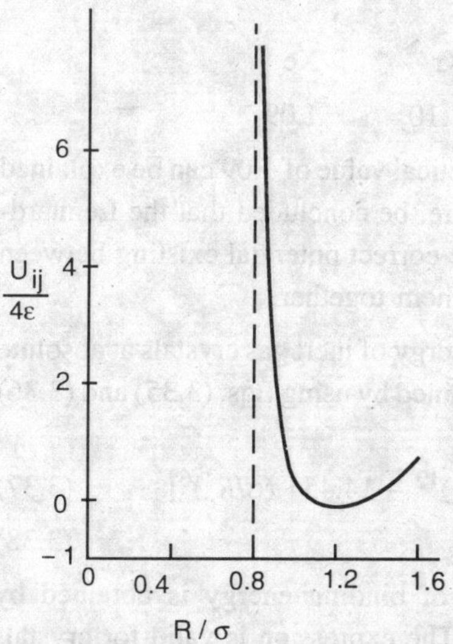

Fig. 3.9. *A plot of Lennard - Jones potential versus R/σ. The minimum occurs for R/σ = 1.122*

From Eq. (3.26), the expression for lattice energy or the total binding energy of the crystal becomes

$$U_{tot} = \tfrac{1}{2}N(4\varepsilon)\sum_j{}'[(\sigma/r_{ij})^{12} - (\sigma/r_{ij})^6] \qquad (3.33)$$

It may be noted that Eq. (3.33) has been written without considering the kinetic energy of the atoms of the crystal. Using Eq. (3.8) in (3.33), we obtain

$$U_{tot} = \tfrac{1}{2}N(4\varepsilon)\sum_j{}'\left[\left(\frac{\sigma}{p_{ij}R}\right)^{12} - \left(\frac{\sigma}{p_{ij}R}\right)^6\right] \qquad (3.34)$$

For *fcc* structures, we have

$$\sum_j{}' p_{ij}^{\;-12} = 12.131 \text{ and } \sum_i{}' p_{ij}^{\;-6} = 14.454 \qquad (3.35)$$

Since there are 12 nearest neighbours of an atom in *fcc* structure, the major contribution to the interaction energy arises from the nearest neighbours. The equilibrium distance R_o is obtained from the condition that at $R = R_o$, $dU_{tot}/dR = 0$. Therefore, from Eq. (3.34), we get

$$-2N\varepsilon[12 \times 12.131\,(\sigma^{12}/R_o^{\;13}) - 6 \times 14.454\,(\sigma^6/R_o^{\;7})] = 0$$

which gives

$$R_o/\sigma = 1.09 \qquad (3.36)$$

The factors of 12 and 6 correspond to the number of first and second nearest neighbours of an atom in an *fcc* structure respectively. This value of R_o/σ matches closely with the independently determined values for molecular crystals of inert gases as given below :

	Ne	Ar	Kr	Xe
R_o/σ	1.14	1.11	1.10	1.09

The slight variation from the theoretical value of 1.09 can be explained from the quantum effects. It can, therefore, be concluded that the Lennard-Jones potential given by Eq. (3.32) is the correct potential existing between the atoms of the inert gas which binds them together.

The final expression for binding energy of inert gas crystals at absolute zero temperature and zero pressure is obtained by using Eqs. (3.35) and (3.36) in Eq. (3.34), i.e.,

$$U_{tot}(R_o) = \tfrac{1}{2}N(4\varepsilon)\,[12.131(\sigma/R_o)^{12} - 14.454\,(\sigma/R_o)^6] \qquad (3.37)$$

$$= -2.15\,(4N\varepsilon) \qquad (3.38)$$

It may be noted that this value of binding energy is obtained by neglecting the kinetic energy of atoms. The expression is valid for crystals of all the inert gases. A more correct expression can be obtained by considering the kinetic energy effect and applying the quantum mechanical corrections. Based on these considerations, Bernardes obtained correction factors which reduce the binding energy to 28, 10, 6 and 4 per cent of the value given by Eq. (3.38) for Ne, Ar, Kr and Xe respectively.

SOLVED EXAMPLES

Example 3.1. The potential energy of a system of two atoms is given by the expression :

$$U = -A/r^2 + B/r^{10}$$

A stable molecule is formed with release of 8.0 eV of energy when the interatomic distance is 2.8 Å. Calculate A and B. Determine the force needed to dissociate this molecule into atoms and the interatomic distance at which the dissociation occurs.

Solution. The potential energy of the system is

$$U = -A/r^2 + B/r^{10} \qquad (3.39)$$

The force between the atoms at an interatomic distance r is

$$F = -dU/dr = -\,2A/r^3 + 10B/r^{11} \qquad (3.40)$$

At equilibrium distance r_o, $F = 0$. Therefore, Eq. (3.40) gives

$$A = 5B/r_o^8 \quad (3.41)$$
$$= 5B/(2.8 \times 10^{-10})^8$$
$$= 1.32 \times 10^{77}B \quad (3.42)$$

The dissociation energy or bond energy, U_o, is the energy released during bond formation. It is obtained from Eq.(3.39) by putting $r = r_o$, i.e.,

$$U_o = -A/r_o^2 + B/r_o^{10} = -\frac{A}{r_o^2}\left(1 - \frac{B}{Ar_o^8}\right)$$

Using Eq. (3.41), we get

$$U_o = -(A/r_o^2)(1 - 1/5) = -(4/5)(A/r_o^2)$$

Now $\quad U_o = -8.0 \text{ eV} = -8.0 \times 1.6 \times 10^{-19}\text{J}$

where the negative sign signifies the release of energy.

$\therefore \quad A = (5/4)(2.8 \times 10^{-10})^2 (8.0 \times 1.6 \times 10^{-19}) = 1.256 \times 10^{-37} \text{ Jm}^2$

Therefore, from Eq. (3.42), we obtain

$$B = 9.52 \times 10^{-115} \text{ Jm}^{10}$$

Referring to Fig. 3.1, it is apparent that the magnitude of the force becomes maximum at a certain critical interatomic distance r_c = OA which is greater that r_o. This gives the force needed to dissociate the molecule. The distance r_c is calculated from the condition that

$$\left.\frac{dF}{dr}\right|_{r=r_c} = 0$$

or, from Eq. (3.41),

$$6A/r_c^4 - 110B/r_c^{12} = 0$$

It gives

$$r_c = (110B/6A)^{1/8} \quad (3.43)$$

$$= \left(\frac{110 \times 9.52 \times 10^{-115}}{6 \times 1.256 \times 10^{-37}}\right)^{1/8}$$

$$= 3.30 \times 10^{-10} \text{ m}$$

Putting $r = r_c$ in Eq. (3.40), the force required to dissociate the molecule is

$$F = -(2/r_c^3)(A - 5B/r_c^8)$$

Using Eq. (3.43), it becomes

$$F = -(2/r_c^3)\,[A-5B\,(6A/110B)]$$

$$= -(2/r_c^3)\,(8A/11)$$

$$= -\frac{2}{\left(3.3\times10^{-10}\right)^3}\left(\frac{8\times1.256\times10^{-37}}{11}\right)$$

$$= -5.08\times10^{-9}\,N$$

The negative sign means that the force existing at $r = r_c$ is the force of attraction. Thus force needed to dissociate the molecule is 5.08×10^{-9} *N*.

Example 3.2. The lattice energy of KCl crystal containing *N* molecules of KCl is given by

$$U = -N(\alpha q^2/R - B/R^n)$$

Determine the repulsive exponent *n* using the following data:

Nearest neighbour equilibrium distance, $R_o = 3.147$ Å

Compressibility of KCl, $K = 5.747 \times 10^{-11} m^2/N$

The Madelung constant, $\alpha = 1.748$

Solution. The expression for energy in SI units is

$$U = -N\left(\frac{\alpha q^2}{4\pi\varepsilon_o R} - \frac{B}{R^n}\right)$$

Differentiating, we get

$$\frac{dU}{dR} = \frac{N\alpha q^2}{4\pi\varepsilon_o R^2} - \frac{nNB}{R^{n+1}} \qquad (3.44)$$

For $R = R_o$, $dU/dR = 0$, which gives

$$B = \frac{\alpha q^2}{4\pi\varepsilon_o n}\,R_o^{\,n-1} \qquad (3.45)$$

The structure of KCl is identical to that of NaCl. Therefore, the expression (3.23) is also valid for the case of KCl. Since the compressibility is reciprocal to the bulk modulus, Eq.(3.23) becomes

$$\frac{1}{K} = \left(\frac{1}{18NR_o}\right)\left.\frac{d^2U}{dR^2}\right|_{R=R_o} \qquad (3.46)$$

From Eq. (3.44) we have

$$\left.\frac{d^2U}{dR^2}\right|_{R=R_o} = -\frac{2N\alpha q^2}{4\pi\varepsilon_o R_o{}^3} + \frac{n(n+1)NB}{R_o{}^{n+2}}$$

$$= \frac{2N\alpha q^2}{4\pi\varepsilon_o R_o{}^3}(n-1) \qquad \text{[Using Eq. (3.45)]}$$

Hence from Eq. (3.46), we obtain

$$n = 1 + \frac{18R_o^4(4\pi\varepsilon_0)}{K\alpha q^2}$$

$$= 1 + \frac{18\times\left(3.147\times10^{-10}\right)^4}{5.747\times10^{-11}\times1.748\times\left(1.6\times10^{-19}\right)^2\times9\times10^9} = 1+7.63 = 8.63$$

SUMMARY

1. There are two types of forces existing amongst the atoms of a crystal : the long range Coulombic forces which are attractive in nature and the short range repulsive forces.

2. The total potential energy of a system of atoms is expressed as

$$U = -A/r^m + B/r^n$$

where r is the interatomic distance and A, B, m and n are constants. Generally, $n > m$. The first term represents the energy of attraction and the second one represents the energy of repulsion.

3. A stable chemical bond is formed at an equilibrium interatomic distance r_o where the forces of attraction balance the forces of repulsion and the total energy U becomes the minimum. This minimum energy is called the binding energy, bond energy, energy of cohesion or dissociation energy of a molecule and is given by

$$U_o = (-A/r_o{}^m)(1-m/n)$$

4. Ionic, covalent and metallic bonds are strong bonds with bond energies ranging from 1 to 10 eV/bond. These are called primary bonds. The bonds like hydrogen bond and van der Waals forces are weak bonds with bond energies ranging from 0.01 to 0.5 eV/bond only. These are called secondary bonds.

5. An ionic bond between two atoms is formed by the transfer of electrons from one atom to the other. It is non-directional.

6. A covalent or valence bond results from the localized sharing of an equal number of valence electrons of the neighbouring atoms. It is a directional bond.

7. In metallic bond, the sharing of electrons is partial and delocalized. A metal may be considered as an array of positive ions which are held together in a free electron cloud. Hence the metallic bond is non-directional.

8. The van der Waals forces or dispersion bonds result from the dipole induced dipole interactions amongst the neighbouring inert gas atoms or the molecules of some other gases. These are non-directional The crystals having these types of forces are called molecular crystals.

9. Hydrogen bond is an electrostatic attraction between the neighbouring molecular dipoles each containing a hydrogen atom. In case the molecules do not contain hydrogen atoms, the bond is termed dipole bond. These are directional bonds.

10. The lattice energy or the total binding energy of an ionic crystal is given by

$$U_{tot} = NU_i = -\ (N\alpha q^2/R_o)\ [1 - \rho/R_o]$$

where U_i is the cohesive energy of the ith ion, N is the number of ionic molecules, r_o is the nearest neighbour equilibrium distance, q is the electronic charge, ρ is the range of repulsive interaction and α is the Madelung constant. The term within the parentheses represents the Madelung energy.

11. The range ρ can be determined from the bulk modulus B of ionic crystal. For NaCl, the two are related as

$$B = \frac{\alpha q^2}{18R_o^4}\ (R_o/\rho - 2)$$

VERY SHORT QUESTIONS

1. What is cohesive energy?
2. Give one example each of primary and secondary bonds.
3. What are short-range forces?
4. What are ionic crystals?
5. What are covalent crystals? Give an example.
6. What is the maximum number of covalent bonds an atom of the nth group can form?

7. What are molecular crystals? Give two examples.
8. What is Van der Waals' interaction? Is it present in all types of solids?
9. What is dispersion bond?
10. What is the origin of the repulsive interaction in inert gas crystals?
11. Give properties of molecular solids.
12. How does hydrogen bond differ from dipole bond?
13. What is Madelung energy?
14. What is Madelung constant? How is it expressed mathematically?
15. What is the significance of the Madelung constant?

SHORT QUESTIONS

1. What are primary and secondary bonds? Give examples.
2. What limits the coordination of a cation and an anion in an ionic crystal?
3. Why are closed-packed structures experienced mostly in metals and not in ionic and covalent solids?
4. Why do covalent crystals usually have lower packing efficiency than the ionic crystals?
5. Why do alkali metals exhibit *bcc* structure? Why does iron at room temperature exhibit *bcc* structure?
6. What is hydrogen bond? How is it different from a dipole bond? Describe the role of hydrogen bond during formation of ice.
7. Why do inert gases get liquefied and solidifed at very low temperatures?
8. Evaluate the Madelung constant for an infinitely long one-dimensional ionic crystal consisting of singly charged alternate positive and negative ions. Take a negative ion in the crystal as the reference ion.
9. The madelung constant of CsCl is greater than that of NaCl but still the NaCl structure is more stable than the CsCl structure. Explain.
10. Calculate the cohesive energy of crystalline NaCl.

LONG QUESTIONS

1. Describe the nature and origin of various forces existing between the atoms of a crystal. Explain the formation of a stable bond using the potential enrgy versus interatomic distance curve.
2. Derive an expression for binding energy for an ionic crystal and obtain the expression for the Madelung constant. Evaluate the Madelung constant for a linear ionic crystal.

3. What type of interaction is responsible for binding among atoms of the inert gas crystals? Explain the origin of both attractive and repulsive parts of the interaction.
4. Define cohesive energy and determine its value for crystals of inert gases.
5. Give the physical characteristics of inert gas crystals. Discuss the form of the potential energy between two inert gas atoms.
6. Consider a line of alternate positive and negative ions each carrying a charge q. The repulsive potential energy between the nearest neighbours is given by A/r^n. Show that, for a total of $2N$ ions, the equilibrium potential energy of the system is

$$U_0 = -\left[2\, Nq^2\, ln(2/r_0)\right]\,(1-1/n)$$

PROBLEMS

1. The potential energy of a system of two atoms is given by

$$U = -\alpha/r^4 + \beta/r^{12}$$

Calculate the amount of energy released when the atoms form a stable bond. Determine the bond length.

$[(4\alpha^3/27\beta)^{1/2},\ (3\beta/\alpha)^{1/8}]$

2. The interaction energy of a system of two atoms is given by

$$U = -A/r^6 + B/r^{12}$$

The atoms form a stable bond with bond length of 3Å and bond energy of 1.8 eV. Calculate A and B. Compute the forces required to break the molecule and the critical interatomic distance for which it occurs. Also, calculate the force required to reduce the interatomic distance by 5 per cent of the value at equilibrium.

(4.19×10^{-76}Jm^6, 1.53×10^{-133}Jm^{12}, 2.56×10^{-9}N, 3.33×10^{-10}m, 5.97×10^{-9} N)

3. The cohesive energy and the nearest neighbour distance for a LiF molecule are 1.68×10^{-18}J and 2.014Å respectively. The structure of LiF is the same as that of NaCl. Calculate the bulk modulus of LiF using the following expression for the potential energy :

$$U = -\,\alpha q^2/R + B/R^n$$

The value of α may be taken as 1.75.

(7.15×10^{10}Nm^{-2})

CHAPTER - IV
LATTICE VIBRATIONS

4.1 INTRODUCTION

A lattice may be regarded as a regular arrangement of atoms which are joined together by elastic springs as shown in Fig. 4.1 for a two-dimensional case. The motion of any single atom is, therefore, shared by all the atoms, i.e., the motion of the atom is coupled. The lattice may vibrate freely in its normal modes due to its internal energy or may experience forced vibrations under the effect of dynamical external forces which may be mechanical or electromagnetic in nature. The vibrations of the former type yield information about the thermal properties of solids, such as specific heat and thermal conductivity and are described in this chapter. The latter type of vibrations are associated with acoustical and some optical properties of solids and are beyond the scope of this book. The characteristics of elastic vibrational motion of a crystal lattice can be easily investigated by considering a one-dimensional lattice, i.e., a lattice consisting of linear chains or lines of atoms, and the results are generalized for two and three-dimensional lattices without giving quantitative details.

Fig. 4.1. *Two-dimensional model of a lattice comprising atoms attached to one another by elastic springs.*

4.2 VIBRATIONS OF ONE-DIMENSIONAL MONOATOMIC LATTICE

Consider a one-dimensional chain of atoms where each atom has mass m and is attached to other atoms by massless springs. Such a discrete arrangement of atoms constitutes a non-homogeneous medium which is distinguished from a homogeneous medium wherein the line of atoms is continuous without any breaks in between. Consider the equilibrium state of the atoms when these are situated at equally spaced sites represented by...$n-1$, n, $n+1$,.. as shown in Fig. 4.2. If the atoms are placed on the x-axis with interatomic spacing a, the x-coordinates of the corresponding atoms are given by...$(n-1)a$, na, $(n+1)a$,.. In the state of vibratory motion along the x-axis, the atoms will execute periodic motion about their equilibrium positions and become sources of elastic waves which propagate through the medium. Let, at any instant of time, the displacements of nth, $(n-1)$th, $(n-2)$th,... atoms from their mean positions be u_n, u_{n-1}, u_{n-2}.... respectively. Assuming the springs to be ideally elastic, the force between any two atoms will be linear, i.e., the force required to produce an atomic displacement is proportional to the displacement itself. Let the force experienced by an atom be mainly due to the nearest neighbours. If u is the displacement of a spring with spring constant β, the force exerted by a spring on an atom is given by

$$F = \beta u$$

Since the nth atom is attached to $(n-1)$th and $(n+1)$th atoms by two springs, it experiences two opposite forces each one of which is proportional to the net displacement of the corresponding spring. The net force on the nth atom is

$$\begin{aligned} F &= \beta(u_{n+1} - u_n) - \beta(u_n - u_{n-1}) \\ &= \beta(u_{n+1} + u_{n-1} - 2u_n) \end{aligned} \quad (4.1)$$

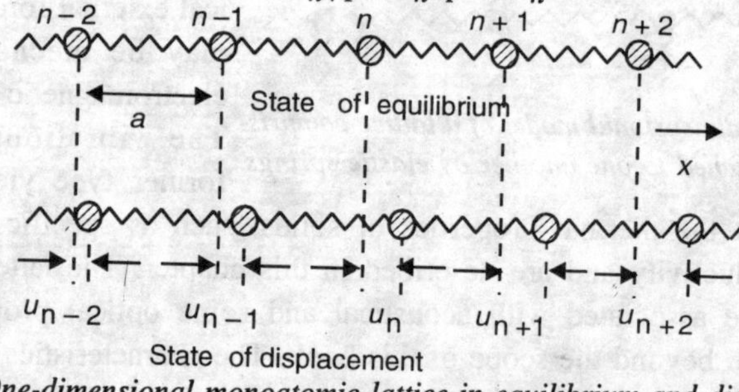

Fig. 4.2. *One-dimensional monoatomic lattice in equilibrium and disturbed states.*

Using the Newton's second law, the equation of motion is written as

$$m\frac{d^2u_n}{dt^2} = \beta(u_{n+1} + u_{n-1} - 2u_n) \quad (4.2)$$

where $\frac{d^2u_n}{dt^2}$ represents the acceleration of the nth atom. We seek the periodic solution to this wave equation as

$$u_n = u_o exp\ i\ (\omega t - Kna) \tag{4.3}$$

where na represents the x-coordinate of the nth atom in the equilibrium state, $K = 2\pi/\lambda$ is the wave vector or propagation vector and ω is the angular frequency of the wave. Similarly, for $(n+1)$th and $(n-1)$th atoms, the corresponding expressions are

$$u_{n+1} = u_o \exp\ [i\{\omega t - K(n + 1)a\}]$$
$$u_{n-1} = u_o \exp\ [i\{\omega t - K(n - 1)a\}] \tag{4.4}$$

From Eqs. (4.2), (4.3) and (4.4), we obtain

$$- m\omega^2 = \beta(e^{iKa} - 2 + e^{-iKa})$$
$$= \beta[e^{iKa/2} - e^{-iKa/2}]^2 \tag{4.5}$$

Since $$\sin x = \frac{e^{ix} - e^{ix}}{2i}$$

or $$\sin^2 x = -\frac{1}{4}\left(e^{ix} - e^{-ix}\right)^2,$$

the Eq. (4.5) becomes

$$-m\omega^2 = -\ 4\beta\sin^2\left(\frac{Ka}{2}\right)$$

or $$\omega = \pm\sqrt{\frac{4\beta}{m}}\ \sin\left(\frac{Ka}{2}\right) \tag{4.6}$$

If c and ρ denote the longitudinal stiffness and the mass per unit length of the line respectively, then

$$c = \beta a$$

and $$\rho = m/a$$

It may be noted here that a line of length a contains a massless spring and an atom of mass m. Therefore, Eq. (4.6) becomes

$$\omega = \pm\frac{2}{a}\sqrt{\frac{c}{\rho}}\ \sin\left(\frac{Ka}{2}\right)$$

$$= \pm\ \frac{2}{a} v_s \sin\left(\frac{Ka}{2}\right) \tag{4.7}$$

where $$v_s = \sqrt{c/\rho} \tag{4.8}$$

is a constant for a given lattice and has dimensions of velocity. It is normally referred to as the velocity of sound waves in solids. Since the frequency ω should always be positive irrespective of the sign of K, we always take the magnitude on the right hand side of Eq. (4.6) or (4.7). Thus, we have

$$\omega = \sqrt{\frac{4\beta}{m}}\left|\sin\left(\frac{Ka}{2}\right)\right| = \frac{2}{a}v_s\left|\sin\left(\frac{Ka}{2}\right)\right| \tag{4.9}$$

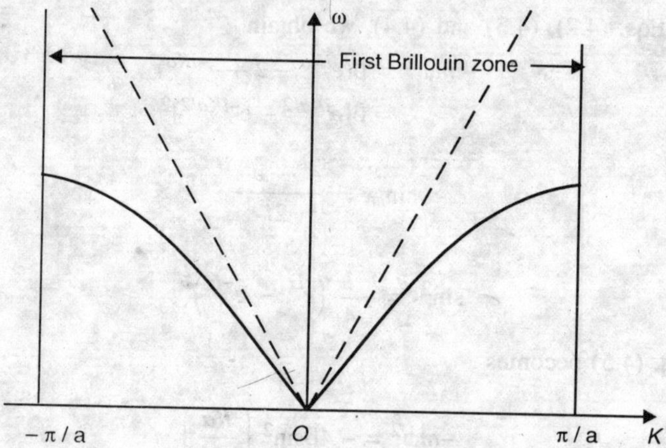

Fig. 4.3. *Dispersion relation for a one-dimensional monoatomic lattice.*

The solutions of the type (4.3) are possible only if ω is related to K as in Eq. (4.9). The relation (4.9) is called *dispersion relation* and is plotted in Fig. 4.3. The following important results are obtained from this relation :

(i) At low frequencies, $K \to 0$.

$$\therefore \qquad \sin\left(\frac{Ka}{2}\right) \to \frac{Ka}{2},$$

and Eq. (4.9) gives

$$\omega = \frac{2}{a}v_s\frac{Ka}{2}$$

or

$$\omega = v_s K \tag{4.10}$$

We introduce here the terms *phase velocity* and *group velocity*. The phase velocity of the waves is defined as the rate of advance of a point of constant phase along the direction of wave propagation and is given by

$$v_p = \omega/K \tag{4.11}$$

The group velocity is defined as the velocity of a group of waves or its envelope and represents the velocity with which the waves transmit energy along the direction of propagation. It is expressed as

$$v_g = \frac{d\omega}{dK} \tag{4.12}$$

It is now obvious from Eq. (4.10) that, in the long wavelength limit, the phase velocity is the same as the group velocity, each being equal to v_s. An exactly similar result is obtained for a homogeneous and continuous line. It, therefore, follows that for long wavelengths the atomic nature of solids is of little importance as far as the dynamical properties of the system are concerned. This is apparently due to insensitiveness of the discrete medium to the waves of long wavelengths. Thus a large number of atoms undergo all types of displacements as on a homogeneous line. This is illustrated in Fig. 4.4a.

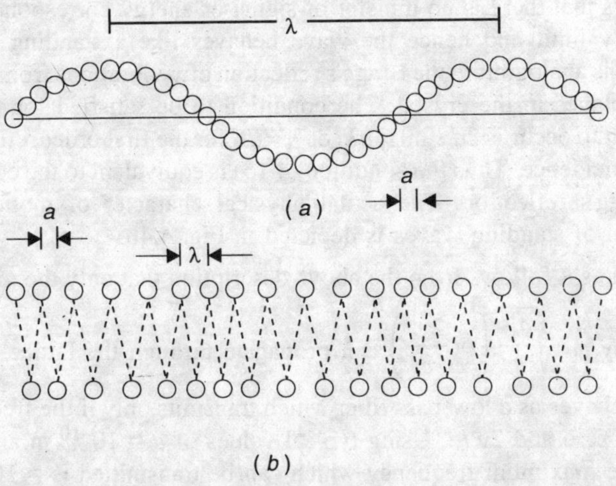

Fig. 4.4. *A linear line of atoms constituting a wave (a) Long wavelength case when the motion approaches that of a homogeneous line (b) Standing wave formation at* $\omega = \sqrt{4\beta/m}$; $\lambda = 2a$. *The displacements are shown transverse for the sake of simplicity, but actually these are longitudinal.*

(ii) At higher frequencies, phase and group velocities are different and are obtained from Eq. (4.9) as

$$v_p = \frac{\omega}{K} = \frac{2v_s}{Ka} \sin \frac{Ka}{2} \tag{4.13}$$

$$v_g = \frac{d\omega}{dK} = v_s \cos \frac{Ka}{2} \tag{4.14}$$

Thus both v_p and v_g are functions of frequency. This is referred to as the *phenomenon of dispersion* and the medium is called the *dispersive medium.* A similar phenomenon of dispersion is encountered when light passes through a medium whose refractive index is a function of frequency. For very long wavelengths, Eqs. (4.13) and (4.14) obviously give $v_p = v_g = v_s$, i.e., dispersion effects are negligible and the medium behaves like a homogeneous continuous medium. The dotted curve in Fig. 4.3 represents the dispersion relation for a continuous string.

(iii) At frequency $\omega = \sqrt{\dfrac{4\beta}{m}}$, which represents the maximum angular frequency of vibrations, Eq. (4.9) gives

$$K = \pi/a \quad \text{or} \quad \lambda = 2a \tag{4.15}$$

and from Eqs. (4.13) and (4.14), we obtain

$$v_p = 2v_s / \pi \; ; \; v_g = 0 \tag{4.16}$$

It follows that there is no transfer of signal or energy corresponding to this frequency limit and hence the wave behaves like a standing wave, The situation is analogous to the Bragg's reflection of x-vibrations from successive atomic planes in the crystal. The condition to be satisfied for the Bragg's reflection to occur is $2d \sin\theta = n\lambda$, or $\lambda = 2d$ for the first order reflection with normal incidence. Thus the condition (4.15) is equivalent to the condition for the Bragg's reflection. The actual physical character of motion for this condition of standing waves is depicted in Fig. 4.4b.

Thus it follows from the above description that only the vibrations of frequency $\omega < \sqrt{\dfrac{4\beta}{m}} \left(\text{or} \dfrac{2v_s}{a}\right)$ can propagate through the lattice. Hence the lattice behaves as a low-pass filter which transmits only if the frequency lies between zero and $2v_s/a$. Using typical values of $a \simeq 10^{-10}$ m and $v_s \simeq 10^4$ ms^{-1}, the maximum frequency which can be transmitted is $\simeq 10^{14}$ s^{-1}.

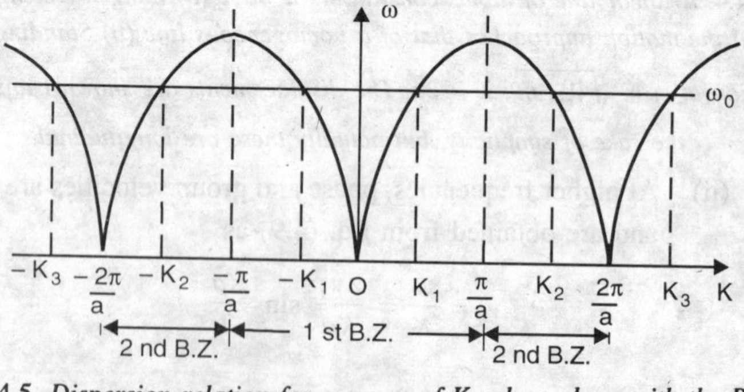

Fig. 4.5. *Dispersion relation for a range of K-values along with the Brillouin zones. A number of K-values correspond to the same frequency* ω_0.

(iv) Now consider the vibrational motion of the lattice corresponding to any frequency $\omega_o < \sqrt{\frac{4\beta}{m}}$ which lies in the vibrational range. A plot of the dispersion relation (4.9) for the values of K lying even beyond $\pm \pi/a$ is shown in Fig. 4.5 which is simply an extension of Fig. 4.3. A number of K-values, K_1, K_2, K_3, etc., are associated with the frequency ω_o which means that the lattice can propagate a number of wavelengths corresponding to these K-values at the same frequency ω_o. It physically means that the pattern of atomic displacements associated with a certain value of K, say K_1, in the range $-\frac{\pi}{a} < K < \frac{\pi}{a}$ and frequency ω_o can also be associated with several other values of K and hence several other wavelengths. This is depicted in Fig. 4.6 where a set of atomic displacements is shown which corresponds to two waves of the same frequency ω_o but of different wavelengths.

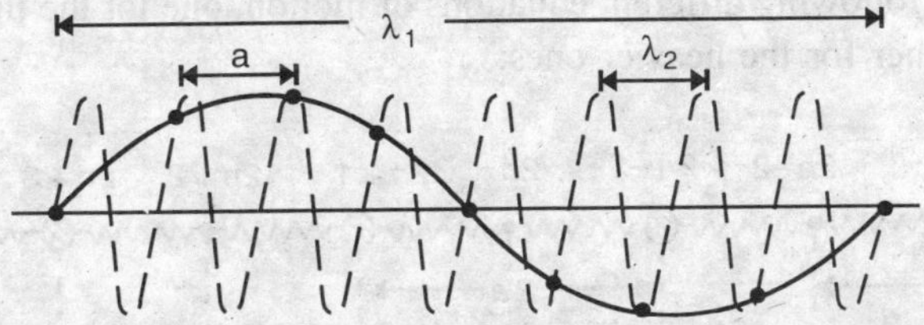

Fig. 4.6. *A set of atomic displacements represented by two sinusoidal waves of different wavelengths. The frequencies of the waves are same in case of a lattice (as in present case) but different in a continuum.*

The region of K-values where $-\frac{\pi}{a} < K < \frac{\pi}{a}$ is known as the *first Brillouin zone* and is of utmost importance in dealing with periodic structures. The region for which $-\frac{2\pi}{a} < K < -\frac{\pi}{a}$ and $\frac{\pi}{a} < K < \frac{2\pi}{a}$ is the *second Brillouin zone*. The third zone corresponds to the regions $-\frac{3\pi}{a} < K < -\frac{2\pi}{a}$ and $-\frac{2\pi}{a} < K < \frac{3\pi}{a}$, and so on. Note that the right half of the second Brillouin zone is similar to the left half region of the first zone. Similarly, the regions of third and higher zones match with those of the first zone. Thus the character of the possible solutions in the second and higher zones is the same as the character of the possible solutions in the first zone except with a difference in K-values. Hence any arrangement of atomic positions representing a sinusoidal wave of wavelength lying in a higher zone can be represented or reduced to correspond to a sinusoidal wave for which K or λ lies in the first zone. It is, therefore, not desirable to consider vibrations contained in regions other than the first Brillouin zone. A K-value corresponding to the second or a higher zone can be made to lie in the first zone by subtracting a suitable integral multiple of $2\pi/a$ from it. The new value of K will give identical results except with a difference of wavelength.

4.3 VIBRATIONS OF ONE-DIMENSIONAL DIATOMIC LATTICE

Consider a one-dimensional (linear) primitive lattice with basis consisting of two atoms of masses m and M ($m<M$) which are placed alternatively along the x-axis with an interatomic distance equal to a. In a state of equilibrium, let the atoms be located at sites represented by ...$2n-2$, $2n-1$, $2n$, $2n+1$, $2n+2$, ...as shown in Fig. 4.7. Also, let u_{2n} be the displacement of an atom corresponding to the $2n$th site at any time during the vibratory motion of atoms. Using the assumptions similar to the monoatomic case, we obtain the following different equations of motion, one for the lighter atoms and the other for the heavier ones:

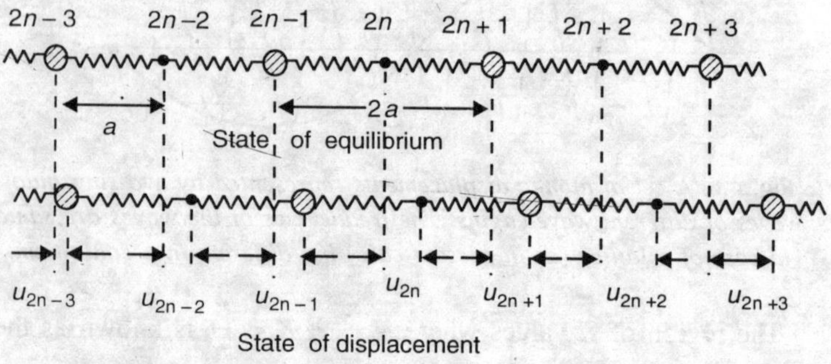

Fig. 4.7. *Linear diatomic lattice in the equilibrium and disturbed states.*

$$\left.\begin{aligned} F_{2n} &= m\frac{d^2u_{2n}}{dt^2} = \beta\,(u_{2n+1} + u_{2n-1} - 2u_{2n}) \\ F_{2n+1} &= M\frac{d^2u_{2n+1}}{dt^2} = \beta\,(u_{2n+2} + u_{2n} - 2u_{2n+1}) \end{aligned}\right] \quad (4.17)$$

We seek the travelling solutions of the type

$$\left.\begin{aligned} u_{2n} &= A \exp i\,[\omega t - 2Kna] \\ u_{2n+1} &= B \exp i\,[\omega t - (2n+1)\,Ka] \end{aligned}\right] \quad (4.18)$$

where K is a wave vector of a particular mode of vibration. It may be noted that the vibrational frequency of both types of atoms is assumed to be the same because both types of atoms participate in the same wave motion. The amplitudes A and B may, however, be different because of different masses of the atoms. Writing similar expressions for u_{2n-1} and u_{2n+2} and substituting these as well as Eqs. (4.18) into Eqs. (4.17), we obtain

$$\left.\begin{aligned} -m\omega^2 A &= \beta B(e^{iKa}+e^{-iKa}) - 2\beta A \\ -M\omega^2 B &= \beta A(e^{iKa}+e^{-iKa}) - 2\beta B \end{aligned}\right] \tag{4.19}$$

Since $e^{iKa} + e^{-iKa} = 2\cos Ka$, Eqs. (4.19) yield

$$\left.\begin{aligned} (2\beta - \omega^2 m)A - (2\beta\cos Ka)B &= 0 \\ (-2\beta\cos Ka)A + (2\beta - \omega^2 M)B &= 0 \end{aligned}\right] \tag{4.20}$$

This set of homogeneous linear equations would give rise to nonzero solutions for A and B only if

$$\begin{vmatrix} 2\beta - \omega^2 m & -2\beta\cos Ka \\ -2\beta\cos Ka & 2\beta - \omega^2 M \end{vmatrix} = 0$$

or

$$(2\beta - M\omega^2)(2\beta - m\omega^2) - 4\beta^2\cos^2 Ka = 0$$

or

$$\omega^4 - \frac{2\beta(m+M)}{mM}\omega^2 + \frac{4\beta^2\sin^2 Ka}{mM} = 0$$

It gives

$$\omega^2 = \beta\left(\frac{1}{m} + \frac{1}{M}\right) \pm \beta\sqrt{\left(\frac{1}{m} + \frac{1}{M}\right)^2 - \frac{4\sin^2 Ka}{mM}} \tag{4.21}$$

This is the dispersion relation for a linear diatomic lattice. From Eq. (4.6), the corresponding relation for the monoatomic lattice is

$$\omega^2 = \frac{4\beta}{m}\sin^2\frac{Ka}{2}$$

Considering only the positive values of ω, we find that, in the monoatomic case, there is only one value of ω for a single value of K whereas in the diatomic case there are two values of ω. These two values are written as ω_+ and ω_- and are expressed as

$$\left.\begin{aligned} \omega_+^2 &= \beta\left(\frac{1}{m} + \frac{1}{M}\right) \pm \beta\sqrt{\left(\frac{1}{m} + \frac{1}{M}\right)^2 - \frac{4\sin^2 Ka}{mM}} \\ \omega_-^2 &= \beta\left(\frac{1}{m} + \frac{1}{M}\right) - \beta\sqrt{\left(\frac{1}{m} + \frac{1}{M}\right)^2 - \frac{4\sin^2 Ka}{mM}} \end{aligned}\right] \tag{4.22}$$

Considering the expression for ω_+, we find that for $K \to 0$, sin Ka is negligible and, therefore, we get

$$\omega_+ = \sqrt{2\beta\left(\frac{1}{m} + \frac{1}{M}\right)} \tag{4.23}$$

For $K \to \dfrac{\pi}{2a}$, sin$Ka \to 1$, and we have

$$\omega^2_+ = \beta\left(\frac{1}{m} + \frac{1}{M}\right) + \beta\sqrt{\left(\frac{1}{m} + \frac{1}{M}\right)^2 - \frac{4}{mM}}$$

$$= \beta\left(\frac{1}{m} + \frac{1}{M}\right) + \beta\left(\frac{1}{m} - \frac{1}{M}\right)$$

or
$$\omega_+ = \sqrt{\frac{2\beta}{m}} \tag{4.24}$$

Now consider the expression for ω_- in Eq. (4.22). For $K \to 0$, we write sinKa $\cong Ka$ (it is not neglected in order to obtain a non-zero value of ω). Therefore, we get

$$\omega_-^2 = \beta\left(\frac{1}{m} + \frac{1}{M}\right) - \beta\left(\frac{1}{m} + \frac{1}{M}\right)\sqrt{1 - \frac{4K^2a^2}{mM}\left(\frac{mM}{m + M}\right)^2}$$

$$= \beta\left(\frac{1}{m} + \frac{1}{M}\right)\left[1 - \sqrt{1 - \frac{mM}{(m + M)^2}4K^2a^2}\right]$$

$$= \beta\left(\frac{1}{m} + \frac{1}{M}\right)\left[1 - 1 + \frac{mM}{(m + M)^2}2K^2a^2 + \cdots\right]$$

(Using binomial theorem)

$$\cong \frac{2\beta}{m + M}K^2a^2$$

$\therefore$
$$\omega_- = Ka\sqrt{\frac{2\beta}{m + M}} \tag{4.25}$$

For $K \to \pi/2a$, we get solution similar to (4.24) with m replaced by M, i.e.,

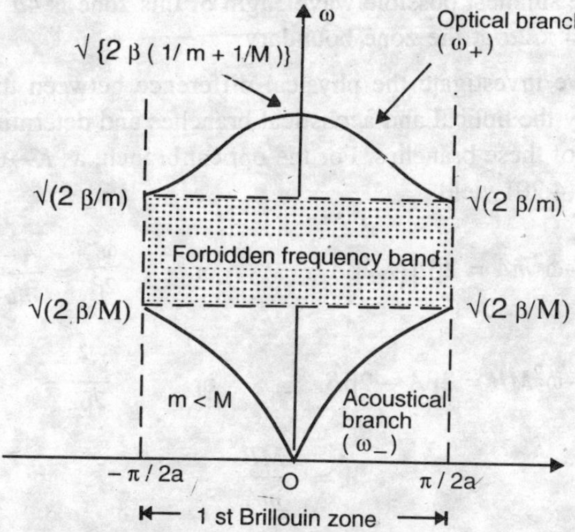

Fig. 4.8. *Dispersion relations for linear diatomic lattice showing acoustical and optical modes.*

$$\omega_- = \sqrt{\frac{2\beta}{M}} \tag{4.26}$$

The plots of dispersion relations (4.22) along with the solutions (4.23) through (4.26) are shown in Fig. 4.8. The following points should be observed :

(i) The allowed frequency range of propagation is split into two branches — an upper branch called the *optical branch* and a lower branch called the *acoustical branch*. The acoustical branch resembles the dispersion relationship curve for a monoatomic lattice, whereas the optical branch represents an entirely different type of wave motion.

(ii) There exists a band of frequencies between these two branches for which the wave-like solutions of the type (4.18) are not possible. It means that it is not possible to excite vibrations in a lattice at a frequency which lies inside this band. This band is called *forbidden band.* The width of this band depends on the mass ratio M/m. The larger the ratio M/m, the greater the width of the forbidden band. The existence of the forbidden band is a characteristic feature of elastic waves in case of diatomic lattices. If $M = m$, the optical and acoustical branches coincide at $K = \pm \pi/2a$ and the forbidden band disappears.

(iii) The first Brillouin zone is defined by K-values which lie in the range

$$-\frac{\pi}{2a} < K < \frac{\pi}{2a}$$

Therefore, the smallest possible wavelength of this zone is $4a$ which corresponds to $K = \pi/2a$ at the zone boundary.

Now we investigate the physical difference between the vibrations represented by the optical and acoustical branches and determine the origin of the names of these branches. For the optical branch, as $K \to 0$, $\cos Ka \to 1$, and the Eqs. (4.20) yield

$$-\omega^2 mA = 2\beta B - 2\beta A \qquad \text{or} \qquad \frac{\omega^2}{2\beta} = \frac{A - B}{mA}$$

and

$$-\omega^2 MB = 2\beta A - 2\beta B \qquad \text{or} \qquad \frac{\omega^2}{2\beta} = -\frac{A - B}{MB}$$

$\therefore$

$$1 = -\frac{MB}{mA}$$

or

$$\frac{A}{B} = -\frac{M}{m} \tag{4.27}$$

This indicates that the two atoms move in opposite directions and their amplitudes are inversely proportional to their masses so that the centre of mass of the unit cell remains unchanged. Such a mode of vibration is shown in Fig. 4.9a.

For $M = m$,

$$\frac{A}{B} = -1$$

Thus, even if their masses are equal, the atoms always move in opposite directions.

Now consider the acoustical branch. As $K \to 0$,

$$\cos Ka \to 1 - \frac{K^2 a^2}{2}$$

We have included the second order term on the right hand side as it is the significant term in the present case. The Eqs. (4.20) give

$$-\omega^2 mA = 2\beta B \left(1 - \frac{K^2 a^2}{2}\right) - 2\beta A$$

$$-\omega^2 MB = 2\beta A \left(1 - \frac{K^2 a^2}{2}\right) - 2\beta B$$

Adding these, we get

$$-\omega^2 (mA + MB) = -K^2a^2 \beta (A + B)$$

Using Eq. (4.25), it becomes

$$-K^2a^2 \frac{2\beta}{m+M}(mA + MB) = -K^2a^2\beta (A + B)$$

or $$2(mA + MB) = (m + M)(A + B)$$

or $$mA + MB = MA + mB$$

or $$\frac{A}{B} = +1 \tag{4.28}$$

This means that the two atoms of different masses move in the same direction with the same amplitude as shown in Fig. 4.9b and there is a movement of their centres of masses as well. We, therefore, conclude that, in the optical case, the neighbouring atoms move in opposite directions whereas in the acoustical case, they move in the same direction. These characteristic features of optical and acoustical branches hold for other values of K as well.

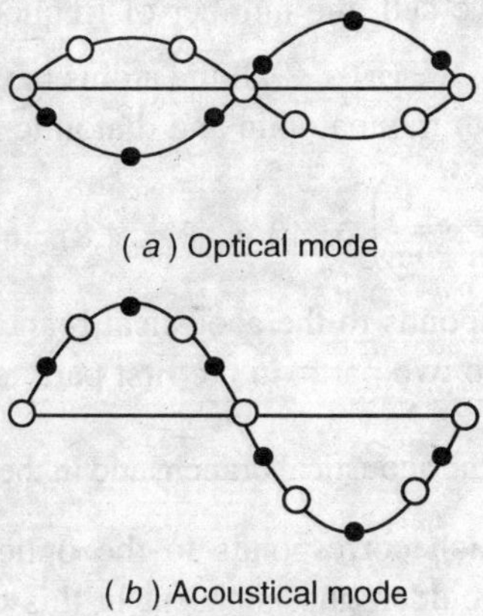

Fig. 4.9. *Transverse optical and transverse acoustical waves in a linear diatomic lattice in the limit $K \to 0$.*

The vibrations of the acoustical branch can be excited by a force which makes the atoms in a crystal move in the same direction. This type of effect may be produced, for example, by directing a beam of sound waves on the surface of a crystal. These vibrations are, therefore, known as *acoustical vibrations*. Monoatomic crystals respond to such vibrations.

The vibrations of the optical branch can be excited by a force which makes the two neighbouring atoms move in opposite directions. Optical radiations induce this type of vibrations in crystals. Hence these vibrations are called *optical vibrations.* Ionic crystals comprising two types of oppositely charged ions respond to such vibrations. The existence of atoms with different charges or different masses is not an essential criterion to induce optical vibrations in a crystal. The mere presence of two atoms per primitive cell is the pre-requisite for the generation of this type of vibrations. The

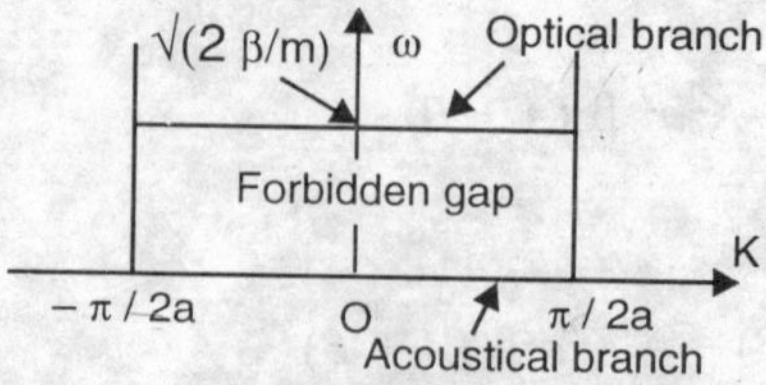

Fig. 4.10. *Dispersion relations for linear diatomic lattice for M → ∞. Acoustical branch coincides with the K-axis and optical branch becomes parallel to the K-axis.*

diamond and hcp lattices having two atoms per primitive cell exhibit optical branches.

Some interesting facts pertaining to diatomic lattices are given as follows :

(i) In diatomic lattices, the allowed frequencies constitute two frequency bands, one corresponding to the acoustical mode and the other to the optical mode of vibration. For polyatomic lattices having *N* atoms per primitive cell, the number of frequency bands is *N*.

(ii) If all the atoms have the same mass m, the frequency range for both monoatomic and diatomic lattices is the same; ω^2 varies from 0 to $2\beta\times\left(\frac{1}{m}+\frac{1}{m}\right)$ or $4\beta/m$ (Fig. 4.8). In monoatomic lattices, the whole range corresponds to the acoustical branch. In diatomic lattices, the range is divided into two parts; in the first part, ω^2 varies from 0 to $2\beta/m$ which corresponds to the acoustical branch, and in the second part, ω^2 lies in the range $\frac{2\beta}{m}<\omega^2<\frac{4\beta}{m}$ which corresponds to the optical branch. The forbidden gap is absent in this case.

(iii) As the heavier mass, *M*, increases, the optical branch flattens and the acoustical branch yields downwards. For $M\to\infty$, the acoustical branch coincides with the *K*-axis and the optical branch becomes parallel to the *K*-axis and represents a single frequency $\sqrt{\frac{2\beta}{m}}$ as shown in Fig. 4.10. The physical meaning of this is that every atom in the lattice behaves as if it is completely free and isolated from its neighbours. It oscillates with its natural frequency $\sqrt{\frac{2\beta}{m}}$. Einstein used this model of lattice vibrations to develop the theory of specific heat of solids. It may be further noted that the allowed frequency bands get compressed into levels. An identical

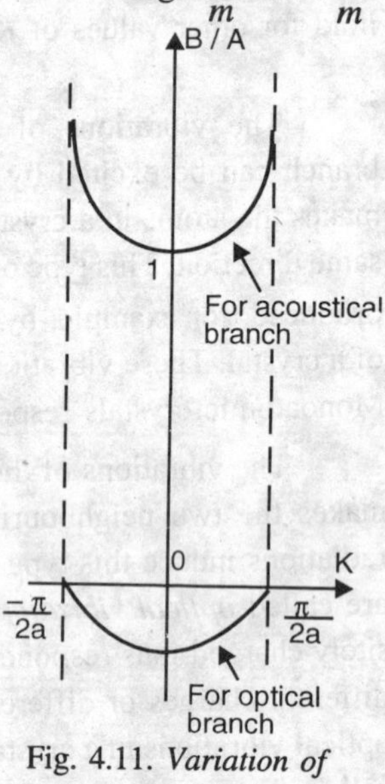

Fig. 4.11. *Variation of amplitude ratio B/A versus K for acoustical and optical branches.*

situation arises for the energies of electrons. We get energy bands when the atoms are bound in a solid and the energy levels when the atoms are free.

(iv) As the lighter mass m decreases, the optical branch moves upwards while the acoustical branch is not affected. As $m \to 0$, the optical branch disappears altogether. This is expected as, for m approaching zero, the lattice becomes monoatomic with a lattice constant $2a$.

(v) The variation of amplitude ratio B/A with wave vector K for both optical and acoustical branches is shown in Fig. 4.11. For $K = \pm\frac{\pi}{2a}$ the ratio B/A is zero for optical branch and infinite for acoustical branch. This indicates that, in the acoustical branch, all the lighter atoms with mass m are at rest ($A = 0$) and, in the optical branch, all the heavier atoms with mass M are at rest ($B = 0$), Thus at the edge of the Brillouin zone, only one of the sublattices is oscillating — the sublattice of heavier atoms in the acoustical branch and the one of lighter atoms in the optical branch. The two modes have different frequencies if masses are unequal and represent standing waves with a phase difference of $\pi/2$. For $K < \pi/2a$, both the sublattices vibrate.

Significance of the Theory

The theory indicates the existence of frequency bands in the diatomic and polyatomic lattices. The bands are separated from one another by forbidden gaps. The crystals cannot propagate the vibrations of all possible frequencies. Only those frequencies are allowed which lie in the allowed bands.

4.4 PHONONS

We know that the energy of an electromagnetic wave is quantized and this quantum of energy is called a photon. Similarly, the energy of a lattice vibration or an elastic wave is also quantized and a quantum of this energy is known as *phonon.* All types of lattice vibrations in crystals comprise phonons — thermal vibrations are thermally excited phonons, sound waves are acoustical phonons and excitations of the optical branch generate optical phonons. Most of the concepts which apply to photons are also valid for phonons. For example, the concept of wave particle duality holds good for phonons. Also, the energy of a phonon is given by $\hbar\omega$. where ω is the angular frequency of a mode of vibration. If n is the number of phonons in a particular mode of vibration, the total energy of that mode is written as

$$\varepsilon = n\hbar\omega \tag{4.29}$$

where n can be zero or a positive integer. Since the number of phonons may change with temperature, the average number of phonons in a vibrational mode is given by

$$\bar{n} = \frac{1}{\exp\left(\frac{\hbar\omega}{k_B T}\right) - 1} \tag{4.30}$$

where k_B is the Boltzmann's constant and T the absolute temperature of the crystal. Thus the number of phonons can be increased or decreased by raising or lowering the temperature respectively. The frequency of phonon waves may vary from 10^4 to 10^{12} cps, i.e., the vibrational spectrum of phonon waves occupies a wide frequency range. The phonons, being indistinguishable particles, require Bose-Einstein distribution function to describe their distribution in the allowed energy states of the system. Due to particle nature of phonons, the interaction of a phonon with another phonon or an electron may be considered as a scattering collision between the two particles.

There is no direct experimental evidence of the quantization of lattice energy. However, the following experimental facts forcefully suggest the existence of phonons :

(i) The lattice heat capacity approaches zero as the temperature approaches zero. This can be explained only if the lattice vibrations are quantized in terms of phonons. This will be described later.

(ii) The crystals scatter x-rays and neutrons inelastically. The change in momentum and energy during this process can be associated with gain or loss of one or more phonons. The properties of these phonons can be determined by measuring the momentum and energy of the scattered x-rays or neutrons.

4.5 MOMENTUM OF PHONONS

Physically, a lattice phonon does not carry any momentum, but it interacts with other particles and fields as if it has a momentum $\hbar\mathbf{K}$, where $\mathbf{K}$ represents the wave vector of the phonon. Also, from the de Broglie relation

$$p = \frac{h}{\lambda} = \hbar K,$$

it is apparent that a phonon of wavelength λ carries a momentum $\hbar\mathbf{K}$. The quantity $\hbar\mathbf{K}$ is sometimes called the *crystal momentum*. The physical significance of $\hbar\mathbf{K}$ is provided by the momentum conservation laws in crystals.

The wave vector conservation law for elastic scattering or the Bragg's diffraction of x-ray photons from crystals is given by

$$\mathbf{k'} = \mathbf{k} + \mathbf{G} \tag{4.31}$$

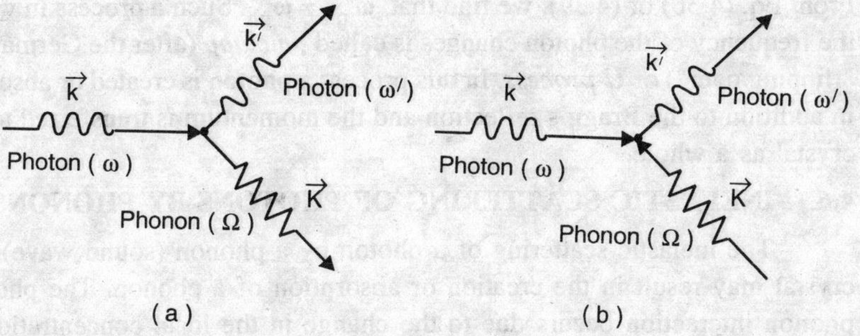

Fig. 4.12. *Inelastic scattering of incident photon of wave vector* **k** *to produce scattered photon of wave vector* **k'** *along with the emission (a) or absorption (b) of a phonon of wave vector* **K**.

where **k'** and **k** represent the wave vectors for the scattered and incident photons respectively and **G** is the reciprocal lattice vector. The conservation of momentum and energy yields.

$$\hbar \mathbf{k}' = \hbar \mathbf{k} + \hbar \mathbf{G} \tag{4.32}$$

and

$$\hbar \omega_{ph} = \hbar \omega'_{ph} \tag{4.33}$$

where ω_{ph} and ω'_{ph} are the frequencies of incident and scattered photons respectively. In this process the crystal as a whole recoils with a momentum $-\hbar\mathbf{G}$ and the frequency of the incident photons remains unchanged, i.e., $\omega_{ph} = \omega'_{ph}$. This also follows from the relation $|\mathbf{k}'| = |\mathbf{k}|$ which holds good for x-ray diffraction. Such a process in which the frequency of the incident photon is the same as that of the scattered photon is called *normal* or *N-process.*

Consider now the case of inelastic scattering of the photon which proceeds with the emission of a phonon of wave vector **K** as shown in Fig. 4.12a. The wave vector conservation law then takes the form

$$\mathbf{k}' + \mathbf{K} = \mathbf{k} + \mathbf{G} \tag{4.34}$$

Accordingly, the momentum and energy conservation laws can be written as

$$\hbar \mathbf{k}' + \hbar \mathbf{K} = \hbar \mathbf{k} + \hbar \mathbf{G} \tag{4.35}$$

and

$$\hbar \omega'_{ph} + \hbar \omega = \hbar \omega_{ph} \tag{4.36}$$

where ω is the frequency of the phonon generated. In case a phonon is absorbed in the scattering process, as shown in Fig. 4.12b, the wave vector conservation law gives

$$\mathbf{k}' = \mathbf{k} + \mathbf{K} + \mathbf{G} \tag{4.37}$$

and the momentum and energy conservation laws become

$$\hbar \mathbf{k}' = \hbar \mathbf{k} + \hbar \mathbf{K} + \hbar \mathbf{G} \tag{4.38}$$

and $$\hbar\omega'_{ph} = \hbar\omega_{ph} + \hbar\omega \tag{4.39}$$

From Eq. (4.36) or (4.39), we find that $\omega'_{ph} \neq \omega_{ph}$. Such a process in which the frequency of the photon changes is called *umclapp* (after the German for "flipping over") or *U-process*. In this process a photon is created or absorbed in addition to the Bragg's reflection and the momentum is transferred to the crystal as a whole.

4.6 INELASTIC SCATTERING OF PHOTONS BY PHONONS

The inelastic scattering of a photon by a phonon (sound wave) in a crystal may result in the creation or absorption of a phonon. The photon-phonon interaction occurs due to the change in the local concentration of atoms, and hence in the refractive index of the crystal. Conversely, the electric field of the light wave induces mechanical vibrations in the medium and hence affects its elastic properties.

Consider a photon of frequency $\upsilon = \dfrac{\omega_{ph}}{2\pi}$ and wave vector **k** propagating through a crystal which is viewed as a continuum of refractive index n. Then, we have

$$\omega_{ph} = \frac{ck}{n} \text{ or } \lambda\upsilon = \frac{c}{n} \tag{4.40}$$

where c is the velocity of light. Let this photon interact with a phonon. As a result of this interaction, the wave vector and frequency of the photon change to **k**' and υ' respectively. If a phonon of wave vector **K** and angular frequency ω is created in this process (Fig. 4.12a), the conservation of energy and momentum gives

$$\hbar\omega_{ph} = \hbar\omega'_{ph} + \hbar\omega \tag{4.41}$$

$$\hbar\mathbf{k} = \hbar\mathbf{k}' + \hbar\mathbf{K} \tag{4.42}$$

The Eq. (4.42) has been written without taking into account the phenomenon of Bragg's diffraction along with that of scattering. Taking v_s as the velocity of sound (phonon) and assuming it to be constant, we can write

$$\omega = v_s k \tag{4.43}$$

The wavelength of the emitted phonon is comparable to the wavelength of the incident photon whereas its energy is only a small fraction of the incident photon energy. From Eqs. (4.40) and (4.43), we find that, since $v_s \ll c$ or c/n,

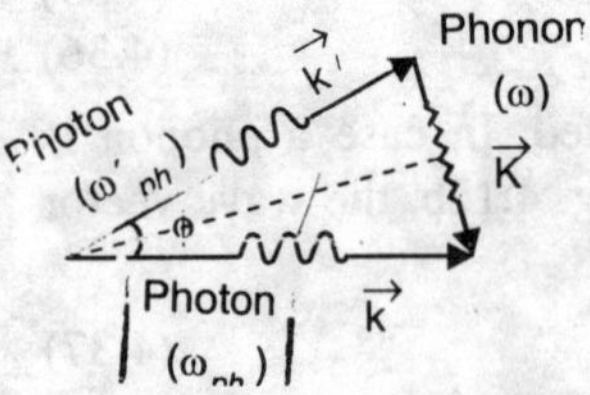

Fig. 4.13. *Selection rule diagram for the emission of a phonon when k = k'*

$\omega_{ph} >> \omega$. It, therefore, follows from Eqs. (4.41) and (4.42) that

$$\omega_{ph} \simeq \omega'_{ph}$$

and

$$k \simeq k'$$

The wave vectors **k**, **k'**, and **K** are related to each other as shown in Fig. 4.13. It also follows from Fig. 4.13 that, for **k** = **k'**, the triangle becomes isosceles and we have

$$K = 2k \sin\frac{\phi}{2} \tag{4.44}$$

where ϕ is the angle between **k** and **k'**. Thus a phonon is produced when a photon is scattered inelastically at an angle $\frac{\phi}{2}$ from the direction of incidence. The frequency of the phonon is given by

$$\omega = v_s K$$

$$= 2\, v_s k \sin\frac{\phi}{2}$$

$$= 2\, \frac{v_s\, \omega_{ph} n}{c} \sin\frac{\phi}{2} \tag{4.45}$$

The phonons have been generated in quartz and sapphire in the microwave frequency range by scattering the visible light produced from an intense laser source. The observed shift in frequency of the photon agrees well with the shift calculated from Eq. (4.45) using the value of the velocity of sound determined by ultrasonic methods at low frequencies.

4.7 SPECIFIC HEAT

The specific heat of a substance is defined as the heat required to raise the temperature of one gram molecule of the substance through 1°C, i.e.,

$$C = \frac{dQ}{dT} \tag{4.46}$$

where dQ is the amount of heat added to a system to raise its temperature from T to $T + dT$. The quantity of heat required to increase the temperature of a body is different under different conditions; accordingly, one can have various types of specific heats. For example, the specific heat at constant pressure, C_p, is generally different from the specific heat at constant volume, C_v. The former is always greater than the latter.

According to the first law of thermodynamics, the heat added to a system is used up in two ways; firstly, to increase the internal energy of the system, thereby raising its temperature, and secondly, to do some work on the system to increase its volume against an external pressure. It is the latter quantity which may have different values under different conditions. If the system expands against a constant pressure, then the first law can be written as

$$dQ = dE + pdV \tag{4.47}$$

The first term on the right hand side represents the change in internal energy of the system and the second one represents the work done to change the volume by an amount dV at a pressure p. If heat is added to the system at constant volume, the second term in Eq. (4.43) vanishes and the specific heat at constant volume may be expressed as

$$C_v = \left(\frac{dQ}{dT}\right)_v = \left(\frac{\partial E}{\partial T}\right)_v \tag{4.48}$$

Similarly, one can express the specific heat at constant pressure as

$$C_p = \left(\frac{dQ}{dT}\right)_p$$

In gases, there is a large difference in C_p and C_v. However, in solids, due to a small change in volume, C_p is almost the same as C_v particularly at low temperatures. For this reason, the term 'specific heat of solids' is commonly used in case of solids. However, it strictly means the specific heat at constant volume and is given by Eq. (4.48). Thus, in solids, most of the heat supplied is used up in increasing the internal energy. The increase in internal energy of a solid may occur in two ways :

(i) The atoms, which ordinarily vibrate freely about their equilibrium positions, are set into rigorous vibrations. This is manifested by a rise in temperature.

(ii) The free electrons in case of metals and semiconductors get thermally excited to higher energy states.

The first contribution arises from the atomic vibrations and may be called the *lattice specific heat.* The second contribution arises from the electronic system and is a relatively small contribution. Thus, in general, the specific heat of a solid may be expressed as

$$C_{solid} = C_{lat} + C_{elec} \tag{4.49}$$

For further discussion on this chapter, it is assumed that no free electrons are present and the specific heat of a crystal is only due to the excitation of thermal vibrations in the lattice. i.e., only the lattice specific heat is to be considered.

Occasionally, the term 'heat capacity' is used instead of 'specific heat' and is defined as the heat required to raise the temperature of the complete mass (or volume) of a solid through 1°C. Thus the specific heat is the heat capacity per gram (or per unit volume). The experimental facts about the heat capacity of some representative inorganic solids are given as follows:

(i) The heat capacity of most of the solids at room temperature is close to $3Nk_B$, where N is the number of atoms in the solid and k_B is the Boltzmann's constant. For one mole of atoms, the heat capacity is given by $3N_a k_B$, where N_a is the Avogadro'a number. Its value is 25 J/mole–K or 6 cal/mole–K approximately.

(ii) At lower temperatures ($T \to 0\,\text{K}$), the heat capacity decreases sharply and follows T^3-law for insulators and T-law for metals. If metal becomes a superconductor, the decrease is even faster.

(iii) In magnetic solids, the heat capacity increases near the Curie temperature when the magnetic moments become ordered. This is also true for alloys exhibiting order-disorder transformations.

We discuss below the theoretical explanation of the facts regarding the lattice heat capacity. The discussion is primarily applicable to insulators.

4.8 CLASSICAL THEORY OF LATTICE HEAT CAPACITY

A crystal consists of atoms which are arranged in a periodic manner and are bound together by strong binding forces. Each atom is free to vibrate about its equilibrium position and constitutes a three-dimensional harmonic oscillator. The effect of imparting thermal energy to a solid is to increase its internal energy in the form of vibrational energy of these harmonic oscillators. Thus, in the classical theory, it is assumed that each atom of a crystal acts as a three-dimensional harmonic oscillator and all the atoms vibrate independent of one another. Further, a system of N vibrating atoms or N independent three-dimensional harmonic oscillators is equivalent to a system of $3N$ identical and independent one-dimensional harmonic oscillators. This is because each vibrating atom has three independent vibrational degrees of freedom and, according to the principle of equipartition of energy, the vibrational energy is distributed equally among all the three degrees of freedom. Thus each vibrational degree of freedom can be regarded as a one-dimensional harmonic oscillator.

We assume that each harmonic oscillator vibrates with its natural frequency ω_0. However, the energies of these osciallators may be different because they may vibrate with different amplitudes. Also, each oscillator may acquire any value of energy ranging continuously from zero to infinity. In order to determine the heat capicity of a solid, we find out the average thermal energy of single one-dimensional harmonic oscillator, multiply it by $3N$ and then use Eq. (4.46) or (4.48).

The energy ε of a one-dimensional harmonic oscillator of mass m and angular frequency ω_o is given by

$$\varepsilon = \frac{p^2}{2m} + V(x)$$

$$= \frac{p^2}{2m} + \frac{1}{2} m\omega_o{}^2 x^2 \tag{4.50}$$

where $\frac{p^2}{2m}$ represents the kinetic energy, p being the momentum, and $V(x)$ is the potential energy at a displacement x from the mean position. Assuming that the distribution of oscillators in energy obeys the Maxwell-Boltzmann distribution law, the average energy of each harmonic oscillator is given by

$$\bar{\varepsilon} = \frac{\int \varepsilon \exp\left(-\frac{\varepsilon}{k_B T}\right) d\varepsilon}{\int \exp\left(-\frac{\varepsilon}{k_B T}\right) d\varepsilon}$$

Using Eq. (4.50), we get

$$\bar{\varepsilon} = \frac{\iint\limits_{p\,x} \left(\frac{p^2}{2m} + \frac{1}{2} m\omega_o^2 x^2\right) \exp\left(-\frac{p^2}{2mk_B T}\right) \exp\left(-\frac{m\omega_o^2 x^2}{2k_B T}\right) dp dx}{\iint\limits_{p\,x} \exp\left(-\frac{p^2}{2mk_B T}\right) \exp\left(-\frac{m\omega_o^2 x^2}{2k_B T}\right) dp dx}$$

$$= \frac{\iint\limits_{p\,x} \left(\frac{p^2}{2m}\right) \exp\left(-\frac{p^2}{2mk_B T}\right) \exp\left(-\frac{m\omega_o^2 x^2}{2k_B T}\right) dp dx}{\iint\limits_{p\,x} \exp\left(-\frac{p^2}{2mk_B T}\right) \exp\left(-\frac{m\omega_o^2 x^2}{2k_B T}\right) dp dx}$$

$$+ \frac{\iint\limits_{p\,x} \left(\frac{m\omega_o^2 x^2}{2}\right) \exp\left(-\frac{p^2}{2mk_B T}\right) \exp\left(-\frac{m\omega_o^2 x^2}{2k_B T}\right) dp dx}{\iint\limits_{p\,x} \exp\left(-\frac{p^2}{2mk_B T}\right) \exp\left(-\frac{m\omega_o^2 x^2}{2k_B T}\right) dp dx}$$

$$= \frac{\int\limits_p \left(\frac{p^2}{2m}\right) \exp\left(-\frac{p^2}{2mk_BT}\right) dp}{\int\limits_p \exp\left(-\frac{p^2}{2mk_BT}\right) dp} + \frac{\int\limits_x \left(\frac{m\omega_o^2 x^2}{2}\right) \exp\left(-\frac{m\omega_o^2 x^2}{2k_BT}\right) dx}{\int\limits_x \exp\left(-\frac{m\omega_o^2 x^2}{2k_BT}\right) dx}$$

Now $$\int_0^\infty u^2 \exp(-au^2)\, du = \frac{1}{4}\sqrt{\frac{\pi}{a^3}}$$

and $$\int_0^\infty \exp(-au^2)\, du = \frac{1}{2}\sqrt{\frac{\pi}{a}}$$

$$\therefore \quad \bar{\varepsilon} = \frac{1}{2m} \cdot \frac{\frac{1}{4}\left[\pi(2mk_BT)^3\right]^{1/2}}{\frac{1}{2}\left[\pi(2mk_BT)\right]^{1/2}} + \frac{1}{2} m\omega_o^2 \frac{\frac{1}{4}\left[\pi\left(\frac{2k_BT}{m\omega_o^2}\right)^3\right]^{1/2}}{\frac{1}{2}\left[\pi\left(\frac{2k_BT}{m\omega_o^2}\right)\right]^{1/2}}$$

$$= \frac{1}{2}k_BT + \frac{1}{2}k_BT$$

$$= k_BT \tag{4.51}$$

Thus the total vibrational energy of a crystal containing N identical atoms or $3N$ one-dimensional harmonic oscillators becomes

$$E = 3N\bar{\varepsilon} = 3Nk_BT \tag{4.52}$$

It is important to note that the total vibrational energy of a crystal is independent of the type of frequency distribution of oscillators assumed in this model. Also, the energy E depends only on the temperature provided the volume remains constant.

Using Eq. (4.48), the heat capacity of a solid consisting of N atoms is

$$C_v = \left(\frac{\partial E}{\partial T}\right)_v = 3Nk_B \tag{4.53}$$

The molar heat capacity is, therefore, given by

$$C_{vm} = 3N_a k_B = 3R = 5.96 \text{ cal/mole–K} \tag{4.54}$$

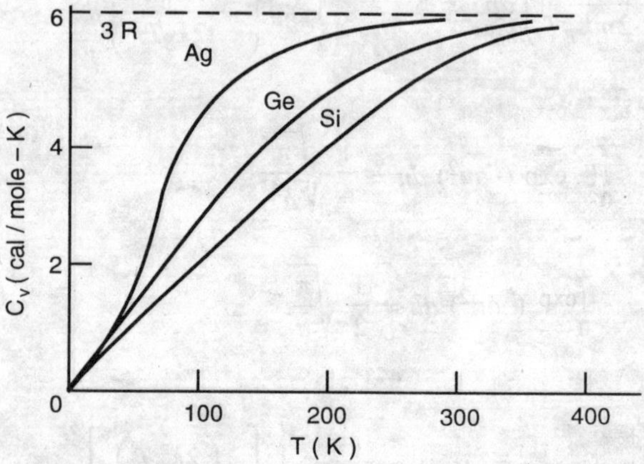

Fig. 4.14. *Heat capacity of silver, germanium and silicon as a function of temperature.*

Thus, according to the classical theory, the molar heat capacity of all the solids is constant and is independent of temperature and frequency. This is called the *Dulong and Petit's law.* The result is in good agreement with the observed heat capacity for a number of solids including metals at high temperatures and often at room temperature too. The theory, however, fails to account for the value of heat capacity at low temperatures. As shown in Fig. 4.14, the heat capacity approaches zero for Ag, Ge and Si as T approaches zero. This discrepancy was resolved first by Einstein and then by Debye by using the quantum theory.

4.9 EINSTEIN'S THEORY OF LATTICE HEAT CAPACITY

Einstein, in 1911, attempted to resolve the discrepancies of the classical theory of specific heat by applying the Planck's quantum theory. Einstein retained all the assumptions of the classical theory as such except replacing the classical harmonic oscillators by quantum harmonic oscillators which can have only discrete energy values. The salient features of the Einstein's theory are listed below :

(i) A crystal consists of atoms which may be regarded as identical and independent harmonic oscillators.

(ii) A solid consisting of N atoms is equivalent to $3N$ one-dimensional harmonic oscillators.

(iii) All the oscillators vibrate with the same natural frequency due to the identical environment of each.

(iv) The oscillators are quantum oscillators and have discrete energy spectrum rather than the continuous one as for classical oscillators.

(v) Any number of oscillators may be present in the same quantum state.

(vi) The atomic oscillators form an assembly of systems which are distinguishable or identifiable due to their location at separate and distinct lattice sites and hence obey the Maxwell-Boltzmann distribution of energies.

Considering the Planck's quantum theory, the discrete energy values of an oscillator with frequency υ are given by

$$\epsilon_n = nh\upsilon = n\hbar\omega_o \tag{4.55}$$

where n = 0, 1, 2, 3, ... is called the quantum number. Einstein later used the wave mechanical result which gives the energy levels of the harmonic oscillator as

$$\varepsilon_n = \left(n + \frac{1}{2}\right) h\upsilon = \left(n + \frac{1}{2}\right) \hbar\omega_o \tag{4.56}$$

where $\frac{1}{2}\hbar\omega_o$ is the temperature independent *zero point energy* contribution to the internal energy of the oscillator. To calculate the average energy of an oscillator, we replace integration by summation in the expression for the Maxwell-Boltzmann distribution of energy and obtain

$$\bar{\varepsilon} = \frac{\sum\limits_{n=0}^{\infty} \varepsilon_n \exp\left(-\frac{\varepsilon_n}{k_B T}\right)}{\sum\limits_{n=0}^{\infty} \exp\left(-\frac{\varepsilon_n}{k_B T}\right)} \tag{4.57}$$

Using Eq. (4.56), we get

$$\bar{\varepsilon} = \frac{\sum\limits_{n=0}^{\infty}\left(n+\frac{1}{2}\right)\hbar\omega_o \exp\left[-\left(n+\frac{1}{2}\right)\frac{\hbar\omega_o}{k_B T}\right]}{\sum\limits_{n=0}^{\infty} \exp\left[-\left(n+\frac{1}{2}\right)\frac{\hbar\omega_o}{k_B T}\right]}$$

$$= \frac{\hbar\omega_o \sum\limits_{n=0}^{\infty}\left(n+\frac{1}{2}\right) \exp\left[\left(n+\frac{1}{2}\right)x\right]}{\sum\limits_{n=0}^{\infty} \exp\left[\left(n+\frac{1}{2}\right)x\right]} \tag{4.58}$$

where $$x = -\frac{\hbar\omega_o}{k_B T}$$

$$\therefore \quad \bar{\varepsilon} = \frac{\hbar\omega_o\left(\frac{1}{2}e^{x/2} + \frac{3}{2}e^{3x/2} + \frac{5}{2}e^{5x/2} + \text{----}\right)}{\left(e^{x/2} + e^{3x/2} + e^{5x/2} + \text{----}\right)}$$

$$= \hbar\omega_o \frac{d}{dx} \ln\left[e^{x/2} + e^{3x/2} + e^{5x/2} + \text{----}\right]$$

$$= \hbar\omega_o \frac{d}{dx} \ln\left[e^{x/2}\left(1 + e^{x} + e^{2x} + \text{---}\right)\right]$$

Now $$\ln(1 + e^x + e^{2x} + \text{---}) = -\ln(1 - e^x)$$

$$\therefore \quad \bar{\varepsilon} = \hbar\omega_o \frac{d}{dx}\left[\frac{x}{2} - \ln\left(1 - e^x\right)\right]$$

$$= \hbar\omega_o\left[\frac{1}{2} + \frac{e^x}{1-e^x}\right] = \hbar\omega_o\left[\frac{1}{2} + \frac{1}{e^{-x}-1}\right]$$

$$= \hbar\omega_o\left[\frac{1}{2} + \frac{1}{e^{\hbar\omega_o/k_B T} - 1}\right]$$

$$= \frac{1}{2}\hbar\omega_o + \frac{\hbar\omega_o}{\exp\left(\frac{\hbar\omega_o}{k_B T}\right) - 1} \tag{4.59}$$

If Eq. (4.55) is used in Eq. (4.57), the expression for the average energy becomes

$$\bar{\varepsilon} = \frac{\hbar\omega_o}{\exp\left(\frac{\hbar\omega_o}{k_B T}\right) - 1} \tag{4.60}$$

The expression (4.59) is identical to (4.60) except with the difference of the energy $\frac{1}{2}\hbar\omega_o$ which is the temperature independent zero point energy as, for $T = 0$, $\bar{\varepsilon} = \frac{1}{2}\hbar\omega_o$. Thus, according to quantum mechanics, the atoms are not at rest even at 0 K and each atom possesses the vibrational energy of $\frac{1}{2}\hbar\omega_o$. This energy, however, has no contribution to C_v. It may be noted that the expression (4.59) also contains the frequency of the oscillator, in contrast to the corresponding expression (4.51) of the classical theory.

The expression for the internal energy of the crystal becomes

$$E = 3N\bar{\varepsilon}$$

$$= \frac{3}{2} N\hbar\omega_o + \frac{3N\hbar\omega_o}{e^{\hbar\omega_o/k_BT} - 1} \tag{4.61}$$

$$\therefore \quad C_v = \left(\frac{\partial E}{\partial T}\right)_v = 3Nk_B \left(\frac{\hbar\omega_o}{k_BT}\right)^2 \frac{e^{\hbar\omega_o/k_BT}}{\left(e^{\hbar\omega_o/k_BT} - 1\right)^2} \tag{4.62}$$

Let
$$\theta_E = \frac{\hbar\omega_o}{k_B} \tag{4.63}$$

where θ_E is the characteristic temperature known as the *Einstein temperature*

$$\therefore \quad C_v = 3Nk_B \left(\frac{\theta_E}{T}\right)^2 \frac{e^{\theta_E/T}}{\left(e^{\theta_E/T} - 1\right)^2} \tag{4.64}$$

We now consider the following cases :

(i) High temperature behaviour

For temperatures such that

$$k_BT >> \hbar\omega_o \quad \text{or} \quad T >> \theta_E,$$

we can write

$$e^{\hbar\omega_o/K_BT} - 1 \cong 1 + \frac{\hbar\omega_o}{k_BT} + \text{-----} - 1 = \frac{\hbar\omega_o}{k_BT} \tag{4.65}$$

and from Eq. (4.59), we get

$$\bar{\varepsilon} = \frac{1}{2}\hbar\omega_o + k_BT \cong k_BT$$

Thus at high temperatures, the average vibrational energy is the same as that obtained from the classical theory. Also, the Eq. (4.62) becomes

$$C_v = 3Nk_B \left(\frac{\hbar\omega_o}{k_BT}\right)^2 \frac{\left(1 + \frac{\hbar\omega_o}{k_BT}\right)}{\left(\frac{\hbar\omega_o}{k_BT}\right)^2} \qquad \text{[Using Eq. (4.65)]}$$

$$= 3Nk_B\left(1 + \frac{\hbar\omega_o}{k_BT}\right)$$

For large T, $\frac{\hbar\omega_o}{k_BT} \rightarrow 0$ and we get

$$C_v = 3Nk_B$$
$$= 3R \qquad \text{(for } N = N_a\text{)}$$

which is the Dulong and Petits's law as obtained from classical theory.

(ii) *Low temperature behaviour*

For temperatures such that

$$k_BT << \hbar\omega_o \text{ or } T << \theta_F,$$

we can write

$$e^{\hbar\omega_o/k_BT} - 1 \cong e^{\hbar\omega_o/k_BT}$$

Therefore, Eq. (4.59) becomes

$$\bar{\varepsilon} = \frac{1}{2}\hbar\omega_o + \hbar\omega_o e^{-\hbar\omega_o/k_BT} \tag{4.66}$$

It shows that, at low temperatures, the average vibrational energy decreases exponentially with decrease in temperature. The expression (4.62) for C_v becomes

$$C_v = 3Nk_B\left(\frac{\hbar\omega_o}{k_BT}\right)^2 e^{-\hbar\omega_o/k_BT} \tag{4.67}$$

or, in terms of θ_E, it becomes

$$C_v = 3Nk_B\left(\frac{\theta_E}{T}\right)^2 e^{-\theta_E/T} \tag{4.68}$$

Thus, for $T << \theta_E$, the heat capacity is proportional to $e^{-\theta_E/T}$ which is the dominating factor. However, experimentally it is found to vary as T^3 for most of the solids.

A theoretical plot of C_v versus T/θ_E using Eq. (4.64) for diamond using $\theta_E = 1320$ K is shown in Fig. 4.15. Some experimental points are also shown for comparison. It is seen that the Einstein curve fits the experimental points well over a wide range of temperature except for very low (liquid helium) temperatures where it falls more rapidly than the observed decrease in C_v which follows T^3 law.

The Einstein temperature defined by Eq. (4.63), which distinguishes between the low and high temperature behaviours, can be determined if ω_0 is known. As described in Secs. 4.2 and 4.3, the natural vibrational frequency can be determined from the atomic mass and the observed elastic constants of the crystal and, for monoatomic crystals, it is equal to $\sqrt{4\beta / m}$. For typical metallic elements, the value of θ_E varies from 100 to 200K.

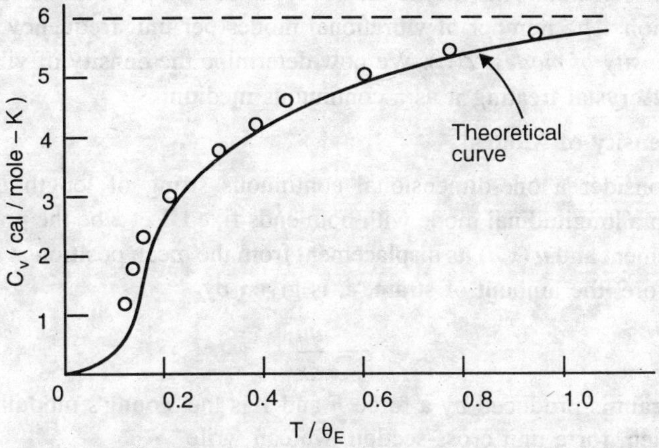

Fig. 4.15. *Comparison of theoretical heat capacity of diamond calculated from Einstein's model with the experimentally observed values.*

Although the Einstein's model provides a much better explanation for the variation of heat capacity with temperature than the classical theory, it fails to account for the values of specific heat at very low temperatures. The discrepancy arises due to the over-simplified assumption of the Einstein's model in which the atomic oscillators are considered to vibrate independently at the same frequency. These oscillators, in fact, are coupled together and there may be a number of possible vibrational frequencies rather than a single frequency ω_0. This fact is accounted for in the Debye's model which is described below.

4.10 DEBYE'S MODEL OF LATTICE HEAT CAPACITY

The Einstein's theory assumed that the atoms of a crystal vibrate totally independent of one another. The vibrational motion of the crystal as a whole was considered to be the same as the vibrational motion of a single atom and, therefore, all the atomic vibrations of the crystal were assigned a common frequency ω_0 which is the natural frequency of vibration of a single atom. This assumption is over-simplified since the atoms are bound together in a crystal and form a system of coupled harmonic oscillators which cannot vibrate independently. This fact was taken into consideration by Debye who, in 1912, put forward his model of lattice heat capacity. In this model, the vibrational motion of the crystal as a whole was considered to be equivalent to the vibrational motion of a system of coupled harmonic oscillators which can propagate a range of frequencies rather than a single frequency. Debye proposed that crystals can propagate elastic waves of wavelengths ranging from low frequencies (sound waves) to high frequencies corresponding to infrared absorption. This means that a crystal can have a number of modes of vibration. The number of vibrational modes per unit frequency range is called *density of modes,* $Z(\upsilon)$. We now determine the density of vibrational modes of crystal treating it as a continuous medium.

4.10.1 Density of Modes

Consider a one-dimensional continuous string of length L. Let it vibrate in a longitudinal mode with both ends fixed. Let x be the position of a line element and $u\,(x, t)$ its displacement from the mean position at any time t. Therefore, the amount of strain, e, is given by

$$e = \frac{du}{dx}$$

If this strain is produced by a force F and Y is the Young's modulus of the string, then, for a unit cross-section, we can write

$$\frac{F}{e} = Y$$

Now consider an element of length Δx. If the strain at one of its ends is $e(x)$, then its value at the other end is

$$e(x) + \frac{\partial e}{\partial x}\,\Delta x = e(x) + \frac{\partial^2 u}{\partial x^2}\,\Delta x$$

The force at the two ends of the string is

$$e(x)Y \qquad \text{and} \qquad [e(x) + \frac{\partial^2 u}{\partial x^2}\,\Delta x]Y$$

Therefore, the net force on the element is

$$\left(\frac{\partial^2 u}{\partial x^2}\right)\Delta x Y$$

The force on the line element Δx can also be written as

$$\rho\Delta x\left(\frac{\partial^2 u}{\partial t^2}\right)$$

where ρ is the density of the string. Therefore, we have

$$\left(\frac{\partial^2 u}{\partial x^2}\right) \Delta x Y = \rho \Delta x \left(\frac{\partial^2 u}{\partial t^2}\right) \text{ or } \frac{\partial^2 u}{\partial x^2} = \frac{\rho}{Y} \frac{\partial^2 u}{\partial t^2}$$

or

$$\frac{\partial^2 u}{\partial x^2} = \frac{1}{v_s^2} \frac{\partial^2 u}{\partial t^2} \tag{4.69}$$

where $v_s = \sqrt{\frac{Y}{\rho}}$ represents the velocity of propagation of the wave along the string. It is independent of frequency. The Eq. (4.69) is the well known one-dimensional wave equation. Since the string is fixed at both the ends, the solutions of Eq. (4.69) should correspond to standing waves. These types of solutions are given by

$$u(x, t) = A \sin\left(\frac{n\pi}{L} x\right) \cos 2\pi \upsilon_n t \tag{4.70}$$

where n is a positive integer ≥ 1. Using Eq. (4.70) in (4.69), we get

$$\lambda_n = \frac{2L}{n}$$

and

$$\upsilon_n = \frac{v_s}{\lambda_n} = \frac{n v_s}{2L} \tag{4.71}$$

i.e.,

$$\upsilon_1 = \frac{v_s}{2L}, \upsilon_2 = \frac{v_s}{L}, \upsilon_3 = \frac{3v_s}{2L}, \text{ etc.} \tag{4.72}$$

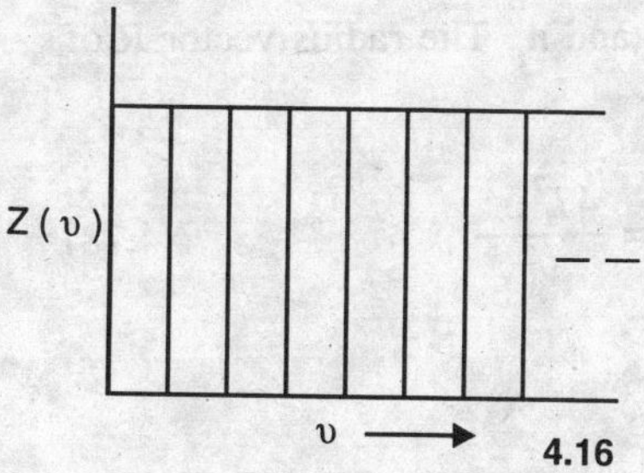

Fig. 4.16. *Frequency spectrum of a continuous string.*

This shows that the frequency of the string can have discrete values only and is an integral multiple of $\frac{v_s}{2L}$. It implies that the frequency spectrum of a continuous string is discrete and contains an infinite number of parallel lines equidistant from each other as shown in Fig. 4.16.

From Eqs. (4.71), we have

$$n = \frac{2L}{v_s} \upsilon_n$$

$$\therefore \quad dn = \frac{2L}{v_s} d\upsilon \tag{4.73}$$

This gives the number of possible modes of vibration in the frequency interval $d\upsilon$.

Considering now the three-dimensional case; the wave equation (4.69) can be written as

$$\frac{\partial^2 u}{\partial x^2} + \frac{\partial^2 u}{\partial y^2} + \frac{\partial^2 u}{\partial z^2} = \frac{1}{v_s^2}\frac{\partial^2 u}{\partial t^2} \tag{4.74}$$

The three-dimensional continuous medium can be taken as a cube of side L whose faces are fixed. In analogy with Eq. (4.70), the standing wave solutions of the wave equation (4.74) are

$$u(x, y, z, t) = A\sin\left(\frac{n_x\pi}{L}x\right)\sin\left(\frac{n_y\pi}{L}y\right)\sin\left(\frac{n_z\pi}{L}z\right)\cos 2\pi\upsilon t \tag{4.75}$$

where n_x, n_y, n_z are positive integers ≥ 1. Substituting this solution into (4.74) and simplifying, we obtain

$$\frac{\pi^2}{L^2}(n_x^2 + n_y^2 + n_z^2) = \frac{4\pi^2\upsilon^2}{v_s^2} \tag{4.76}$$

or

$$n_x^2 + n_y^2 + n_z^2 = \frac{4L^2\upsilon^2}{v_s^2} \tag{4.77}$$

This equation gives the possible modes of vibration. The integers n_x, n_y and n_z determine the possible frequencies or wavelengths. In order to determine the number of possible modes of vibration, $Z(\upsilon)d\upsilon$, present in the frequency range υ and $\upsilon + d\upsilon$, we consider a network of points in the space defined by three positive integral coordinates n_x, n_y and n_z. The radius vector R of any point from the origin is given by

$$R^2 = n_x^2 + n_y^2 + n_z^2 = \frac{4L^2\upsilon^2}{v_s^2} \tag{4.78}$$

This is the equation of a sphere of volume

$$V' = \frac{4}{3}\pi R^3$$

Differentiating it, we get

$$dV' = 4\pi R^2 dR$$

The number of modes present in the frequency range υ and $\upsilon + d\upsilon$ should be the same as number of points lying in the volume interval V' and $V' + dV'$ or in the range R and $R + dR$ of the radius vector. Since each point occupies, on an average, a unit volume in the space of integers, the number of points present in the volume dV' of the spherical shell is numerically equal to the volume of the shell, i.e.,

$$dn = 4\pi R^2 dR$$

But since a mode of vibration is always determined by the positive values of n_x, n_y and n_z only, we must consider the number of points lying in the octant defined by these positive integers only. Thus the number of possible modes of vibraton is

$$Z(\upsilon)d\upsilon = \frac{1}{8}\left(4\pi R^2 dR\right)$$

$$= \frac{1}{8}4\pi \frac{4L^2\upsilon^2}{v_s^2}\frac{2L}{v_s}d\upsilon = \frac{4\pi L^3\upsilon^2}{v_s^3}\,d\upsilon$$

$$= \left(\frac{4\pi V}{v_s^3}\right)\upsilon^2 d\upsilon \tag{4.79}$$

where $V = L^3$ is the volume of the solid cube. For a perfect continuum, the possible frequencies vary between zero and infinity and the number of possible vibrational modes increases with square of the frequency as shown in Fig. 4.17. This type of situation arises in case of electromagnetic waves travelling in a box. Thus Eq. (4.76) is fundamental to the theory of black body radiations.

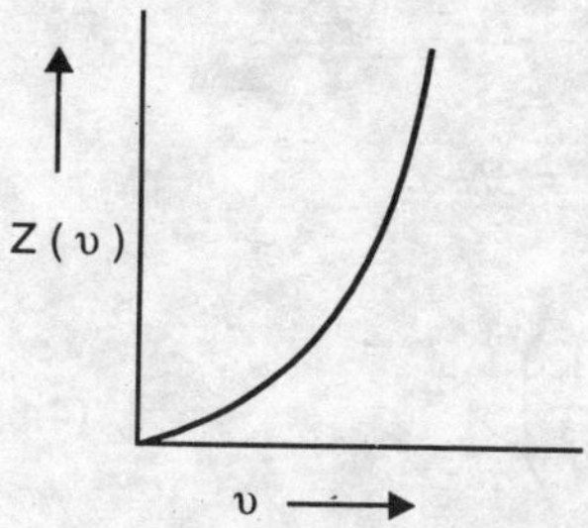

Fig. 4.17. *Frequency spectrum of a three-dimensional continuum.*

In general, the elastic waves propagating in a solid are of two types — transverse waves and longitudinal waves. The velocity of propagation, v_t, of transverse waves is generally different from the velocity of propagation, v_l, of longitudinal waves. Also, for each frequency or direction of propagation, the transverse waves have two vibrational modes perpendicular to the direction of propagation whereas the longitudinal waves have only one mode which lies along the direction of propagation. For such a case, the total number of vibrational modes is expressed as

$$Z(\upsilon)d\upsilon = 4\pi V\left(\frac{2}{v_t^3} + \frac{1}{v_l^3}\right)\upsilon^2 d\upsilon \tag{4.80}$$

4.10.2 The Debye Approximation

If the interatomic distance is small as compared to the wavelength of elastic waves, the crystal can be regarded as a continuum from the point of view of the wave. Based on this idea, Debye assumed that the continuum model is applicable to all possible vibrational modes of the crystal. Further, the fact that the crystal consists of a finite number (N) of atoms is taken into account by limiting the total number of vibrational modes to $3N$. This puts an upper limit to the frequency of the elastic waves which can propagate through the crystal. This highest frequency propagating through a crystal is called the *Debye frequency*, υ_D. It is common to transverse and longitudinal modes of vibrations. Hence the frequency spectrum of a continuous medium is cut off at the Debye frequency as shown in Fig. 4.18. For the total number of vibrational modes with frequencies ranging from zero to υ_D, we can write

$$\int_0^{\upsilon_D} Z(\upsilon)\, d\upsilon = 3N \tag{4.81}$$

Using Eq. (4.80), we obtain

$$\int_0^{\upsilon_D} 4\pi V\left(\frac{2}{v_t^3}+\frac{1}{v_l^3}\right)\upsilon^2 d\upsilon = 3N$$

or

$$4\pi V\left(\frac{2}{v_t^3}+\frac{1}{v_l^3}\right)\frac{\upsilon_D^3}{3} = 3N$$

or

$$\upsilon_D^3 = \frac{9N}{4\pi V}\left(\frac{2}{v_t^3}+\frac{1}{v_l^3}\right)^{-1} \tag{4.82}$$

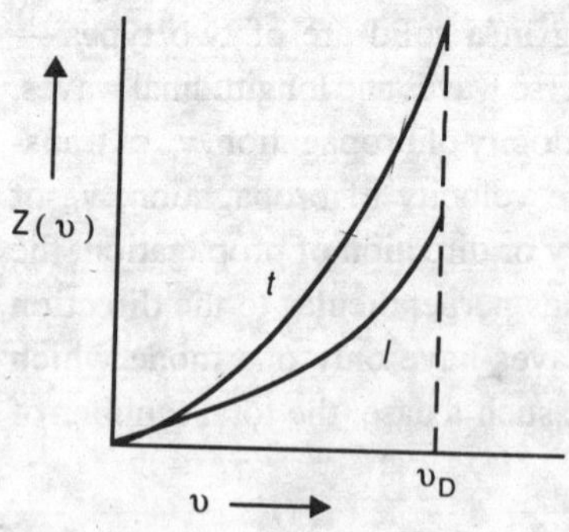

Fig. 4.18. *Frequency spectrum of transverse (t) and longitudinal (l) modes in a continuum showing cut-off at the Debye frequency.*

This equation can be used to determine υ_D. Taking velocity of sound of the order of 10^3 ms $^{-1}$ and $N/V \cong 10^{28}$m^{-3}, we get $\upsilon_D \cong 10^{13}$ s^{-1}. This corresponds to the wavelength of about 1 Å which is of the same order as interatomic distance. This violates the basic assumption of Debye and hence puts a question mark on the validity of the continuum theory, particularly in the high frequency region. Furthermore, the velocities v_t and v_l have been assumed to be independent of wavelength which may not be always true.

The vibrational energy of a crystal is determined by using the Planck's theory. The average energy of an oscillator having frequency υ at a temperature T is given by

$$\bar{\varepsilon} = \frac{h\upsilon}{e^{h\upsilon/k_BT} - 1} \tag{4.83}$$

We can associate a harmonic oscillator of the same frequency with each vibrational mode. Thus the vibrational energy of the crystal is given by

$$E = \int_0^{\upsilon_D} \bar{\varepsilon}\, Z(\upsilon)\, d\upsilon$$

Using Eqs. (4.80) and (4.83) we get

$$E = \int_0^{\upsilon_D} 4\pi V\left(\frac{2}{v_t^3} + \frac{1}{v_l^3}\right) \frac{h\upsilon^3}{e^{h\upsilon/k_BT} - 1}\, d\upsilon$$

$$= 4\pi hV\left(\frac{2}{v_t^3} + \frac{1}{v_l^3}\right) \int_0^{\upsilon_D} \frac{\upsilon^3 d\upsilon}{e^{h\upsilon/k_BT} - 1}$$

From Eq. (4.82), we obtain

$$4\pi V\left(\frac{2}{v_t^3} + \frac{1}{v_l^3}\right) = \frac{9N}{\upsilon_D^3}$$

$$\therefore \qquad E = \frac{9Nh}{\upsilon_D^3} \int_0^{\upsilon_D} \frac{\upsilon^3 d\upsilon}{e^{h\upsilon/k_BT} - 1} \tag{4.84}$$

Putting $\dfrac{h\upsilon}{k_BT} = x$ and $\dfrac{h\upsilon_D}{k_BT} = x_m$, we get

$$\upsilon = \frac{xk_BT}{h}$$

or $$d\upsilon = \frac{k_BT}{h}\, dx$$

Therefore, Eq. (4.84) becomes

$$E = \frac{9Nh}{\upsilon_D^3}\left(\frac{k_BT}{h}\right)^4 \int_0^{x_m} \frac{x^3 dx}{e^x - 1}$$

$$= 9N\left(\frac{k_BT}{h\upsilon_D}\right)^3 k_BT \int_0^{x_m} \frac{x^3 dx}{e^x - 1} \tag{4.85}$$

As in the Einstein's theory, we introduce here a characteristic temperature, θ_D, called the *Debye temperature* defined as

$$\theta_D = \frac{h\upsilon_D}{k_B} \tag{4.86}$$

$$\therefore \qquad x_m = \frac{\theta_D}{T}$$

Equation (4.85) then becomes

$$E = 9Nk_BT\left(\frac{T}{\theta_D}\right)^3 \int_0^{\theta_D/T} \frac{x^3\, dx}{e^x - 1} \tag{4.87}$$

The specific heat is given by

$$C_v = \left(\frac{\partial E}{\partial T}\right)_v = 9Nk_B\left(\frac{T}{\theta_D}\right)^3 \int_0^{\theta_D/T} \frac{e^x x^4}{\left(e^x - 1\right)^2} dx \tag{4.88}$$

$$= 3R\left(\frac{\theta_D}{T}\right) F_D \qquad (\text{for } N = N_a)$$

The function F_D is called the Debye function and is expressed as

$$F_D = 3\left(\frac{T}{\theta_D}\right)^4 \int_0^{\theta_D/T} \frac{e^x x^4}{\left(e^x - 1\right)^2} dx \tag{4.89}$$

We now consider the high and low temperature cases.

(i) High temperature case

For $T >> \theta_D$, x is small compared with unity for the complete range of integration. Therefore, we can write

$$e^x - 1 \cong x$$

Equation (4.87) then becomes

$$E = 9Nk_BT\left(\frac{T}{\theta_D}\right)^3 \int_0^{\theta_D/T} x^2 dx$$

$$= 3Nk_BT \qquad (4.90)$$

$$\therefore \quad C_v = \frac{\partial E}{\partial T} = 3Nk_B$$

$$= 3R \quad (\text{for } N = N_a)$$

Thus at high temperatures, the Debye's theory also obeys the Dulong and Petit's law as obeyed by the classical theory and the Einstein's theory. This, in fact, means that the quantum considerations carry almost no significance at high temperatures.

(ii) Low temperature case

$$\text{For } T << \theta_D,\ x_m = \frac{\theta_D}{T} \to \infty$$

Therefore, Eq. (4.87) becomes

$$E = 9Nk_BT\left(\frac{T}{\theta_D}\right)^3 \int_0^{\infty} \frac{x^3 dx}{e^x - 1}$$

Now

$$\int_0^{\infty} \frac{x^3 dx}{e^x - 1} = \frac{\pi^4}{15}$$

$$\therefore \quad E = 9Nk_BT\left(\frac{T}{\theta_D}\right)^3 \frac{\pi^4}{15}$$

$$= \frac{3}{5}\pi^4 Nk_B \frac{T^4}{\theta_D^3} \qquad (4.91)$$

This shows that the vibrational energy is proportional to T^4 which is analogous to the Stefen's law of black body radiation. Thus it follows that both phonons and photons obey the same statistics except with the difference that photons obey T^4 law at all temperatures whereas phonons do so only at low temperatures.

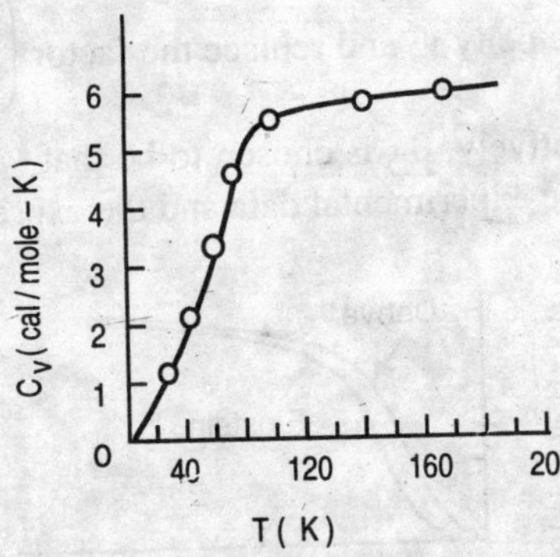

Fig. 4.19. *Comparison of Debye heat capacity obtained from Eq. (4.87) using θ_D = 225K with the experimentally observed values (circles) for silver.*

The expression for specific heat is obtained as

$$C_v = \left(\frac{\partial E}{\partial T}\right)_v = \frac{12}{5}\pi^4 Nk_B\left(\frac{T}{\theta_D}\right)^3$$

$$= \frac{12}{5}\pi^4 R\left(\frac{T}{\theta_D}\right)^3 \qquad (\text{for } N = N_a) \qquad (4.92)$$

Thus, at very low temperatures, the specific heat is proportional to T^3. This is called the *Debye T^3 law* and holds for $T \leq \frac{\theta_D}{10}$.

There is an excellent matching of the experimentally determined values of the specific heats at various temperatures with those calculated theoretically using the Debye's model as shown in Fig. 4.19 for silver with θ_D = 225K. This also proves the validity of the Debye's approximation at sufficiently low temperatures.

A comparison of results of the specific heat obtained from the Einstein and Debye models is made in Fig. 4.20. The Debye model yields somewhat higher values of specific heat as compared to the Einstein model. This is because of the fact that the Debye model takes into account the low frequency modes which, at low temperatures, have higher vibrational energy and hence larger specific heat. The Debye temperature θ_D can be obtained from Eqs. (4.82) and (4.86) provided the velocities of sound in transverse and longitudinal modes are known. For isotropic medium, one can use the average velocity v_s and replace the factor $\left(\frac{2}{v_t^3} + \frac{1}{v_l^3}\right)$ by $3/v_s^3$ in Eq. (4.80). Alternatively, θ_D is chosen to be that value which produces the best fit between the experimental data and the expression (4.88). There is, in general, a good agreement between the theoretical and experimental results.

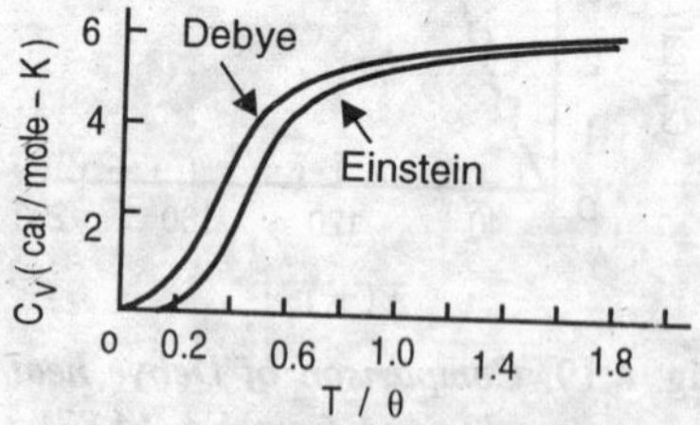

Fig. 4.20 Comparison of specific heats obtained from the Einstein and Debye models [Eqs. (4.63) and (4.87)].

Accurate measurements have indicated that the temperatures at which the T^3 law holds are quite low. According to this model, the law should hold for termperatures below 0.1 θ_D, whereas it may be strictly valid for $T \leq \theta_D/50$. The specific heat is, however, relatively insensitive to the variations in the density of modes.

4.10.3 Limitations of the Debye Model

(i) The Debye's continuum model is valid for long wavelengths only, i.e., only low frequencies are active in the solid.

(ii) The total number of vibrational modes are assumed to be 3*N*. This is difficult to justify as the solid is considered to be an elastic continuum which should possess infinite frequencies.

(iii) The cut off frequency is assumed to be the same for both longitudinal and transverse waves. This is again difficult to justify because different velocities of transverse and longitudinal waves should imply different values of cut off frequency for these waves.

(iv) According to the Debye's theory, θ_D is independent of temperature, whereas actually it is found to vary up to an extent of 10% or even more.

(v) The theory does not take into account the actual crystalline nature of the solid. The theory cannot be applied to crystals comprising more than one type of atoms.

(vi) The theory completely ignores the interaction among the atoms and the contribution of electrons to the specific heat.

SOLVED EXAMPLES

Example 4.1. The visible light of wavelength 5000 Å undergoes scattering from a crystal of refractive index 1.5. Calculate the maximum frequency of the phonon generated and the fractional change in frequency of the incident radiation, given the velocity of sound in the crystal as 5000 ms^{-1}.

Solution. The frequency of the phonon emitted is given by

$$\omega = \frac{2v_s\omega_{ph}n}{c}\sin\frac{\phi}{2}$$

Velocity of sound in the crystal, $v_s = 5000\ ms^{-1}$

Refractive index of the crystal, $n = 1.5$

The frequency of incident radiation, ω_{ph}, is calculated as :

$$\omega_{ph} = 2\pi\upsilon = 2\pi\frac{c}{\lambda}$$

$$= \frac{2\pi \times 3 \times 10^8}{5000 \times 10^{-10}}$$

$$= 3.77 \times 10^{15} \text{ rad/s.}$$

For ω to be maximum,

$$\sin \frac{\phi}{2} = 1$$

where φ is the angle of scattering.

$$\therefore \qquad \omega = \frac{2v_s\omega_{ph}n}{c}$$

$$= \frac{2 \times 5000 \times 3.77 \times 10^{15} \times 1.5}{3 \times 10^8}$$

$$= 1.86 \times 10^{11} \text{ rad/s.}$$

If ω'_{ph} represents the frequency of the scattered photon, then we have

$$\omega_{ph} - \omega'_{ph} = \omega$$

Therefore, the fractional change in frequency of the incident photon is

$$\frac{\omega_{ph} - \omega'_{ph}}{\omega_{ph}} = \frac{\omega}{\omega_{ph}} = \frac{1.86 \times 10^{11}}{3.77 \times 10^{15}} = 5 \times 10^{-5}$$

Example 4.2. The radii of Na^+ and Cl^- ions are 0.98 and 1.81Å respectively. The Young's modulus of NaCl in [100] direction is 5×10^{10} Nm^{-2}. Assuming that the extension in [100] direction produces negligible contraction in the perpendicular directions, calculate the wavelength at which the electromagnetic radiation is strongly reflected by NaCl crystal. Atomic masses of Na and Cl are 23 and 35.5 amu respectively.

Solution. The frequency of radiation strongly reflected by the ionic crystals is given by

$$\omega^2 = 2\beta\left(\frac{1}{m} + \frac{1}{M}\right)$$

where m and M denote the masses of the constituent ions and β is the force constant. Since the extension in [100] direction produces negligible contraction in the perpendicular directions, we can write

$$\beta = aY$$

where a is the interatomic distance and Y is the Young's modulus

$$\therefore \qquad \omega^2 = 2aY\left(\frac{1}{m} + \frac{1}{M}\right)$$

The interatomic distance along [100] direction of NaCl is the sum of the radii of Na^+ and Cl^- ions.

$$\therefore \quad \omega^2 = 2(0.98 + 1.81) \times 10^{-10} \times 5 \times 10^{10} \times \left(\frac{1}{23} + \frac{1}{35.5}\right) \times \frac{1}{1.67 \times 10^{-27}}$$

or $\quad \omega = 3.46 \times 10^{13}$ rad/s

Wavelength of e.m radiation which is strongly reflected by NaCl is

$$\lambda = \frac{c}{\upsilon} = \frac{2\pi c}{\omega} = \frac{2\pi \times 3 \times 10^{10}}{3.46 \times 10^{13}} = 5.45 \times 10^{-3} \text{ m.}$$

Example 4.3. Gold has the same structure as copper. The velocity of sound in gold is 2100 ms^{-1} and that in copper is 3800 ms^{-1}. If the Debye temperature of copper is 348 K, determine the Debye temperature of gold. The densities of gold and copper are 1.93×10^4 kgm^{-3} and 8960 kgm^{-3} and their atomic weights are 197.0 and 63.54 amu respectively.

Solution. The Debye temperature, θ_D, is defined as

$$\theta_D = \frac{h\upsilon_D}{k_B}$$

where υ_D is the Debye frequency and is given by the relation

$$\upsilon_D^3 = \frac{9N}{4\pi V}\left(\frac{1}{v_l^3} + \frac{2}{v_t^3}\right)^{-1}$$

Here N is the number of atoms present in volume V of the crystal and v_l and v_t represent the longitudinal and transverse velocities of sound waves in the crystal respectively. Replacing v_l and v_t by v_s, the mean velocity of sound, we get

$$\upsilon_D = v_s\left(\frac{3N}{4\pi V}\right)^{1/3}$$

$$\therefore \quad \theta_D = \frac{h}{k_B} v_s \left(\frac{3N}{4\pi V}\right)^{1/3}$$

Considering one mole of atoms, we can write

$$\theta_D = \frac{hv_s}{k_B}\left(\frac{3N_a\rho}{4\pi M}\right)^{1/3}$$

where N_a is the Avogadro's number, M is the mass of one mole of atoms and ρ is the density of crystal. The individual expressions for θ_D for copper and gold are

$$(\theta_D)_{Cu} = \frac{h}{k_B}(v_s)_{Cu}\left(\frac{3N_a\rho_{Cu}}{4\pi M_{Cu}}\right)^{1/3}$$

and

$$(\theta_D)_{Au} = \frac{h}{k_B}(v_s)_{Au}\left(\frac{3N_a\rho_{Au}}{4\pi M_{Au}}\right)^{1/3}$$

From the above equations, we obtain

$$(\theta_D)_{Au} = (\theta_D)_{Cu}\frac{(v_s)_{Au}}{(v_s)_{Cu}}\left[\frac{M_{Cu}}{M_{Au}}\frac{\rho_{Au}}{\rho_{Cu}}\right]^{1/3}$$

$$= 348\left(\frac{2100}{3800}\right)\left[\frac{63.54\times 1.93\times 10^4}{197.0\times 8960}\right]^{1/3}$$

$$= 170 \text{ K}$$

Example 4.4. The Debye temperature of diamond is 2000 K. Calculate the mean velocity of sound in diamond, given the density and atomic mass of diamond as 3500 kgm^{-3} and 12 amu respectively. If the interatomic spacing is 1.54 Å, estimate the frequency of the dominant mode of lattice vibration.

Solution. As shown in the previous example, the Debye temperature may be expressed by the relation.

$$\theta_D = \frac{h}{k_B}v_s\left(\frac{3N_a\rho}{4\pi M}\right)^{1/3}$$

which gives

$$v_s = \frac{k_B\theta_D}{h}\left(\frac{4\pi M}{3N_a\rho}\right)^{1/3}$$

Here, we have

$$\theta_D = 2000 \text{ K}$$
$$M = 12 \text{ kg}$$
$$\rho = 3500 \text{ kgm}^{-3}$$

$$\therefore \quad v_s = \frac{1.38\times 10^{-23}\times 2000}{6.626\times 10^{-34}}\left(\frac{4\pi\times 12}{3\times 6.023\times 10^{26}\times 3500}\right)^{1/3}$$

$$= 1.2\times 10^4 \text{ ms}^{-1}.$$

The order of magnitude of the dominant phonon wavelength at a temperature T (T << θ_D) is estimated from the relation

$$\lambda_d \approx \left(\frac{\theta_D}{T}\right)a$$

where a is the interatomic spacing.

$$\therefore \quad \lambda_d = \frac{2000}{298} \times 1.54 \times 10^{-10} = 1.03 \times 10^{-9}\, m$$

The frequency of the corresponding lattice vibration is

$$\upsilon_d = v_s / \lambda_d = \frac{1.2 \times 10^4}{1.03 \times 10^{-9}} = 1.16 \times 10^{13} \text{ Hz}$$

SUMMARY

1. The atoms of a crystal experience coupled vibrations. A crystal may oscillate in normal mode due to its internal energy or may undergo forced vibrations due to the effect of some external forces which may be mechanical or electromagnetic. The vibrations of the former type are associated with thermal properties and those of the latter type are associated with acoustical and some optical properties of solids.

2. The dispersion relation for a one-dimensional monoatomic lattice is

$$\omega = \omega_m \left|\sin \frac{Ka}{2}\right| = \sqrt{\frac{4\beta}{m}} \left|\sin \frac{Ka}{2}\right|$$

where K is the propagation vector of the wave, a is the interatomic distance and $\omega_m = \sqrt{4\beta/m}$ is the maximum angular frequency of vibration, β and m being the interatomic force constant and mass of each atom respectively.

3. In a monoatomic lattice, the dispersion effects are negligible at low frequencies ($K \to 0$) and the lattice behaves as a continuum. For frequencies $< \sqrt{4\beta/m}$, the lattice behaves as a dispersive medium. It acts as a low-pass filter and can propagate a number of wavelengths at the same frequency. For frequency $\omega = \sqrt{4\beta/m}$, the standing waves are formed and there is no transfer of signal or energy through the lattice.

4. The K-values associated with vibrations of monoatomic lattice can be grouped into Brillouin zones. The lattice waves can be described by K-

values which lie within the first Brillouin zone.

5. The dispersion curves for a one-dimensional diatomic lattice comprise two branches — the upper one the optical branch and the lower one the acoustical branch. The latter resembles the dispersion curve for monoatomic lattice. The two branches are separated by a forbidden band whose width increases with difference in masses of the two types of atoms. This band consists of frequencies the vibrations corresponding to which are not allowed. Thus the lattice acts as a band-pass filter.

6. In the optical branch, the neighbouring atoms move in opposite directions, whereas in the acoustical branch, they move in the same direction. The optical vibrations are excited by a force that distinguishes between the two types of charged ions, e.g., the response of ionic crystals to electromagnetic radiation.

7. In a lattice containing N atoms per primitive cell, the number of frequency bands is N.

8. A quantum of energy of a lattice vibrations is called a phonon. Phonons may be emitted or absorbed during inelastic scattering of photons from crystals. The energy of a phonon is $\hbar\omega$ where ω is the angular frequency of the lattice vibration.

9. The average number of phonons in a mode of lattice vibrations at termperature T is

$$\bar{n} = \frac{1}{\exp\left(\frac{\hbar\omega}{k_B T}\right) - 1}$$

10. According to the Dulong and Petit's law, the molar heat capacity of all solids at all temperatures is $3N_a k_B$ where N_a is the Avogadro's number and k_B is the Boltzmann constant. The law approximately holds for most of the solids at or above room temperature and completely fails as T approaches 0 K where C_v tends to vanish.

11. Einstein considered the atoms of a crystal as identical and independent three-dimensional quantum harmonic oscillators, all vibrating with the same frequency ω_0 and obtained the following expression for the molar heat capacity :

$$C_v = 3R\left(\frac{\theta_E}{T}\right)^2 \frac{e^{\theta_E/T}}{\left(e^{\theta_E/T} - 1\right)^2}$$

where θ_E is the Einstein temperature which is defined as

$$\theta_E = \frac{\hbar\omega_0}{k_B}$$

The Einstein model does not explain well the temperature variation of specific heat near 0 K.

12. Debye explained the variation of specific heat over the complete range of temperature by considering the atom of a crystal as coupled harmonic oscillators and obtained the following expression for molar specific heat

$$\overline{C}_v = 9R\left(\frac{T}{\theta_D}\right)^3 \int_0^{\theta_D/T} \frac{e^x x^4}{\left(e^x - 1\right)^2} dx$$

where

$$x = \frac{h\upsilon}{k_B T}$$

and

$$\theta_D = \frac{h\upsilon_D}{k_B}$$

is the Debye temperature, υ being the frequency of an oscillator and υ_D the maximum frequency of a vibrational mode. According to this model, $C_v \propto T^3$ at low temperatures. This is called Debye T^3 law and is in good agreement with experimental observations. At high temperatures, C_v approaches the classical value $3R$.

VERY SHORT QUESTIONS

1. Define phase velocity and group velocity?
2. Give dispersion relation for one-dimensional monoatomic lattice.
3. What are Brillouin zones?
4. Does a linear diatomic crystal have a constant forbidden gap?
5. What is meant by optical mode of wave propagation in a linear diatomic crystal?
6. What is meant by acoustical mode of wave propogation in a diatomic crystal?
7. What is a phonon? Give an evidence for the existence of phonons.

8. Plot phonon dispersion curve for a diatomic lattice ($m_1 > m_2$).
9. What is the normal scattering process of a photon?
10. What is umclapp scattering process of a photon?
11. List the major contributions to the lattice heat capacity.
12. What is the value of heat capacity for most of the solids at room temperature?
13. State Dulong and Petit's law.
14. What is Einstein temperature?
15. Give drawbacks of the Debye model.
16. Define the Debye temperature.

SHORT QUESTIONS

1. Show that for one-dimensional monoatomic lattice, the phase velocity is equal to the group velocity at low frequencies.
2. Obtain dispersion relation for the lattice waves in a monoatomic linear lattice of mass m, spacing a and and the nearest neighbour interaction c. Hence find
   (*a*) The maximum frequency that can be propagated through the lattice.
   (*b*) The allowed values of the phonon wave vector.
3. Describe the concept of Brillouin zones as applicable to one-dimensional monoatomic lattice.
4. Discuss the effect of variations of relative masses of the two types of atoms and the wavelength of elastic waves of the forbidden gap for a linear diatomic lattice.
5. Describe the movement of atoms in the optical and acoustical modes of wave propagation for a linear diatomic crystal.
6. Differentiate between optical and acoustical branches of diatomic linear lattice. Why are these branches named so?
7. Name different branches of the dispersion relation curve. Identify individual modes of vibration for KBr?
8. What are phonons? Express the laws of conservation of energy and momentum in the case of inelastic scattering of a photon by a phonon.
9. Show that, in the long wavelength limit, the velocity of sound is independent of the frequency.

10. What are normal and umclapp processes? Explain with the help of vector diagrams.
11. Give laws of conservation of energy and momentum for the normal and umclapp scattering of a photon.
12. Why specific heat at constant pressure is greater than the specific heat at constant volume?
13. Enlist the salient features of the Einstein's theory of lattice heat capacity. How is it different from the classical theory?
14. What is Debye temperature? What is its significance? If a solid has the Debye temperature of 2000°C, what can you say about its room temperature specific heat?

LONG QUESTIONS

1. Derive dispersion relationship for a one-dimensional atomic crystal and discuss the nature of acoustic and optical modes. Show that the group velocity vanishes at the zone boundary. Give pysical interpretation of the result.
2. Deduce vibrational modes of a finite one-dimensional monoatomic lattice. How does this knowledge help in calculating the specific heat?
3. Describe inelastic scattering of photons by phonons. Obtain an expression for the frequency of phonons generated when a photon is scattered inelastically at an angle θ
4. Describe the classical theory of lattice heat capacity and obtain the value of molar heat capacity for metals.
5. Describe the Einstein model of lattice heat capacity. Discuss the successes and failures of this model.
6. Derive an expression for the lattice heat capacity of a solid following Einstein model. Discuss the assumptions and predictions of this model and comapre it with experimental observations.
7. How does the Debye model differ from the Einstein model of lattice heat capacity? Discuss the consequences of this difference explaining the low temperature behaviour of specific heat in each case.
8. Discuss the Debye model of lattice heat capacity. What is Debye T^3 law?
9. Calculate the density of normal modes for a linear chain of atoms assuming the dispersion relation

$$\omega = \omega_m \left|\sin \tfrac{1}{2} Ka\right|$$

where a is the spacing and ω_m is the maximum frequency. What will be its value in Debye approximation?

10. What are the assumptions of the Debye model of lattice specific heat? Discuss its predictions and limitations as compared with Einstein model.

PROBLEMS

1. The visible light of wavelength 4000 Å undergoes scattering from a diamond crystal of refractive index 2.42. Calculate the maximum frequency of the phonon generated and the fractional change in the frequency of the incident radiation, given the velocity of sound in diamond as 1.2×10^4 ms^{-1}. (9.12×10^{11} rad/s, 1.9×10^{-4})
2. Compute the cut-off frequency for a linear monoatomic lattice if the velocity of sound and the interatomic spacing in the lattice are 3×10^3 ms^{-1} and 3×10^{-10}m respectively. (2×10^{13}Hz)
3. A system consists of 10^{25} atoms which act as simple harmonic oscillators each having a frequency of 10^{12}Hz. Ignoring the zero point energy, calculate the mean thermal energy of the system at 2.5K, 25K, 250K and 2500K. What conclusions can you draw from your answers ? [9.07×10^{-5}J, 3.41×10^{3}J, 9.39×10^{4}J, 1.03×10^{6}J]
4. Show that the zero point energy of a solid according to the Debye model is $\frac{9}{8}R\theta_D$.
5. Using the Debye approximation, show that the heat capacity of a linear monoatomic lattice at a temperature $T << \theta_D$ is proportional to T / θ_D. The effective Debye temperature in one dimension may be expressed as

$$\theta_D = \frac{\hbar\omega}{k_B} = \frac{\pi\hbar v_s}{k_B a}$$

where v_s is the effective velocity of sound and a is the interatomic spacing.

6. NaCl has the same structure as KCl. The Debye temperature of NaCl and KCl are 281 K and 230 K respectively. If the lattice heat capacity of NaCl at 5K is 1.6×10^{-2} Jmol^{-1}K^{-1}, estimate the heat capacity of KCl at 5K and 3K. (2.9×10^{-2}, 6.26×10^{-3}Jmol^{-1}K^{-1})

CHAPTER – V

FREE ELECTRON THEORY OF METALS

5.1 DRUDE-LORENTZ'S CLASSICAL THEORY (FREE ELECTRON GAS MODEL)

Drude, in 1900, postulated that the metals consist of positive ion cores with the valence electrons moving freely among these cores. The electrons are, however, bound to move within the metal due to electrostatic attraction between the positive ion cores and the electrons. The potential field of these ion cores, which is responsible for such an interaction, is assumed to be constant throughout the metal and the mutual repulsion among the electrons is neglected. The behaviour of free electrons moving inside the metals is considered to be similar to that of atoms or molecules in perfect gas. These free electrons are, therefore, also referred to as *free electron gas* and the theory is accordingly named as *free electron gas model*. The free electron gas, however, differs from an ordinary gas in some respects. Firstly, the free electron gas is negatively charged whereas the molecules of an ordinary gas are mostly neutral. Secondly, the concentration of electrons in an electron gas is quite large as compared to the concentration of molecules in an ordinary gas. The valence electrons are also called the conduction electrons and obey the Pauli's exclusion principle. These electrons are responsible for conduction of electricity through metals. Since the conduction electrons move in a uniform electrostatic field of ion cores, their potential energy remains constant and is normally taken as zero, i.e., the existence of ion cores is ignored. Thus the total energy of a conduction electron is equal to its kinetic energy. Also, since the movement of conduction electrons is restricted to within the crystal only, the potential energy of a stationary electron inside a metal is less than the potential energy of an identical electron just outside it. This energy difference, V_o, serves as a potential barrier and stops the inner electrons from leaving the surface of the metal. Thus, in free electron gas model, the movement of free electrons in a metal is equivalent to the movement of a free electron gas inside a 'potential energy box' which, in one-dimensional case, is represented by a line as shown in Fig. 5.1.

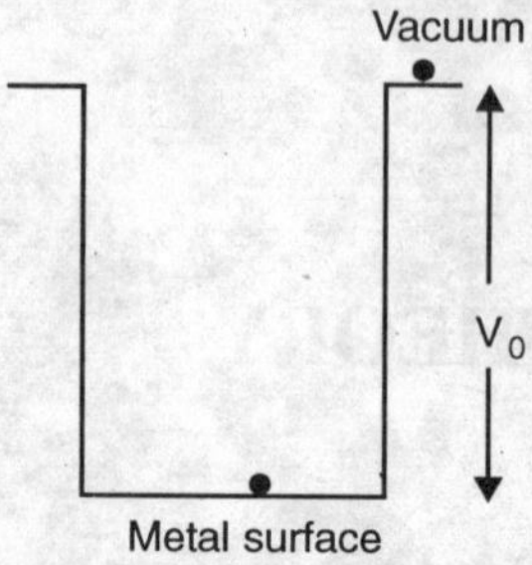

Fig. 5.1. *Metallic surface bounded by potential barrier* V_0 *which represents the difference in potential energy of a stationary electron present at the surface of the metal and just outside it (in vacuum).*

Based on the Drude's considerations that the electron gas behaves as a perfect gas, H.A. Lorentz postulated in 1909 that the electrons constituting the electron gas obey Maxwell-Boltzmann statistics under equilibrium conditions. The combined ideas of Drude and Lorentz constitue the *Drude-Lorentz theory*. As the theory is based on the classical ideas of Maxwell-Boltzmann statistics, it is called the classical theory.

The theory has been successfully applied to explain the various properties of metals. For example, it proves the validity of Ohm's law. The free electrons in a metal move in random directions and do not constitute a current until an electric field is applied across the metal which accelerates the electrons in a particular direction. The electrons, however, cannot be accelerated indefinitely. During their motion, the electrons suffer elastic collisions with the metal ions which slow down their speed. This gives rise to a steady state current of magnitude proportional to the voltage applied provided the temperature remains constant. This leads to Ohm's law. Also, as the free electrons can move easily, the metals exhibit high electrical and thermal conductivities. Moreover, since the electrons move freely inside the metals irrespective of the crystal structure, the ratio of the electrical conductivity, σ, to the thermal conductivity, K, should be constant for all metals at a constant temperature, i.e.,

$$\frac{\sigma}{K} = \text{constant}$$

This is called the *Wiedemann-Franz law* and has been realized in practice. The theory also explains the high lustre and complete opacity of metals. The opacity is due to absorption of all the incident electromagnetic radiations by free electrons which are then set into forced oscillations. The electrons return to their normal states by emitting the same amount of energy in all directions, thus producing metallic lustre.

Besides these successes, the theory also met with a number of failures. It correctly predicted the room temperature resistivity of various metals but the temperature dependence of resistivity could not be established accurately. The theory predicted that resistivity varies as $\sqrt{T}$ whereas actually it is found to vary linearly with temperature. The theory failed to explain the heat capacity and paramagnetic susceptibility of the conduction electrons. These properties are based on interactions of free elecurons with the external sources of energy which may be thermal or magnetic in nature. The application of Maxwell-Boltzmann statistics to this theory allows all the free electrons to gain energy which results in much higher values of heat capacity and paramagnetic susceptibility. As we shall describe later, the use of Fermi-Dirac or quantum statistics allows only a fraction of the total number of free electrons to gain energy and the values of heat capacity and paramagnetic susceptibility thus obtained match with the observed values. These are the principal achievements of the Sommerfeld's theory. The classical theory is also unable to account for the occurrence of long mean free paths (~10^8 to 10^9 interatomic spacings, or more than one centimetre) at low temperatures.

5.2 SOMMERFELD'S QUANTUM THEORY

Considering the free nature of valence electrons as assumed in the classical theory, Sommerfeld treated the problem quantum mechanically using the Fermi-Dirac statistics rather than the classical Maxwell-Boltzmann statistics. The possible electronic energy states in the potential energy box and the distribution of electrons in these states are then determined using quantum statistics. We describe the one-dimensional and three-dimensional cases separately.

5.2.1 Free Electron Gas in One-Dimensional Box

Consider an electron of mass m which is bound to move in a one-dimensional crystal of length L. The electron is prevented from leaving the crystal by the presence of a large potential energy barrier at its surfaces. Although the barriers extend over a few atomic layers near the surface, these are taken infinitely large for the sake of simplicity. The problem is identical to that of an electron moving in a one-dimensional potential box which is represented by a line and is bounded by infinite potential energy barriers as shown in Fig. 5.2. The potential energy within the crystal or box is assumed to be zero. Thus, we have

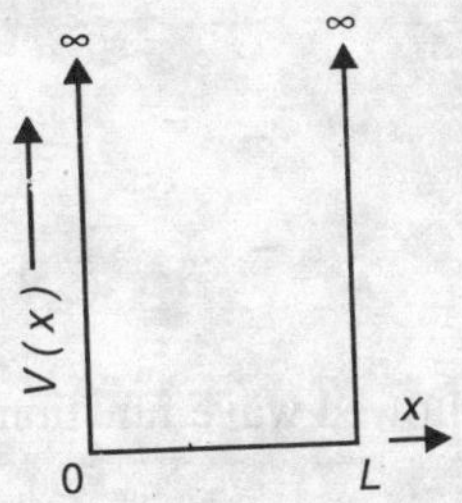

Fig. 5.2 One-dimensional potential box bounded by infinite potential energy barriers.

$$V(x) = 0 \qquad \text{for } 0 < x < L$$

$$V(x) = \infty \qquad \text{for } x \leq 0 \text{ and } x \geq L \tag{5.1}$$

The wave function ψ_n of the electron occupying the nth state is determined from the solution of the Schrodinger equation, i.e.,

$$\frac{d^2\psi_n}{dx^2} + \frac{2m}{\hbar^2}(E_n - V)\,\psi_n = 0 \tag{5.2}$$

where E_n represents the kinetic energy of the electron in the nth state and V is its potential energy. Since $V = 0$ inside the box, Eq. (5.2) becomes

$$\frac{d^2\psi_n}{dx^2} + \frac{2m}{\hbar^2} E_n\psi_n = 0 \tag{5.3}$$

The general solution to this equation is

$$\psi_n(x) = A \sin kx + B \cos kx \tag{5.4}$$

where A and B are arbitrary constants to be determined from the boundary conditions and k is given by

$$k = \sqrt{\frac{2mE_n}{\hbar^2}} \tag{5.5}$$

The boundary conditions are

$$\psi_n(0) = 0 \quad \text{and} \quad \psi_n(L) = 0 \tag{5.6}$$

These conditions are based on the fact that at $x = 0$ and L, $V \to \infty$ and the product $V(x)\,\psi_n(x)$ in Eq. (5.2) also approaches infinity. Thus in order that the wave function $\psi_n(x)$ may be continuous, the kinetic energy E_n must also become infinite which is not feasible. Hence $\psi_n(x)$ must vanish for $x = 0$ and L.

For $x = 0$, Eq. (5.4) gives B = 0 and the solution (5.4) becomes

$$\psi_n(x) = A \sin kx \tag{5.7}$$

Also, since $\psi_n(L) = 0$, Eq. (5.7) yields

$$\sin kL = 0$$

or

$$k = \frac{n\pi}{L} \tag{5.8}$$

where n = 1, 2, 3,....Thus the expression (5.7) for the allowed wave function becomes

$$\psi_n(x) = \text{A} \sin\left(\frac{n\pi}{L}\right)x \tag{5.9}$$

The allowed energy values can be obtained from Eqs. (5.5) and (5.8) as

$$E_n = \frac{\hbar^2}{2m}\left(\frac{n\pi}{L}\right)^2 = \frac{n^2h^2}{8mL^2} \tag{5.10}$$

or

$$E_n \propto n^2$$

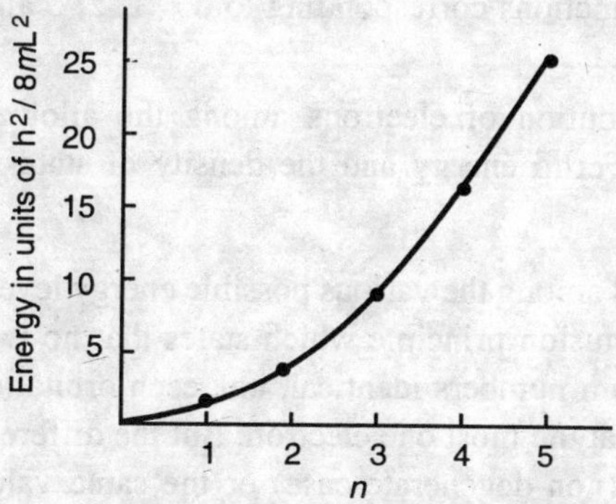

Fig. 5.3. E_n *versus n for a one-dimensional crystal.*

It is apparent from Eqs. (5.9) and (5.10) that the allowed wave functions $\psi_n(x)$ and the allowed energy values E_n exist only for integral values of n. The number n is called the quantum number. Hence the energy spectrum consists of discrete energy levels where the spacing between the levels is determined by the values of n and L. It decreases with increase in L. If L is of the order of a few centimetres, the energy levels form almost a continuum. But if L has atomic dimensions, the spacing between the levels becomes appreciable. The plot of E_n versus n is shown in Fig. 5.3.

The constant A in Eq. (5.9) is determined by using the condition that the probability of finding an electron somewhere on the line is unity, i.e.,

$$\int_0^L \psi_n^*(x)\psi_n(x)dx = 1$$

Using Eq. (5.9), we get

$$A^2 \int_0^L \sin^2\left(\frac{n\pi}{L}x\right)dx = 1$$

or
$$\frac{A^2}{2}\int_0^L \left[1-\cos\left(\frac{2n\pi}{L}x\right)\right]dx = 1$$

or
$$\frac{A^2}{2}\int_0^L dx = 1$$

or
$$A = \sqrt{\frac{2}{L}}$$

Substituting it in Eq. (5.9), we get the normalized wave function as

$$\psi_n(x) = \sqrt{\frac{2}{L}}\sin\left(\frac{n\pi}{L}x\right) \tag{5.11}$$

The energy levels and the wave functions corresponding to n = 1, 2, 3 and 4 are shown in Fig. 5.4.

We now discuss the distribution of electrons among the allowed energy levels and determine the Fermi energy and the density of states.

(i) Fermi Energy

The electrons are distributed among the various possible energy levels in accordance with the Pauli's exclusion principle which states that no two electrons can have all their quantum numbers identical, i.e., each orbital or quantum state can be occupied by at the most one electron. But the different states may have different values (non-degenerate case) or the same value (degenerate case) of energy. In a solid, an electron in a conduction electronic state has the quantum numbers n and m_s where n is the principal quantum number and m_s is the magnetic spin quantum number. Each set of values of n and m_s define a quantum state. For each value of n, m_s can have two possible values, +1/2 or −1/2. This means that each energy level defined by the quantum number n can have two quantum states and hence can accommodate a maximum of two electrons, one with spin up and the other with spin down. In other words, each energy level is doubly degenerate. For example, if there are seven electrons of appropriate spins then, in the ground state, these can be accommodated in four energy levels; the first three levels would have two electrons each with opposite spins and the fourth level would contain the last unpaired electron. These four energy levels are represented by the values 1, 2, 3 and 4 of the quantum number n. Thus, we find that if the total number of electrons to be accommodated is seven, the energy levels with $n < 4$ would be occupied while the level with $n > 4$ would be empty. The topmost filled energy level at 0 K is known as the *Fermi level* and the energy corresponding to this level is called the *Fermi energy*, E_F.

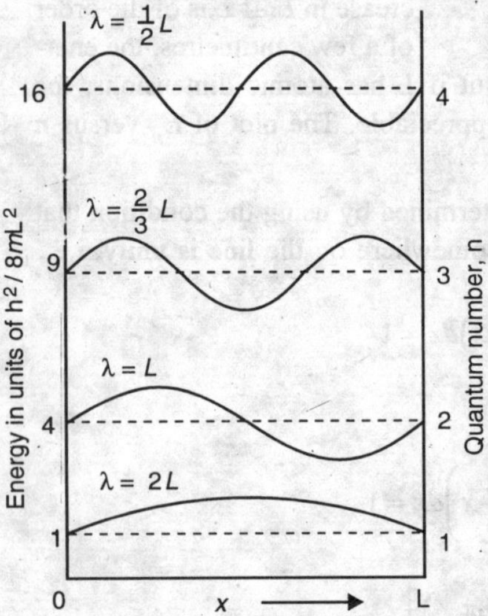

Fig. 5.4. *First four wave functions (solid lines) and the corresponding energy levels (broken lines) of an electron in a one-dimensional crystal.*

If N is the total number of electrons to be accommodated on the line then, for even n, we can write

$$2n_F = N \tag{5.12}$$

where n_F represents the principal quantum number of the Fermi level. Thus, for $n = n_F$, Eq. (5.10) gives

$$E_F = \frac{\hbar^2}{2m}\left(\frac{n_F\pi}{L}\right)^2 = \frac{\hbar^2}{2m}\left(\frac{N\pi}{2L}\right)^2 \tag{5.13}$$

Thus the value of the Fermi energy depends upon the length of the box the number of electrons in the box. For example, taking

$$\frac{N}{L} = 0.5 \text{ electrons/Å or } 5 \times 10^7 \text{ electrons/cm,}$$

the Eq. (5.13) gives

$$E_F = 3.7 \times 10^{-12} \text{ erg} = 2.4 \text{ eV}$$

Thus if we accommodate 5×10^7 electrons on one centimetre length of the line, the energy of the topmost electron would be 2.4 eV.

(ii) Total Energy

The total energy, E_o, of all the N electrons in the ground state is determined by summing up the energies of the individual electrons. For N electrons, the number of filled energy levels is $N/2$ and E_o is given by

$$E_o = 2\sum_{n=1}^{N/2} E_n$$

Here the factor of 2 appears because each level contains two electrons with equal energies. Using Eq. (5.10), we get

$$E_o = 2\frac{\hbar^2}{2m}\left(\frac{\pi}{L}\right)^2 \sum_{n=1}^{N/2} n^2 \tag{5.14}$$

Since

$$\sum_{n=1}^{s} n^2 = \frac{1}{6}s\left(2s^2 + 3s + 1\right)$$

$$\cong \frac{1}{3}s^3 \quad \text{for} \quad s \geq 1$$

$$\therefore \qquad \sum_{n=1}^{N/2} n^2 \cong \frac{1}{3}\left(\frac{N}{2}\right)^3$$

Hence Eq. (5.14) becomes

$$E_o = 2\frac{\hbar^2}{2m}\left(\frac{\pi}{L}\right)^2\frac{1}{3}\left(\frac{N}{2}\right)^3 = \frac{1}{3}\frac{\hbar^2}{2m}\left(\frac{N\pi}{2L}\right)^2 N$$

or, using Eq. (5.13), we obtain

$$E_o = \frac{1}{3}NE_F \tag{5.15}$$

Thus, for one-dimensional crystal, the average kinetic energy in the ground state is one-third of the Fermi energy.

(iii) Density of States

The density of states is defined as the number of electronic states present in a unit energy range. It is denoted by $D(E)$ and is given by

$$D(E) = \frac{dn}{dE} \tag{5.16}$$

where dn represents the number of electronic quantum states present in the energy interval E and $E + dE$. For a free electron gas, since each energy level contains two electronic states, one with spin up and the other with spin down, the actual density of states is twice the value given by Eq. (5.16). Thus, we have

$$D(E) = 2\frac{dn}{dE} \tag{5.17}$$

From. Eq. (5.10), we obtain

$$\frac{dE}{dn} = \frac{\hbar^2}{2m}\left(\frac{\pi}{L}\right)^2 2n = \frac{h^2 n}{4mL^2}$$

Therefore, Eq. (5.17) becomes

$$D(E) = 2\left(\frac{4mL^2}{nh^2}\right) = \frac{8mL^2}{h^2}\frac{1}{n}$$

Again, from Eq. (5.10), we get

$$\frac{1}{n} = \left(\frac{h^2}{8mL^2 E}\right)^{1/2}$$

Hence

$$D(E) = \left(\frac{8mL^2}{h^2 E}\right)^{1/2} = \frac{4L}{h}\left(\frac{m}{2E}\right)^{1/2} \tag{5.18}$$

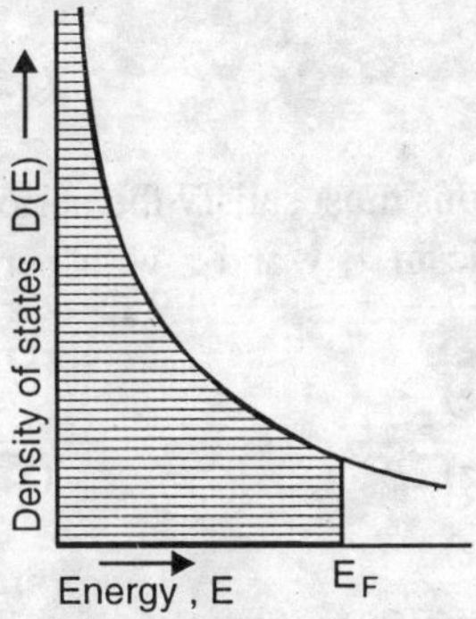

Fig. 5.5. *Variation of density of electronic states with energy for a one-dimensional metallic crystal. At* 0 K, *all the states upto the Fermi level are filled.*

The plot of $D(E)$ versus E is shown in Fig. 5.5. It indicates that all the levels present below the Fermi level are filled and all those present above it are empty. This type of situation, in fact, exists at absolute zero. Thus at 0 K, the Fermi level divides the filled and unfilled levels in the metallic crystals.

5.2.2 Free Electron Gas in Three Dimensions

Consider now the behaviour of the free electron gas in a three-dimensional metallic crystal. The potential of the electrons inside the crystal is constant and may be taken as zero, whereas it has a large value outside the crystal. For simplicity, the three-dimensional crystal may be regarded as a cubical box having length of the edge equal to L. The free particle Schrodinger equation in three dimensions is

$$-\frac{\hbar^2}{2m}\left(\frac{\partial^2}{\partial x^2}+\frac{\partial^2}{\partial y^2}+\frac{\partial^2}{\partial z^2}\right)\psi_k(\mathbf{r}) = E_k\psi_k(\mathbf{r})$$

or

$$\nabla^2\psi_k(\mathbf{r}) + \frac{2m}{\hbar^2}E_k\psi_k(\mathbf{r}) = 0 \tag{5.19}$$

where

$$\nabla^2 \equiv \frac{\partial^2}{\partial x^2}+\frac{\partial^2}{\partial y^2}+\frac{\partial^2}{\partial z^2}$$

is the Laplacian operator and E_k is the total energy (kinetic energy in the present case) of the electron in the k-state. Since the electrons are confined to a cubical box of edge L, the solution to Eq. (5.19) is just an extension of the one-dimensional normalized wave function (5.11), i.e.,

$$\psi_k(\mathbf{r}) = \sqrt{\frac{8}{L^3}}\,\sin\left(\frac{\pi n_x x}{L}\right)\sin\left(\frac{\pi n_y y}{L}\right)\sin\left(\frac{\pi n_z z}{L}\right) \tag{5.20}$$

where n_x, n_y and n_z are positive integers and $\sqrt{8/\mathrm{L}^3}$ is the normalising constant. This equation represents a standing wave solution. It is, however, more convenient to work with the plane travelling wave solution of the type

$$\psi_k(\mathbf{r}) = A\,e^{i\mathbf{k}.\mathbf{r}} = Ae^{i(k_x x + k_y y + k_z z)} \tag{5.21}$$

where
$$k^2 = \frac{2m}{\hbar^2} E_k \tag{5.22}$$

and A is an arbitrary constant. Such wave functions must satisfy the periodic boundary conditions, i.e., they must be periodic in x, y and z with period equal to L. These boundary conditions are

$$\begin{aligned}\psi(x + L, y, z) &= \psi(x, y, z)\\ \psi(x, y + L, z) &= \psi(x, y, z)\\ \psi(x, y, z + L) &= \psi(x, y, z)\end{aligned} \tag{5.23}$$

An application of the first boundary condition in (5.23) to the wave function (5.21) yields

$$\exp[i\{k_x(x + L) + k_y y + k_z z\}] = \exp\,[i\{k_x x + k_y y + k_z z\}]$$

or

$$\exp(ik_x L) = 1$$

or

$$k_x = 0, \pm\frac{2\pi}{L}, \pm\frac{4\pi}{L}, \text{------} \pm\frac{2n\pi}{L} \tag{5.24}$$

Similar results are obtained for k_y and k_z. This implies that any component of k is of the form $\frac{2n\pi}{L}$, where n is a positive or negative integer. These three components of **k** also form the quantum numbers of the problem in addition to the quantum number m_s which represents the spin direction. Thus the state of an electron is specified completely by a set of four quantum numbers k_x, k_y, k_z and m_s.

The allowed eigen values of the state or orbital with wave vector **k** are obtained from Eq. (5.22) as

$$E_k = \frac{\hbar^2 k^2}{2m} = \frac{\hbar^2}{2m}\left(k_x^2 + k_y^2 + k_z^2\right) \tag{5.25}$$

where the magnitude of the wave vector **k** is related to the wavelength λ as

$$k = \frac{2\pi}{\lambda} \tag{5.26}$$

Thus it is easy to see that the energy spectrum consists of discrete energy levels. These levels are generally very close to each other (energy difference ~ 10^{-15} eV) and may be regarded as continuous for most of the purposes.

These levels are also said to be *quasi-continuous*. In fact, the spacing between the levels depends upon the dimensions of the box. For dimensions of the order of a centimetre, the levels are so closely spaced that they almost form a continuum as predicted by the classical mechanics. If, however, the box has atomic dimensions, the energy levels are widely spaced and discrete.

The constant A in Eq. (5.21) can be determined by using the normalization condition, i.e.,

$$\int_0^V \psi^*(\mathbf{r})\,\psi(\mathbf{r})\,dV = 1$$

or
$$\int_0^L\int_0^L\int_0^L A^2 e^{-i\mathbf{k}.\mathbf{r}} e^{i\mathbf{k}.\mathbf{r}}\,dx\,dy\,dz = 1$$

or
$$A = \left(\frac{1}{L^3}\right)^{1/2} = \left(\frac{1}{V}\right)^{1/2}$$

Hence the normalized wave function is written as

$$\psi_k(\mathbf{r}) = \left(\frac{1}{V}\right)^{1/2} e^{i\mathbf{k}.\mathbf{r}} \tag{5.27}$$

The distribution of electrons among the allowed energy levels and the density of states are described in the following sections.

(i) Filling of Energy Levels

As described earlier, the distribution of electrons in the allowed energy levels follows the Pauli's exclusion principle. In the present case, we have a set of four quantum numbers k_x, k_y, k_z and m_s which represents a state of the system. Apparently, each state can be occupied by at the most one electron. It also follows from the expression (5.25) that a set of values of k_x, k_y and k_z defines an energy level. Now since for each set of k_x, k_y and k_z the fourth quantum number can have two values $\pm 1/2$, it follows that each energy level contains two quantum states of orbitals and hence can accommodate a maximum of two electrons, one with spin up and the other with spin down. In other words, each energy level is doubly degenerate. Thus a total of N non-interacting electrons at 0 K can be filled in $N/2$ energy levels. The topmost filled level is the $(N/2)$th level and all the levels lying

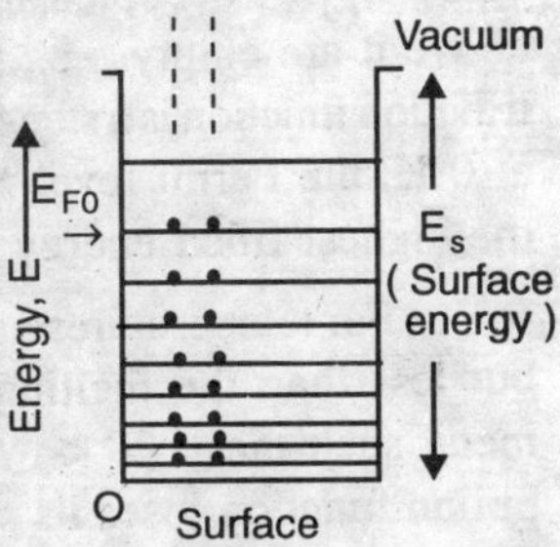

Fig. 5.6. *Sommerfeld's free electron model at* 0 K. *All the levels below* E_{F0} *are filled and all those above* E_{F0} *are empty.*

above it are empty. This level is, therefore, the *Fermi level* as it divides the filled and empty levels at 0 K. The energy of the Fermi level, or the Fermi energy, is denoted by E_{Fo}. The distribution of electrons among the energy levels present inside the box at 0 K is shown in Fig. 5.6. Thus we note that, unlike the classical theory, the Sommerfeld's quantum theory does not allow the condensation of all the electrons into the state of zero energy even at absolute zero. These electrons are rather distributed among the discrete energy levels having energies ranging from 0 to E_{Fo}.

We now discuss the effect of temperature on the occupancy of energy levels. It is apparent that, for temperatures greater than 0 K, the Fermi level may not be the topmost filled level since some of the electrons from the filled energy levels may be excited to the higher levels. Thus some of the levels below E_{Fo} would be empty while some above it would be occupied. The probability that a particular quantum state of energy E is occupied at a temperature T is given by the so called *Fermi function*

$$f(E) = \frac{1}{\exp\left(\dfrac{E - E_F}{k_B T}\right) + 1} \tag{5.28}$$

where E_F is the Fermi energy. The plot of $f(E)$ versus E is shown in Fig. 5.7 for different temperatures.

At absolute zero, the Eq. (5.28) gives

$$\begin{aligned} f(E) &= 1 \quad \text{for} \quad E \le E_{Fo} \\ &= 0 \quad \text{for} \quad E > E_{Fo} \end{aligned} \tag{5.29}$$

This indicates that all the energy states below E_{Fo} are occupied and all the states above it are empty, i.e., the Fermi distribution function is a step function. Thus, at 0 K, the Fermi level E_{Fo} represents the highest filled energy level.

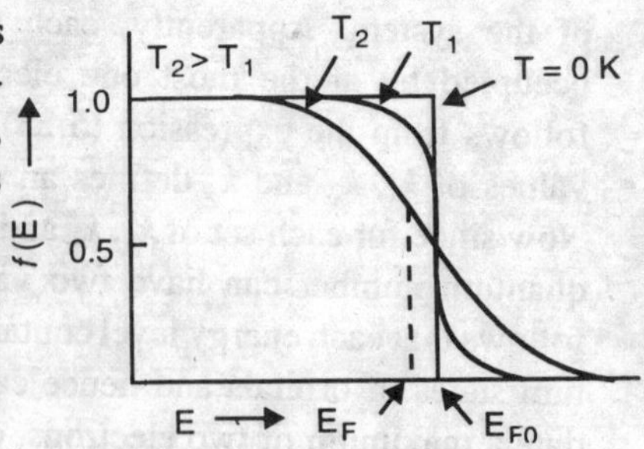

Fig. 5.7. *Schematic representation of the Fermi distribution function for three different temperatures. The variation of the Fermi energy with temperature is also shown.*

For temperatures greater than 0 K but less than the melting point of the metal such that $k_B T << E_F$, the distribution function loses its step character. The probability of occupation, $f(E)$, decreases gradually from 1 to 0 near E_F. This indicates that some of the states below E_F are empty while some others above it are filled. This is because some of the electrons from the energy states below E_F gain thermal energy and get excited to the states above E_F. At $E = E_F$, Eq. (5.28) gives

$$f(E_F) = \frac{1}{2}$$

Thus, for temperatures greater than 0 K, the Fermi level may be defined as the level where the probability of occupation is 1/2. Unlike E_{Fo}, it is not the topmost filled level; instead, it lies between the filled levels and empty levels. Apparently, the position of the Fermi level is not fixed, but changes with temperature. An approximate relationship between E_F and E_{Fo} is given as follows :

$$E_F \cong E_{Fo}\left[1 - \frac{\pi^2}{12}\left(\frac{k_BT}{E_{F_o}}\right)^2\right] \qquad (5.30)$$

However, in case of metals, since the spacing between the levels is quite small ($\sim 10^{-19}$ eV), the highest filled energy level is usually taken as the Fermi level.

For energies below E_F such that $(E_F - E) >> k_BT$*, $f(E) \cong 1$, provided T does not differ much from 0 K. Thus it is only in the vicinity of E_F minus a few k_BT that $f(E)$ becomes less than unity, i.e., some of the states below E_F would be empty and some above E_F would be filled. It, therefore, follows that as the temperature is raised above 0 K, all the electrons would not gain energy as expected classically, but only those which lie within an energy range k_BT below the Fermi level can do so. This is because the electrons present near the Fermi level can jump to the vacant higher energy states after acquiring energy of the order of k_BT, but the electrons present well below the Fermi level cannot do so due to the non-availability of empty states within the range k_BT. Thus, according to quantum mechanics, only a small fraction of the electrons can gain thermal energy and get excited to the energy states. The value of this fraction is $\dfrac{k_BT}{E_F} \cong 0.01$ at room temperature for $E_F = 3.0$ eV.

(ii) Density of Available Electron States, D(E)

As defined earlier, the density of states, $D(E)$, is the total number of available electronic states (or orbitals) per unit energy range at energy E. To obtain the expression for $D(E)$, we consider the linear momentum **p** which, in quantum mechanics, is represented by the operator

---

* According to the classical mechanics, k_BT represents the energy acquired by a particle at a temperature T K. At room temperature, $k_BT \sim 0.025$ eV.

$$\mathbf{p} = -i\hbar\nabla$$

When operated on Eq. (5.21), it gives

$$\mathbf{p}\psi_k(\mathbf{r}) = -i\hbar\nabla\psi_k(\mathbf{r}) = \hbar\mathbf{k}\psi_k(\mathbf{r}) \tag{5.31}$$

This indicates that the plane wave function ψ_k is an eigen function of the linear momentum with the eigen value $\hbar\mathbf{k}$. The particle velocity is given by

$$v = \frac{\hbar k}{m} \tag{5.32}$$

In a system of N free electrons, the occupied states or orbitals in the ground state may be represented by points inside a sphere in the **k**-space. The energy corresponding to the surface of the sphere then represents the Fermi energy, Let $\mathbf{k_F}$ be the wave vector from the origin of the **k**-space to the surface of the sphere as shown in Fig. 5.8. Then, using Eq. (5.25), the Fermi energy is written as

$$E_F = \frac{\hbar^2}{2m}k_F^2 \tag{5.33}$$

It is observed from Eqs. (5.24) and Fig. 5.8 that there is one allowed wave vector or one distinct triplet of quantum numbers k_x, k_y and k_z which corresponds to the volume element $(2\pi/L)^3$ of **k**-space. Thus, in the sphere called the *Fermi sphere* of volume $(4\pi/3)k^3_F$, the total number of electronic states or orbitals is

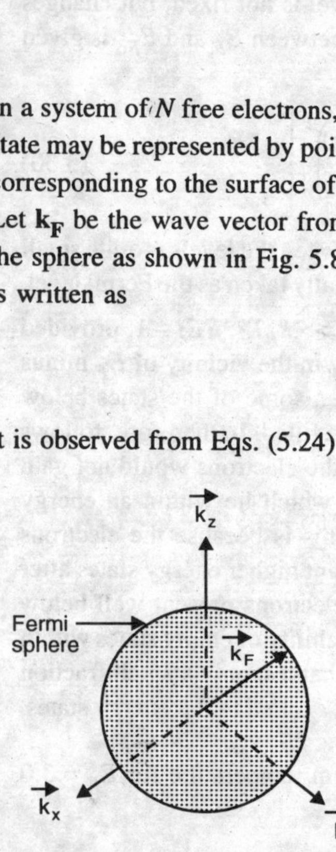

Fig. 5.8. *A sphere of radius k_F in the **k**-space containing points which represent the occupied states of a system of N free electrons in the ground state.*

$$2\,\frac{\frac{4\pi}{3}k_F^3}{\left(\frac{2\pi}{L}\right)^3} = \frac{L^3}{3\pi^2}k_F^3 = \frac{V}{3\pi^2}k_F^3 = N \tag{5.34}$$

where the factor 2 appears because there are two allowed values of m_s, and hence two electronic states, corresponding to each value of k. The number of states has been set equal to the number of electrons, N. From Eq. (5.34), we obtain

$$k_F = \left(\frac{3\pi^2 N}{V}\right)^{1/3} \tag{5.35}$$

This shows that the value of k_F depends on the concentration of electrons, N/V, and is independent of the mass of the electron. The expression for the Fermi energy is obtained by using Eq. (5.35) in (5.33) as

$$E_F = \frac{\hbar^2}{2m}\left(\frac{3\pi^2 N}{V}\right)^{2/3} \tag{5.36}$$

i.e., E_F depends on both the electronic concentration and mass. The total number of electrons is, therefore, given by

$$N = \frac{V}{3\pi^2}\left(\frac{2mE_F}{\hbar^2}\right)^{3/2} \tag{5.37}$$

The electron velocity, v_F, at the Fermi surface is obtained from Eqs. (5.32) and (5.35) as

$$v_F = \frac{\hbar k_F}{m} = \frac{\hbar}{m}\left(\frac{3\pi^2 N}{V}\right)^{1/3} \tag{5.38}$$

The density of states function, $D(E)$, is obtained by using the fact that, in the ground state of the system, all the energy states below E_F are occupied and the total number of states is equal to the total number of electrons, i.e.,

$$\int_0^{E_F} D(E)dE = N \tag{5.39}$$

Substituting the value of N from Eq. (5.37), we get

$$\int_0^{E_F} D(E)dE = \frac{V}{3\pi^2}\left(\frac{2mE_F}{\hbar^2}\right)^{3/2}$$

Expressing the integral in an indefinite form, we obtain

$$\int D(E)dE = \frac{V}{3\pi^2}\left(\frac{2mE}{\hbar^2}\right)^{3/2} = \int \frac{V}{2\pi^2}\left(\frac{2m}{\hbar^2}\right)^{3/2} E^{1/2}\, dE$$

or

$$D(E) = \frac{V}{2\pi^2}\left(\frac{2m}{\hbar^2}\right)^{3/2} E^{1/2} \tag{5.40}$$

The variation of $D(E)$ with E is parabolic and is shown in Fig. 5.9. The value of $D(E)$ also increases with increase in crystal volume; this is to accommodate the total number of electrons N which also increases [Eq. (5.37)] with volume of the crystal. At the Fermi surface, the density of states is obtained by putting $E = E_F$ in Eq. (5.40), i.e.,

$$D(E_F) \equiv \frac{dN}{dE_F} = \frac{V}{2\pi^2}\left(\frac{2m}{\hbar^2}\right)^{3/2} E_F^{1/2} \tag{5.41}$$

This result may also be obtained and expressed in a simple way by taking the natural logarithm of the expression (5.37), i.e.,

$$\ln N = \frac{3}{2}\ln E_F + \text{constant}$$

or

$$\frac{dN}{N} = \frac{3}{2}\frac{dE_F}{E_F} \tag{5.42}$$

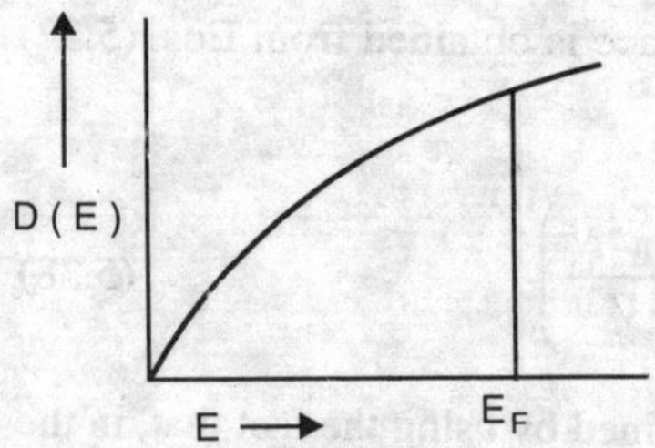

Fig. 5.9. *Variation of density of electronic states with energy for a free electron gas in three dimensions.*

Thus, within a factor of the order of unity, the number of electronic states or orbitals per unit energy range at the Fermi energy is equal to the total number of conduction electrons divided by the Fermi energy.

The density of filled electronic states, $N(E)$, at any temperature T is obtained by multplying the density of states, $D(E)$, at 0 K by the probability of occupation, $f(E)$, of the quantum state E at that temperature, i.e.,

$$N(E) = D(E)\,f(E) \tag{5.43}$$

The function $f(E)$ is the Fermi function and is defined by Eq. (5.28). The number of filled electronic states in the energy range E and $E + dE$ at temperature T is then given by

$$N(E)dE = D(E)\,f(E)dE \tag{5.44}$$

We now discuss some parameters of free electron gas at absolute zero.

(a) The Fermi Energy, E_{Fo}

The total number of electrons can be obtained by integrating Eq. (5.43), i.e.,

$$N = \int_0^\infty N(E)dE = \int_0^\infty D(E)f(E)dE \tag{5.45}$$

Using the expression (5.40), we get

$$N = \frac{V}{2\pi^2}\left(\frac{2m}{\hbar^2}\right)^{3/2}\int_0^\infty E^{1/2} f(E)dE$$

$$= \frac{V}{2\pi^2}\left(\frac{2m}{\hbar^2}\right)^{3/2}\left[\int_0^{E_{Fo}} E^{1/2} f(E)dE + \int_{E_{Fo}}^\infty E^{1/2} f(E)dE\right.$$

Now using Eqs. (5.29), we get

$$N = \frac{V}{2\pi^2}\left(\frac{2m}{\hbar^2}\right)^{3/2}\int_0^{E_{Fo}} E^{1/2} dE$$

$$= \frac{V}{3\pi^2}\left(\frac{2m}{\hbar^2}\right)^{3/2} E_{Fo}^{3/2}$$

or $$E_{Fo} = \frac{\hbar^2}{2m}\left(\frac{3\pi^2 N}{V}\right)^{2/3} \tag{5.46}$$

Putting $N/V = n$, the electron concentration, we obtain

$$E_{Fo} = \frac{\hbar^2}{2m}\left(3\pi^2 n\right)^{2/3} = \frac{h^2}{8m}\left(\frac{3n}{\pi}\right)^{2/3} \tag{5.47}$$

This shows that the Fermi energy at 0 K can be calculated simply from the electron concentration n. We also conclude from the above relationship that the electrons have some definite energy even at absolute zero. This is something entirely different from the classical theory according to which the electrons should have zero energy at absolute zero.

(b) Average Kinetic Energy, $\overline{E}_0$

The average kinetic energy of an electron at absolute zero is calculated as

$$\overline{E}_0 = \frac{1}{N}\int_0^\infty E D(E) f(E) dE$$

Using the expression (5.40) for $D(E)$, we obtain

$$\overline{E}_o = \frac{V}{2\pi^2}\left(\frac{2m}{\hbar^2}\right)^{3/2}\frac{1}{N}\left[\int_0^{E_{Fo}} E E^{1/2} f(E)dE + \int_{E_{Fo}}^\infty E E^{1/2} f(E)dE\right]$$

Now using Eq. (5.29), we get

$$\overline{E}_o = \frac{V}{2\pi^2}\left(\frac{2m}{\hbar^2}\right)^{3/2} \frac{1}{N}\int_0^{E_{Fo}} E^{3/2} dE$$

$$= \frac{2}{5}\,\frac{V}{2\pi^2}\left(\frac{2m}{\hbar^2}\right)^{3/2} \frac{1}{N} E_{Fo}^{5/2} \tag{5.48}$$

Substituting the value of N from Eq. (5.37), we obtain

$$\overline{E}_o = \frac{3}{5} E_{Fo} \tag{5.49}$$

This relates the average kinetic energy of an electron at absolute zero to the Fermi energy at the same temperature.

5.3 APPLICATIONS OF THE FREE ELECTRON GAS MODEL

The electronic properties of metals can be broadly divided into the following two groups :

(i) Static properties

(ii) Transport properties

The static properties are those which can be treated effectively by considering just the energy levels or the distribution of energy levels to which the electrons belong. More specifically, we need to consider the overall change in potential of the electrons without investigating deeply the detailed process producing the transitions. These properties include various electron emission properties, the magnetic properties and the properties like heat capacity and contact potential. These properties arise as a result of the excitations of the electrons by light, thermal energy, strong electric fields, etc.

The transport properties are those which can be treated by considering the detailed response of the electrons to an external field, i.e., by considering the acceleration properties of the electrons. These properties include electrical and thermal conductivities, Hall effect, thermoelectricity, etc.

We consider below the electronic specific heat of metals.

5.3.1 Electronic Specific Heat

The average kinetic energy of a free electron as given by the classical statistical mechanics is

$$\overline{E}_o = \frac{3}{2} k_B T$$

If the metal contains N free electrons, then the total kinetic energy becomes

$$E = N\bar{E}_0 = \frac{3}{2} Nk_BT$$

The electronic specific heat is, therefore, given by

$$C_v = \frac{3}{2} Nk_B \tag{5.50}$$

The measurement of optical reflection coefficient of metals indicates that the number of electrons per atom in metals is of the order of unity. Therefore, considering one gram atom of the metal, the value of electronic specific heat is obtained as (3/2)R or 3 cal $g^{-1}K^{-1}$ approximately. On the other hand, the specific heat associated with the lattice vibrations at high temperatures is about 3R or 6 cal $g^{-1}K^{-1}$. Thus one might conclude that the specific heat of metals should be about 50% greater than the specific heat of insulators. This is contradictory to the experimental observations which indicate that the electronic contribution to the specific heat is very small. At room termperature, the electronic contribution is not more than 0.01 of the value given by Eq. (5.50). Also, this contribution decreases linearly to zero as T approaches zero. It, therefore, appears that all the electrons might not contribute to the specific heat; instead, only a small fraction of the total number of electrons might contribute. This view is supported by the quantum theory.

As described earlier, the quantum mechanics suggests that only those electrons contribute to the specific heat which lie within an energy range k_BT below the Fermi level. This happens because when the electron gas is heated up to a temperature T, these electrons acquire energy of the order of k_BT and jump to the empty higher energy states. The deep lying electrons cannot do so because the unfilled energy states are not available to these electrons for excitation. These electrons, as such, do not contribute to the specific heat. The number of electrons which contribute to the specific heat is of the order $N\left(\frac{k_BT}{E_F}\right)$ or $N\left(\frac{T}{T_F}\right)$ where T_F is called the *Fermi termperature* and is defined by the equation

$$E_F = k_BT_F$$

Using the effective value of N, Eq. (5.50) becomes

$$C_v \cong \frac{3}{2} Nk_B\left(\frac{T}{T_F}\right) \tag{5.51}$$

This indicates that C_v is proportional to T and approaches zero as $T \to 0$. Using typical values of T and T_F as 300 K and 3000 K respectively, the above equation gives the value of C_v as

$$C_v \cong (0.01)\left(\frac{3}{2}\right)Nk_B$$

The specific heat calculated from this equation agrees with the experimental observations. Thus the quantum mechanics modifies the thermal behaviour of free electrons in a simple and satisfying manner.

SOLVED EXAMPLES

Example 5.1. The atomic radius of sodium is 1.86 Å. Calculate the Fermi energy of sodium at absolute zero.

Solution. The Fermi energy at 0 K is given by the relation

$$E_{Fo} = \frac{\hbar^2}{2m}\left(\frac{3\pi^2 N}{V}\right)^{2/3}$$

where N is the number of free (valence) electrons present in volume V of the metal. We consider a unit cell of sodium which is *bcc*.

Radius of the sodium atom, $r = 1.86 \times 10^{-10}$m

$$\text{Volume of the unit cell, } V = a^3 = \left(\frac{4r}{\sqrt{3}}\right)^3$$

$$= \left(\frac{4 \times 1.86 \times 10^{-10}}{\sqrt{3}}\right)^3$$

$$= 7.93 \times 10^{-29} \text{ m}^3$$

Since an atom of sodium has only one valence electron, the number of free electrons in a unit cell of sodium is 2.

$$\therefore \quad \frac{N}{V} = \frac{2}{7.93 \times 10^{-29}} = 2.52 \times 10^{28} \text{ electrons/m}^3.$$

$$\therefore \quad E_{Fo} = \frac{\left(1.05 \times 10^{-34}\right)^2}{2 \times 9.1 \times 10^{-31}}\left(3\pi^2 \times 2.52 \times 10^{28}\right)^{2/3}$$

$$= 4.98 \times 10^{-19} \text{ J}$$

$$= 3.11 \text{ eV}.$$

Example 5.2. Derive pressure versus volume relationship for a free electron gas at 0 K. Hence obtain an expression for the bulk modulus in terms of the average kinetic energy.

Solution. From thermodynamics, we have

$$P = -\frac{\partial E}{\partial V}$$

where E is the internal energy of a system of particles occupying a volume V at pressure P. For a free electron gas containing N electrons with average kinetic energy $\bar{E}_o$ at 0 K, the energy E may be replaced by $N\bar{E}_o$. Therefore, we have

$$P = -N\frac{\partial \bar{E}_o}{\partial V}$$

Using Eq. (5.49), we obtain

$$P = -\frac{3}{5}N\frac{\partial E_{Fo}}{\partial V}$$

Now, from Eq. (5.46), we have

$$E_{Fo} = \frac{\hbar^2}{2m}\left(\frac{3\pi^2 N}{V}\right)^{2/3}$$

which gives

$$\frac{\partial E_{Fo}}{\partial V} = -\frac{2}{3}\frac{\hbar^2}{2m}\left(3\pi^2 N\right)^{2/3}\left(\frac{1}{V}\right)^{5/3}$$

$$\therefore \qquad P = \frac{2}{5}N\frac{\hbar^2}{2m}\left(3\pi^2 N\right)^{2/3}\left(\frac{1}{V}\right)^{5/3} \tag{5.52}$$

or

$$P = \frac{2}{5}\frac{NE_{Fo}}{V} \tag{5.53}$$

This is the pressure versus volume relationship for a free electron gas at 0 K. The bulk modulus is given by the expression

$$B = -V\frac{\partial P}{\partial V}$$

Differentiating (5.52) with respect to V, we get

$$\frac{\partial P}{\partial V} = -\frac{5}{3}\left[\frac{2}{5}\frac{N\hbar^2}{2m}\left(3\pi^2 N\right)^{2/3}\right]\left(\frac{1}{V}\right)^{7/3}$$

$$= -\frac{5}{3}\,\frac{P}{V}$$

$$\therefore \qquad B = \frac{5}{3}P = \frac{2}{3}\,\frac{NE_{Fo}}{V} = \frac{10}{9}\,\frac{N\overline{E}_o}{V} = \frac{10}{9}\frac{E}{V} \qquad (5.54)$$

where we have used Eqs. (5.53) and (5.49).

SUMMARY

1. According to the free electron gas model, the valence electrons of the atoms present in a metal move freely among the positive ion cores just like the molecules of an ideal gas. These electrons are also called the conduction electrons and are responsible for conduction of electricity.

2. The energy spectrum of a free electron gas present in one or three-dimensional box is discrete particularly when the box has atomic dimensions. For laboratory dimensions, the levels almost form a continuum. The allowed eigen values of the states with wave vector **k** are given as

$$E_k = \frac{\hbar^2 k^2}{2m} = \frac{\hbar^2}{2m}\left(k_x^2 + k_y^2 + k_z^2\right)$$

where m is the mass of the electron.

3. The Fermi level is the level which divides the filled and unfilled states. It is the topmost filled level at 0 K. For T > 0 K, the probability of occupation of this level is 1/2. The position of the Fermi level changes with temperature and is expressed by the relation

$$E_F \cong E_{Fo}\left[1 - \frac{\pi^2}{12}\left(\frac{k_B T}{E_{Fo}}\right)^2\right]$$

where E_{Fo} is the Fermi energy at 0 K.

4. The Fermi energy varies with electron concentration n as

$$E_F = \frac{\hbar^2}{2m}\left(3\pi^2 n\right)^{2/3}$$

5. The number of electronic states present in a unit energy range is called the density of states. In a three-dimensional box of volume V, the density of states at energy E is given by

$$D(E) = \frac{V}{2\pi^2}\left(\frac{2m}{\hbar^2}\right)^{3/2} E^{1/2}$$

At the Fermi level, $D(E)$ may be expressed as

$$D(E_F) = \frac{3}{2}\frac{N}{E_F}$$

where N is the total number of electrons.

6. The average kinetic energy of an electron at 0 K is related to the Fermi energy at the same temperature as

$$\overline{E}_o = \frac{3}{5} E_{Fo}$$

7. The free electron gas model explains the electronic specific heat of metals by restricting the number of electrons contributing to specific heat to $N\left(\frac{k_B T}{E_F}\right)$. It yielded the expression for specific heat as

$$C_v \cong \frac{3}{2} N k_B \left(\frac{k_B T}{E_F}\right)$$

VERY SHORT QUESTIONS

1. What do you understand by free electron gas ?
2. What is Wiedemann-Franz law ?
3. Define the Fermi energy for metals. How does it depend on mass of the sample?
4. What is the significance of the Fermi distribution function?
5. Define density of states.
6. What do you understand by a Fermi gas ?
7. What is the Fermi sphere ?
8. Can a metal be associated with two Fermi temperatures ?

SHORT QUESTIONS

1. How does a free electron gas differ from an ordinary gas ?
2. How does the classical free electron theory lead to the Ohm's law?
3. Explain complete opacity and high lustre on the basis of the Drude-Lorentz free electron theory.
4. What are the failures of the Drude-Lorentz free electron theory ?
5. Obtain an expression for the energy eigen values of an electron using the Sommerfeld's free electron theory in one dimension.
6. Derive the three-dimensional wave function for the free electron gas and obtain its eigen values.
7. Write the expression for the Fermi-Dirac distribution function and discuss its behaviour with change in temperature.

8. How does the Fermi energy of a metal depend on temperature, concentration of electrons and the total number of electrons present in the metal ?
9. How does the quantum free electron theory explain the observed small values of electronic specific heat of metals ?

LONG QUESTIONS

1. Describe the free electron gas model of metals. How does it help to explain the lattice heat capacity of metals.
2. Obtain expressions for the Fermi energy, the total energy and the density of states for a free electron gas in one dimension. Show the variation of density of states with energy.
3. What is the Fermi energy? Calculate its value for the free electron gas at 0K and mention its significance.
4. Derive expressions for the Fermi energy and density of states for a free elctron gas in one dimension.
5. Show that for a one-dimensional metallic crystal, the average kinetic energy of electron in the ground state is one-third of the Fermi energy.
6. Show that the average kinetic energy per electron for a three-dimensional free electron gas at 0K is

$$\overline{E}_0 = (3/5)E_{F0}$$

where E_{F0} is the Fermi energy at 0K.
7. Discuss qualitatively why the electronic specific heat is temperature dependent and is much less than that expected from the classical behaviour of free electron gas.

PROBLEMS

1. Calculate the Fermi energy of electrons at 0K for a metal with electron density of $10^{28}m^{-3}$. (1.69eV)
2. Silver (*fcc*) has an atomic radius of 1.44 Å. Assuming silver to be monovalent metal, calculate the value of the Fermi energy, the Fermi temperature and the Fermi velocity.
(5.50 eV, 63850K, 1.39×10^6 ms^{-1})
3. Aluminium metal crystallises in *fcc* structure. If each atom contributes single electron as free electron and the lattice constant is 4Å, calculate (treating conduction electrons as free electron Fermi gas)
(*i*) Fermi energy (E_F) and the Fermi wave vector (k_F),
(*ii*) Total kinetic energy of free electron gas per unit volume at 0K.
Given $\hbar = 1.054 \times 10^{-27}$ erg-s; Electron rest mass = 9.11×10^{-28}g.
(2.28 eV, 7.73×10^9 m^{-1}, 2.14×10^{28} eVm^{-3})
4. The density of potassium is 860 kgm^{-3}. Calculate the Fermi energy and compressibility of potassium at 0K. (Atomic mass of K = 39.1 amu) (2.02 eV, 3.51×10^{-10} $N^{-1}m^2$)
5. If Fermi temperature is of the order of 5×10^4K, provide a rough estimate of electronic specific heat at 100K. (» 0.025J)

CHAPTER -VI

BAND THEORY OF SOLIDS

6.1 INTRODUCTION

The free electron theory described in the previous chapter was successful in explaining the various electronic and thermal properties of metals, such as specific heat, paramagnetism, etc. However, there are various other properties which could not be explained by this theory. For example, the theory could not explain why certain solids have a large number of free electrons and thus behave as good conductors while certain others have hardly any electrons and are, therefore, insulators. It also could not account for the variation of resistivity with temperature for the latter type of materials. Furthermore, the properties of semiconductors could not be explained on the basis of this theory.

The failure of the free electron model is due to the oversimplified assumption that a conduction electron in a metal experiences a constant or zero potential due to the ion cores and hence is free to move about in a crystal; the motion being restrained only by the surface of the crystal. In fact, the potential due to ion cores is not constant and may change with position of the electron in the crystal. Some contribution to potential may also arise because of the other electrons present in the crystal. Thus the actual nature of potential under which an electron moves in a crystal is very complicated. To a reasonable approximation, the ion cores may be considered at rest and the potential experienced by an electron in a crystal is assumed to be periodic with period equal to the lattice constant as shown in Fig. 6.1 for a one-dimensional case. This assumption is based on the fact that the ion cores in a crystal are distributed periodically on the lattice sites. The potential contribution due to all other free electrons may be taken as constant. This type of periodic potential extends up to infinity in all directions except at the surface of the crystal where, due to interrruption in periodicity of the lattice, it has a shape similar to as shown on the right hand edge of the figure. This type of potential irregularity may, however, be ignored.

The periodic potential described above forms the basis of the band theory of solids. The behaviour of an electron in this potential is described by constructing the electron wave functions using one-electron approximation. In this approximation, the total wave function for the system is obtained from a combination of wave functions each one of which involves the coordinates of one electron. In other words, the field experienced by a particular electron is assumed to be the resultant of the field due to fixed nuclei and the average field due to the charge distribution of all other electrons. As we shall discuss later, the motion of an electron in a periodic potential yields the following results :

(a) There exist allowed energy bands separated by forbidden regions or band gaps.

(b) The electronic energy functions E(**k**) are periodic in the wave vector **k**.

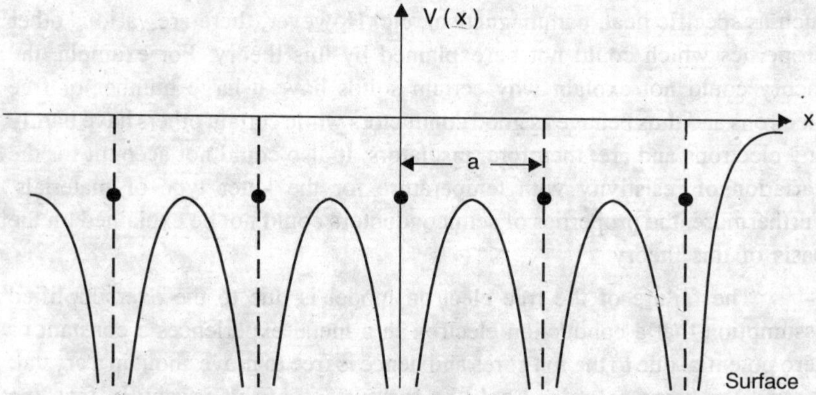

Fig. 6.1. *One-dimensional representation of potential experienced by an electron in a perfectly periodic crystal lattice with lattice constant a. The potential near the surface is shown on the right.*

These results are in contrast with those obtained from the free electron theory (constant or zero potential case) in the sense that, in the free electron theory, E varies with k as

$$E = \frac{\hbar^2 k^2}{2m}$$

i.e., there is no upper limit to the energy and k can have discrete values, which means that the energy levels are discrete and may have any spacing depending on the dimensions of the box. The existence of band gaps is the most important new property which emerges when the free electron model is extended to include the effect of periodic potential of the ion cores. It will be shown that these band gaps result from the interaction of the conduction electron waves with ion cores of the crystal and are of decisive significance to determine whether a solid is a conductor, a semiconductor, or an insulator. Besides this, we shall also encounter some other remarkable properties of electrons in crystals. We first determine the electron wave functions for a periodic lattice.

6.2 THE BLOCH THEOREM

As described in case of the free electron theory, the one-dimensional Schrodinger equation for an electron moving in a constant potential V_o is

$$\frac{d^2\psi}{dx^2} + \frac{2m}{\hbar^2}(E - V_o)\psi = 0 \tag{6.1}$$

The solutions to this equation are plane waves of the type

$$\psi(x) = e^{\pm ikx} \tag{6.2}$$

where

$$(E - V_o) = \frac{\hbar^2 k^2}{2m} = \frac{p^2}{2m} = E_{kin}$$

For an electron moving in a one-dimensional periodic potential, $V(x)$, the Schrodinger equation is written as

$$\frac{d^2\psi}{dx^2} + \frac{2m}{\hbar^2}[E - V(x)]\psi = 0 \tag{6.3}$$

Since the potential is periodic with period equal to the lattice constant, a, we have

$$V(x) = V(x + a)$$

There is an important theorem known as the *Bloch theorem* or the *Floquet's theorem* concerned with the solutions to Eq. (6.3). According to this theorem, the solutions to Eq. (6.3) are plane waves of the type of Eq. (6.2) which are modulated by a function $u_k(x)$ having the same periodicity as that of the lattice. Thus the solutions are of the form

$$\psi(x) = e^{\pm ikx} u_k(x) \tag{6.4}$$

with

$$u_k(x) = u_k(x + a) \tag{6.5}$$

The wave functions of the type of Eq. (6.4) are called the *Bloch functions.*

Proof of the Bloch Theorem

Let $g(x)$ and $f(x)$ be two real and independent solutions to the second order differential equation (6.3). Then the general solution can be written as

$$\psi(x) = A f(x) + B g(x) \tag{6.6}$$

where A and B are arbitrary constants. Since $V(x) = V(x + a)$, the functions $g(x + a$ and $f(x + a)$ must also be the solutions to Eq. (6.3) in addition to $g(x)$ and $f(x)$

However, Eq. (6.3) cannot have more than two independent solutions, the functions $g(x+a)$ and $f(x+a)$ must be expressible in terms of the functions $g(x)$ and $f(x)$, i.e.,

$$\left.\begin{aligned} f(x+a) &= \alpha_1 f(x) + \alpha_2 g(x) \\ g(x+a) &= \beta_1 f(x) + \beta_2 g(x) \end{aligned}\right] \tag{6.7}$$

where α_1, α_2, β_1 and β_2 are the real functions of energy E. Another general solution to Eq. (6.3) as obtained from the functions (6.7) is

$$\begin{aligned} \psi(x + a) &= A\, f(x + a) + B\, g(x + a) \\ &= (A\alpha_1 + B\beta_1)\, f(x) + (A\alpha_2 + B\beta_2)\, g(x) \end{aligned} \tag{6.8}$$

If the constants A and B are chosen such that

$$\left.\begin{aligned} A\alpha_1 + B\beta_1 &= \lambda A \\ \text{and} \qquad A\alpha_2 + B\beta_2 &= \lambda B \end{aligned}\right] \tag{6.9}$$

where λ is a constant, we find from Eqs. (6.6) and (6.8) that the functions $\psi(x)$ has the property

$$\psi(x + a) = \lambda\psi(x) \tag{6.10}$$

The Eqs. (6.9) yield non-zero values for A and B only if the determinant of the coefficients vanishes, i.e. if

$$\begin{vmatrix} (\alpha_1 - \lambda) & \beta_1 \\ \alpha_2 & (\beta_2 - \lambda) \end{vmatrix} = 0 \tag{6.11}$$

or
$$\lambda^2 - (\alpha_1 + \beta_2)\lambda + \alpha_1\beta_2 - \alpha_2\beta_1 = 0 \tag{6.12}$$

We now show that $\alpha_1\beta_2 - \alpha_2\beta_1 = 1$ Since $g(x)$ and $f(x)$ are the real and independent solutions to Eq. (6.3), we get

$$\frac{d^2 g(x)}{dx^2} + \frac{2m}{\hbar^2}\left[E - V(x)\right] g(x) = 0$$

and
$$\frac{d^2 f(x)}{dx^2} + \frac{2m}{\hbar^2}\left[E - V(x)\right] f(x) = 0$$

Multiplying the former by $f(x)$ and the latter by $g(x)$ and subtracting, we obtain

$$f(x)\frac{d^2 g(x)}{dx^2} - g(x)\frac{d^2 f(x)}{dx^2} = 0$$

or
$$f(x)\frac{dg(x)}{dx} - g(x)\frac{df(x)}{dx} = \text{constant}$$

The left hand side is called Wronskian, $W(x)$, of the solutions and is constant in this case. Further, from Eqs. (6.7), we obtain

$$f(x+a)\frac{dg(x+a)}{dx} - g(x+a)\frac{df(x+a)}{dx} = (\alpha_1\beta_2 - \alpha_2\beta_1)\left[f(x)\frac{dg(x)}{dx} - g(x)\frac{df(x)}{dx}\right]$$

or
$$W(x+a) = (\alpha_1\beta_2 - \alpha_2\beta_1)\ W(x)$$

Hence

$$\alpha_1\beta_2 - \alpha_2\beta_1 = 1 \tag{6.13}$$

Thus, Eq. (6.12) becomes

$$\lambda^2 - (\alpha_1 + \beta_2)\lambda + 1 = 0 \tag{6.14}$$

It gives two possible values of λ and hence two functions $\psi_1(x)$ and $\psi_2(x)$ which have the property as given in Eq. (6.10). We consider the following two cases :

Case I *For energy ranges such that* $(\alpha_1 + \beta_2)^2 < 4$

The Eq. (6.14) will have complex roots λ_1 and λ_2. Since $\lambda_1\lambda_2 = 1$ these will be complex conjugates of each other. We may, therefore, write

$$\left.\begin{aligned}\lambda_1 &= e^{ika}\\ \text{and}\quad \lambda_2 &= e^{-ika}\end{aligned}\right] \tag{6.15}$$

where k is real. The corresponding functions $\psi_1(x)$ and $\psi_2(x)$ then have the property

$$\left.\begin{aligned}\psi_1\ (x + a) &= e^{ika}\ \psi_1(x)\\ \text{and}\quad \psi_2\ (x + a) &= e^{-ika}\ \psi_2(x)\end{aligned}\right] \tag{6.16}$$

or, in general,

$$\psi(x+a) = e^{\pm ika}\ \psi(x) \tag{6.17}$$

It is easy to see that a function having such a property is the Bloch function of the type of Eq. (6.4). To check this we replace x by $x+a$ in the Bloch function (6.4) and obtain

$$\psi(x+a) = e^{\pm ik(x+a)}\ u_k(x+a)$$

Using Eq. (6.5), it becomes

$$\psi(x+a) = e^{\pm ika}\ \psi(x)$$

which is the same as (6.17). Thus we find that the function $\psi(x)$, which represents a solution to Eq. (6.3), has the same property as that expressed by Eq. (6.17), Eq. (6.10) with $\lambda = e^{\pm ika}$, or the Bloch function (6.4). Hence the Bloch functions of the type of Eq. (6.4) represent the solutions to the Schrodinger equation (6.13). This proves the Bloch theorem.

Case II *For the energy ranges such that* $(\alpha_1 + \beta_2)^2 > 4$

The Eq. (6.14) will have real roots λ_1 and λ_2 which may be taken as

$$\lambda_1 = e^{\mu a} \quad \text{and} \quad \lambda_2 = e^{-\mu a} \tag{6.18}$$

where μ is real. The corresponding solutions to the Schrodinger equation are

$$\psi_1(x) = e^{\mu x}u(x) \quad \text{and} \quad \psi_2(x) = e^{-\mu x}\, u(x) \tag{6.19}$$

Though mathematically valid, these solutions are not the acceptable electron wave functions because these are not bounded. The ratio $\frac{\psi(x+na)}{\psi(x)}$ approaches infinity as n approaches $\pm\infty$. This is contradictory to the requirement that the wave functions should remain finite at all points in space because of their probabilistic interpretation. In case I, however, $|\psi(x+na)|$ is equal to $|\psi(x)|$ and the ratio of these is always finite.

The roots (6.15) and (6.18) depend on α and β and hence are functions of energy E. The allowed roots (6.15) are associated with allowed energy regions and the disallowed roots (6.18) are associated with the forbidden energy regions. This means that the energy spectrum of an electron moving in a periodic potential consists of allowed and forbidden energy regions or bands. This result will be illustrated using the Kronig-Penny model.

In three dimensions, the Bloch theorem is expressed as

$$\psi_k(\mathbf{r}) = e^{i\mathbf{k}.\mathbf{r}}\, u_k(\mathbf{r}) \tag{6.20}$$

If the lattice potential vanishes, $u_k(\mathbf{r})$ becomes constant and Eq. (6.20) give

$$\psi_k(\mathbf{r}) = e^{i\mathbf{k}.\mathbf{r}}$$

Thus the wave function becomes the one of a free electron.

6.3 THE KRONIG-PENNEY MODEL

This model illustrates the behaviour of electrons in a periodic potential by assuming a relatively simple one-dimensional model of periodic potential as shown in Fig. 6.2. The potential energy of an electron in a linear array of positive nuclei is assumed to have the form of a periodic array of square wells with period of $(a + b)$. At the bottom of a well, i.e. for $0 < x < a$, the electron is assumed to be in the vicinity of a nucleus and the potential energy is taken as zero, whereas outside a well, i.e. for $-b < x < 0$, the potential energy is assumed to be V_0. Although this model employs a very crude approximation to the type of periodic potential existing inside a lattice, yet it is very useful to illustrate various important features of the quantum behaviour of electrons in the periodic lattice. The wave functions

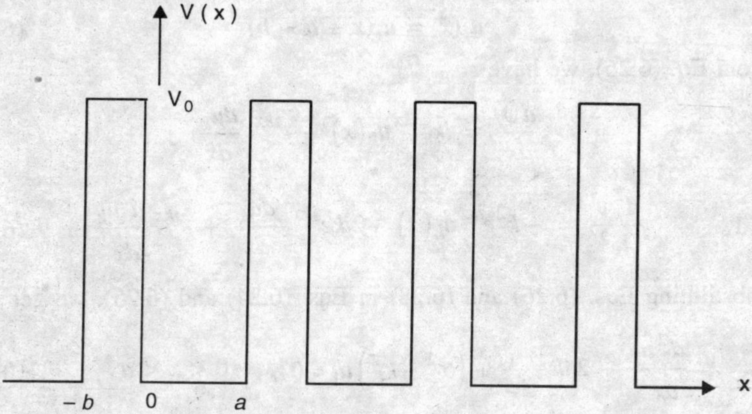

Fig. 6.2. *One-dimensional periodic potential used by Kronig and Penney.*

are obtained by writing the Schrodinger equations for the two regions as

$$\frac{d^2\psi}{dx^2} + \frac{2m}{\hbar^2} E\psi = 0 \qquad \text{for } 0 < x < a \tag{6.21}$$

and

$$\frac{d^2\psi}{dx^2} + \frac{2m}{\hbar^2} (E - V_o)\psi = 0 \qquad \text{for } -b < x < 0 \tag{6.22}$$

Assuming that the energy E of the electrons is less than V_o, we define two real quantities α and β as

$$\alpha^2 = \frac{2mE}{\hbar^2} \quad \text{and} \quad \beta^2 = \frac{2m(V_o - E)}{\hbar^2} \tag{6.23}$$

Therefore, the Eqs. (6.21) and (6.22) become

$$\frac{d^2\psi}{dx^2} + \alpha^2\psi = 0 \qquad \text{for } 0 < x < a \tag{6.24}$$

and

$$\frac{d^2\psi}{dx^2} - \beta^2\psi = 0 \qquad \text{for } -b < x < 0 \tag{6.25}$$

Since the potential is periodic, the wave functions must be of the form of Bloch function (6.4), i.e.,

$$\psi(x) = e^{ikx}\, u_k(x) \tag{6.26}$$

where $u_k(x)$ is the periodic function in x with periodicity of $(a+b)$, i.e.,

$$u_k(x) = u_k(x + a + b) \tag{6.27}$$

From Eq. (6.26), we have

$$\frac{d\psi}{dx} = ike^{ikx}u_k(x) + e^{ikx}\frac{du_k}{dx}$$

and
$$\frac{d^2\psi}{dx^2} = -k^2e^{ikx}u_k(x) + 2ike^{ikx}\frac{du_k}{dx} + e^{ikx}\frac{d^2u_k}{dx^2} \tag{6.28}$$

Substituting Eqs. (6.26) and (6.28) in Eqs. (6.24) and (6.25), we get

$$\frac{d^2u_1}{dx^2} + 2ik\frac{du_1}{dx} + \left(\alpha^2 - k^2\right)u_1 = 0 \qquad 0 < x < a \tag{6.29}$$

and

$$\frac{d^2u_2}{dx^2} + 2ik\frac{du_2}{dx} - \left(\beta^2 + k^2\right)u_2 = 0; \qquad -b < x < 0 \tag{6.30}$$

where u_1 and u_2 represent the values of $u_k(x)$ in the intervals $0 < x < a$. and $-b < x < 0$ respectively. The general solutions to these equations are

$$u_1 = A\,e^{i(\alpha-k)x} + B\,e^{-i(\alpha+k)x}; \qquad 0 < x < a \tag{6.31}$$

and

$$u_2 = C\,e^{(\beta-ik)x} + D\,e^{-(\beta+ik)x}; \qquad -b < x < 0 \tag{6.32}$$

where A, B, C and D are constants which can be determined from the following boundary conditions :

$$u_1(0) = u_2(0)\,; \qquad \left(\frac{du_1}{dx}\right)_{x=0} = \left(\frac{du_2}{dx}\right)_{x=0} \tag{6.33}$$

$$u_1(a) = u_2(-b)\,; \qquad \left(\frac{du_1}{dx}\right)_{x=a} = \left(\frac{du_2}{dx}\right)_{x=-b} \tag{6.34}$$

The conditions (6.33) correspond to the requirements that the wave function ψ and its derivative $\frac{d\psi}{dx}$, and hence u and $\frac{du}{dx}$ must be continuous, whereas the conditions (6.34) correspond to the requirements of the periodicity of $u_k(x)$. Applying these boundary conditions to (6.31) and (6.32) we obtain the following four linear homogeneous equations :

$$\begin{aligned}
A + B &= C + D \\
Ai(\alpha-k) - Bi(\alpha+k) &= C(\beta-ik) - D(\beta+ik) \\
Ae^{i(\alpha-k)a} + Be^{-i(\alpha+k)a} &= Ce^{-(\beta-ik)b} + De^{(\beta+ik)b} \\
Ai(\alpha-k)e^{i(\alpha-k)a} - Bi(\alpha+k)e^{-i(\alpha+k)a} &= C(\beta-ik)e^{-(\beta-ik)b} - Di(\beta+ik)e^{(\beta+ik)b}
\end{aligned} \tag{6.35}$$

These equations are used to determine the constants *A*, *B*, *C* and *D*. A non-zero solution to these equations exists only if the determinant of the coefficients of *A*, *B*, *C* and *D* vanishes, i.e.,

$$\begin{vmatrix} 1 & 1 & 1 & 1 \\ i(\alpha-k) & -i(\alpha+k) & \beta-ik & -(\beta+ik) \\ e^{i(\alpha-k)a} & e^{-i(\alpha+k)a} & e^{-(\beta-ik)b} & e^{(\beta+ik)b} \\ i(\alpha-k)e^{i(\alpha-k)a} & -i(\alpha+k)e^{-i(\alpha+k)a} & (\beta-ik)e^{-(\beta-ik)b} & -(\beta+ik)e^{(\beta+ik)b} \end{vmatrix} = 0$$

On solving this determinant, we obtain

$$\frac{\beta^2+\alpha^2}{2\beta\alpha}\sinh\beta b\,\sin\alpha a+\cosh\beta b\,\cos\alpha a=\cos k(a+b) \tag{6.36}$$

To simplify this equation, Kronig and Penney considered the case when V_o tends to infinity and *b* approaches zero but the product $V_o b$ remains finite, i.e., the potential barriers become delta functions. Under these circumstances, the model is modified to the one of a series of wells separated by infinitely thin potential barriers. The quantity lim $V_o b$ for $V_o \to \infty$ and $b \to 0$ is called the *barrier strength.*

As $b \to 0$, $\sinh \beta b \to \beta b$ and $\cosh \beta b \to 1$. Also, from Eqs. (6.23), we have

$$\frac{\beta^2+\alpha^2}{2\alpha\beta}=\frac{mV_o}{\alpha\beta\hbar^2}$$

Therefore, Eq. (6.36) becomes

$$\left(\frac{mV_o b}{\hbar^2\alpha}\right)\sin\alpha a+\cos\alpha a = \cos ka \tag{6.37}$$

We define a quantity *P* as

$$P=\frac{mV_o ba}{\hbar^2} \tag{6.38}$$

which is a measure of the area $V_o b$ of the potential barrier. Thus increasing *P* has the physical meaning of binding an electron more strongly to a particular potential well. Using Eq. (6.38) in (6.37), we get

$$P\frac{\sin\alpha a}{\alpha a}+\cos\alpha a = \cos ka \tag{6.39}$$

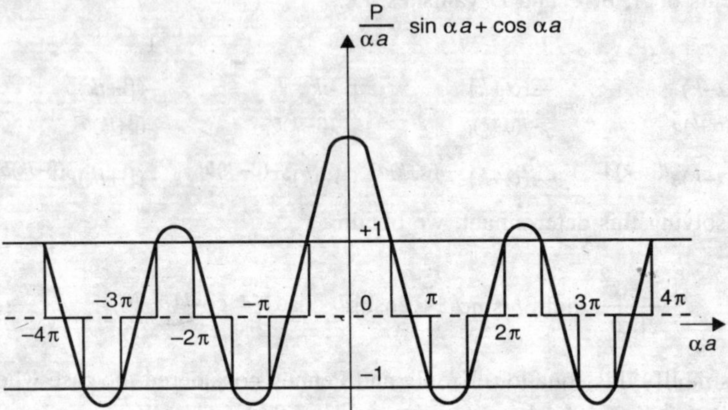

Fig. 6.3. *Plot of the left hand side of Eq. (6.39) as a function of* αa *using* $P = 3\pi/2$. *The solid and broken lines on the abscissa correspond to allowed and forbidden energy regions respectively.*

This is the condition which must be satisfied for the solutions to the wave equation to exit. Since cos*ka* lies between +1 and −1, the left hand side should assume only those values of αa for which its value lies between +1 and −1. Such values of αa, therefore, represent wave like solutions of the type of Eq. (6.26) and are allowed. The other values of αa are not allowed. Figure 6.3 shows a plot of the left hand side versus αa for $P = 3\pi/2$. The vertical axis lying between −1 and +1, as indicated by the horizontal lines, represents the values acceptable to the left hand side. It may be noted that since α^2 is proportional to the energy E, the abscissa is a measure of the energy. The following conclusions may be drawn from Fig. 6.3 :

(i) The energy spectrum of the electrons consists of alternate regions of allowed energy bands (solid lines on abscissa) and forbidden energy bands (broken lines).

(ii) The width of the allowed energy bands increases with αa or the energy.

(iii) The width of a particular allowed energy band decreases with increase in the value of P, i.e., with increase in the binding energy of the electrons. As $P \to \infty$, the allowed energy bands are compressed into energy levels and a line spectrum is resulted. In such a case, the Eq. (6.39) has solutions only if

$$\sin\alpha a = 0$$

or
$$\alpha a = \pm n\pi$$

where n is an integer. Therefore, from Eq. (6.23), we have

$$E = \frac{\alpha^2\hbar^2}{2m} = \frac{\pi^2\hbar^2}{2ma^2}n^2 \tag{6.40}$$

This expression gives the energy levels of a particle in a constant potential box of atomic dimensions. This is what is expected physically as, for large P, the tunnelling through the barrier becomes difficult. In the other extreme case, i.e., when $P \to 0$, Eq. (6.39) becomes

$$\cos \alpha a = \cos ka$$

or
$$\alpha = k$$

Therefore, Eq. (6.23) gives

$$E = \frac{k^2\hbar^2}{2m}$$

which is applicable to completely free electrons for which all the energies are allowed. This is again an expected result as, for P approaching zero, the electrons become free and, therefore, the free electron model involving quasi-continuous energy levels becomes applicable.

These results are summarized in Fig. 6.4 which shows the energy spectrum as a function of P. On the extreme left, i.e. for $P = 0$, the energy spectrum is quasi-continuous and on the extreme right, i.e. for $P = \infty$, a line spectrum is resulted. For any other value of P, the position and width of the allowed and forbidden bands are obtained by drawing a vertical line.

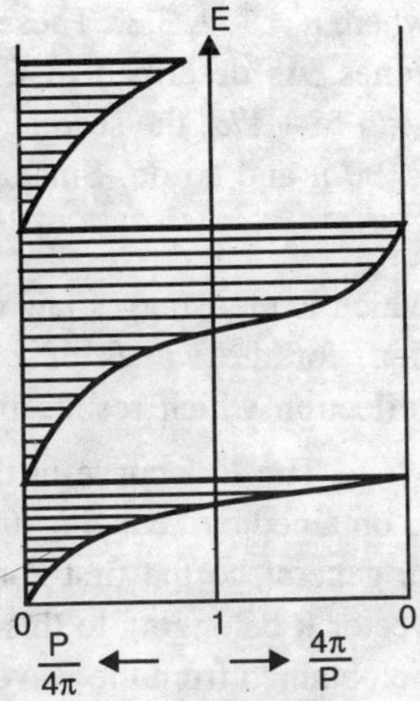

Fig. 6.4. *Allowed (shaded) and forbidden (open) energy ranges as a function of P.*

6.3.1 Energy versus Wave-Vector Relationship — Different Representations

It follows from Eq. (6.39) that cos*ka* takes a specific value for each allowed energy value E. However, cos*ka* is an even periodic function of k and remains unchanged if k is replaced by $k + 2\pi n/a$, where n is an integer. This means that the energy E is also an even periodic function of k with period of $2\pi/a$. This type of periodic representation of energy is shown in Fig. 6.5a for the first few allowed bands. This representation can be obtained by the periodic repetition of the region $-\pi/a < k < \pi/a$, i.e., the first *Brillouin* *zone,* and is known as the *periodic or repeated zone scheme.* In this scheme, k is not uniquely defined as, for each value of E, there exist a number of k values. It is, therefore, generally convenient to introduce the following two schemes :

(*a*) The extended zone scheme

(*b*) The reduced zone scheme

A curve of E versus k in the extended zone scheme is shown by solid lines in Fig. 6.5b. The corresponding parabolic curve (dotted) for free electrons in the constant potential is also shown for comparison. We note that the E–k curve of the extended zone scheme is not continuous and has discontinuities at

$$k = \pm \frac{n\pi}{a} \tag{6.41}$$

where $n = 1, 2, 3, ...$ These values of k define the boundaries of the Brillouin zones. As described in Chapter II, the first Brillouin zone extends from $-\pi/a$ to $+\pi/a$, the second one extends from π/a to $2\pi/a$ and from $-\pi/a$ to $-2\pi/a$, and so on. Since $k = 2\pi/\lambda$, Eq. (6.41) gives

$$n\lambda = 2a \tag{6.42}$$

which is the Bragg's law of reflection for normal incidence. It thus follows that, for the values of k as given by (6.41), the electrons suffer Bragg reflection which results in discontinuities in the E–k curve.

The E–k curve in the reduced zone scheme is shown in Fig. 6.5c It is obtained by 'reducing' the contents of the other zones so as to correspond, in general, to the first zone, i.e., to the region $-\pi/a \leq k \leq \pi/a$. The wave vector **k** belonging to this region is called the *reduced wave vector* and can be obtained from the wave vector **k** belonging to the region outside the first Brillouin zone by using the relation

$$\mathbf{k} = \mathbf{k}' - \mathbf{G} \tag{6.43}$$

where **G** is a suitably chosen reciprocal lattice vector.

6.3.2 Number of Wave Functions in a Band

It has been shown that, in an infinitely long one-dimensional crystal, there are certain allowed energy ranges which have a continuous distribution of energy. We now consider a finite crystal of length L and determine the number of possible wave functions, states or k-values per band. Using the periodic boundary conditions, we have

$$\psi(x + L) = \psi(x) \tag{6.44}$$

Since the wave functions are Bloch functions, it requires

$$e^{ik(x+L)}\, u_k(x+L) = e^{ikx}\, u_k(x)$$

Using Eq. (6.5), we obtain

$$e^{ik(x+L)} = e^{ikx}$$

or

$$e^{ikL} = 1$$

which gives

$$k = \frac{2\pi n}{L} \tag{6.45}$$

with $n = \pm 1, \pm 2, \ldots.,$ or

$$k = \pm\frac{2\pi}{L},\ \pm\frac{4\pi}{L},\ \ldots. \tag{6.46}$$

Therefore, the number of possible wave functions in the range dk is

$$dn = \frac{L}{2\pi} dk \tag{6.47}$$

Considering now the first Brillouin zone, the maximum value of k is π/a where a is the length of the primitive cell. If N is the number of primitive cells in the crystal of length L, then $a = L/N$. Therefore, the maximum value of k in the first Brillouin zone is $\frac{N\pi}{L}$. Thus the series (6.46) is cut off at N/L which means that the total number of allowed k-values in the first Brillouin zone is equal to N. Hence the total number of possible states in any energy band is equal to the number of primitive unit cells N. We further note that since each band can be occupied by a maximum of two electrons, each band can accommodate a maximum of $2N$ electrons. This result is quite important for explaining the distinction between metals, insulators and semiconductors.

6.4 VELOCITY AND EFFECTIVE MASS OF ELECTRON

6.4.1 Velocity of Electron

We consider the velocity of an electron described by a wave vector **k**. According to the wave mechanical theory of particles, a particle moving with a velocity v is equivalent to a wave packet moving with a group velocity v_g. Thus, we have

$$v = v_g = \frac{d\omega}{dk} \tag{6.48}$$

where ω is the angular frequency of the deBroglie waves. Also, the energy E of the particle is given by

$$E = \hbar\omega$$

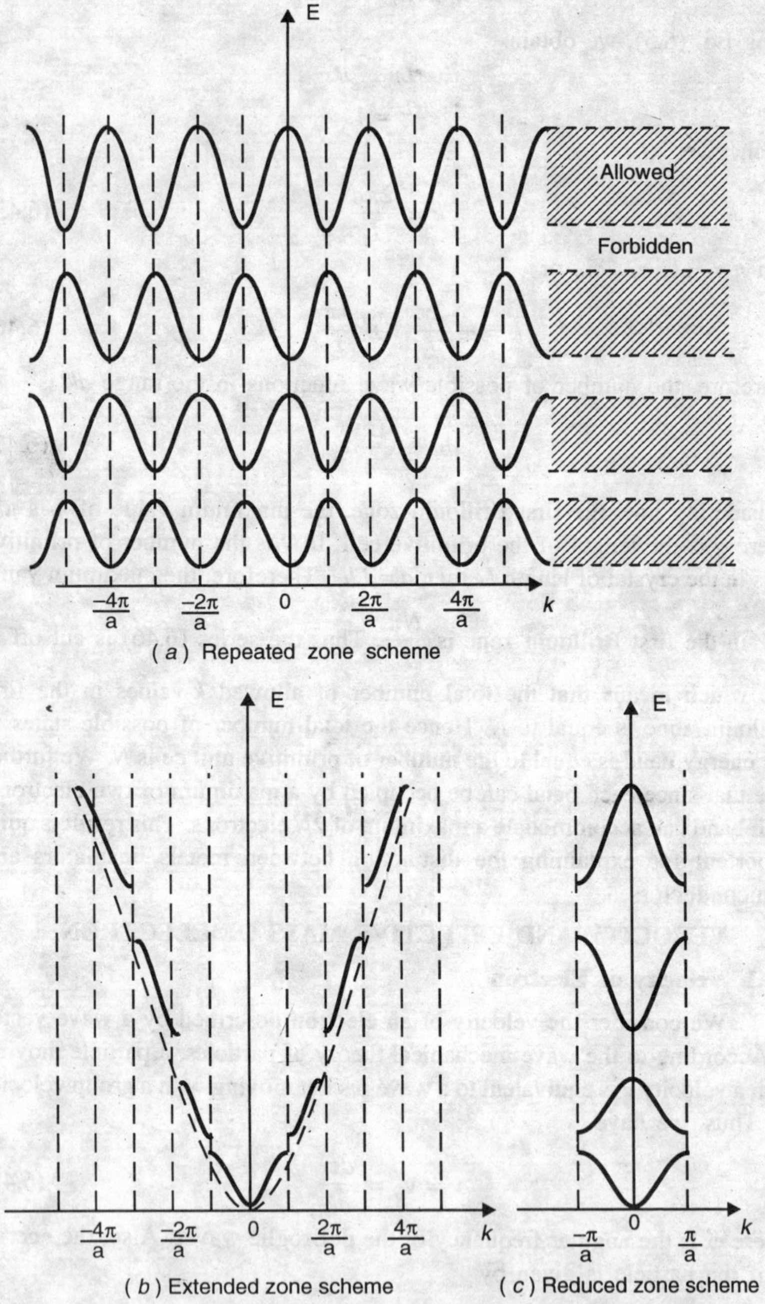

Fig. 6.5. *E versus k curves in three different zone schemes drawn on the same scale. Figure (b) also shows allowed and forbidden energy bands. The dotted parabolic curve in Fig. (b) corresponds to free electrons in a constant potential.*

Therefore, Eq. (6.48) can be written as

$$v = \frac{1}{\hbar}\frac{dE}{dk} \tag{6.49}$$

For free electrons, we have

$$E = \frac{\hbar^2 k^2}{2m}$$

and, therefore, Eq. (6.49) gives

$$v = \frac{\hbar k}{m} = \frac{p}{m}$$

This shows that, for a free electron, v is proportional to k. However, in the band theory, E is generally not proportional to k. The variation of E with k based on the band theory is shown in Fig. 6.6a. This shows that the slope dE/dk of the $E(k)$ curve is not constant but changes with k. Using this curve and employing Eq. (6.49), one can obtain v versus k curve as shown in Fig. 6.6b. This curve indicates that the velocity of the electron is zero for k=0 and $\pm\pi/a$ where the slope dE/dk is zero, i.e., at the top and bottom of the energy band (first Brillouin zone). For $k = k_o$, where k_o corresponds to the inflexion point of the $E(k)$ curve, the absolute value of the velocity attains a maximum value equal to the free electron velocity. Beyond the inflexion point, the velocity decreases with increase in energy which is altogether different from the behaviour of free electrons.

6.4.2 Effective Mass of Electron

Consider now the effect of applied electric field $\mathcal{E}$ on the motion of an electron present in the Brillouin zone. We assume only one electron in the Brillouin zone so that the Pauli's exclusion principle is not violated. If c is the velocity of the electron and the field $\mathcal{E}$ acts on it for a time dt, the increase in energy of the electron is given by

$$dE = e\mathcal{E}dx = e\mathcal{E}v\,dt = \frac{e\mathcal{E}}{\hbar}\left(\frac{dE}{dk}\right)dt \tag{6.50}$$

We can write

$$dE = \frac{dE}{dk}dk$$

$$\therefore \qquad \frac{dE}{dk}dk = \frac{e\mathcal{E}}{\hbar}\left(\frac{dE}{dk}\right)dt$$

or

$$\frac{dk}{dt} = \frac{e\mathcal{E}}{\hbar} \tag{6.51}$$

or
$$\hbar \frac{dk}{dt} = \frac{dp}{dt} = F = e\mathcal{E} \qquad (6.52)$$

where p represents the crystal momentum. It follows from Eq. (6.52) that the rate of change of crystal momentum is equal to the impressed force $e\mathcal{E}$. It is thus an analogue to Newton's second law for an electron in a periodic lattice.

The acceleration a of the electron is obtained by differentiating (6.49) with respect to t, i.e.,

$$a = \frac{dv}{dt} = \frac{1}{\hbar} \frac{d^2E}{dk^2} \left(\frac{dk}{dt} \right)$$

Using Eq. (6.51), we obtain

$$a = \frac{e\mathcal{E}}{\hbar^2} \frac{d^2E}{dk^2} \qquad (6.53)$$

For a free electron of mass m,

$$a = \frac{e\mathcal{E}}{m} \qquad (6.54)$$

A comparison of Eq. (6.53) with (6.54) yields an interesting result of the band theory, i.e., an electron moving in a crystal has an effective mass m* given by

$$m^* = \frac{\hbar^2}{\left(\frac{d^2E}{dk^2} \right)} \qquad (6.55)$$

This indicates that the effective mass of an electron is not constant but depends on the value of d^2E/dk^2, i.e., on the shape of the $E(k)$ curve. The variation of the effective mass with k is shown in Fig. 6.6c. The effective mass is positive in the lower half of the energy band and negative in the upper half. It becomes infinite at the inflexion points of the $E(k)$ curve. Thus the electron behaves as a positively charged particle in the upper half of the band only.

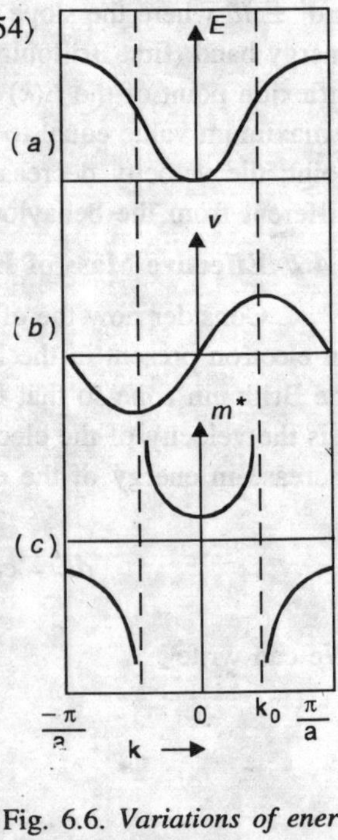

Fig. 6.6. *Variations of energy, velocity and effective mass with k. The dashed lines correspond to inflexion points in the E-k curve.*

It is generally convenient to introduce a factor

$$f_k = \frac{m}{m^*} = \frac{m}{\hbar^2}\left(\frac{d^2E}{dk^2}\right) \tag{6.56}$$

which measures the extent up to which an electron in the k-state behaves as a free electron. If $f_k = 1$, the electron behaves totally as a free electron. If $f_k < 1$, i.e. $m^* > m$, the electron behaves as a heavy particle.

The concept of the negative effective mass may be understood in terms of the Bragg's reflection when k is close to $\pm\pi/a$. Due to the Bragg's reflection, a force applied in one direction leads to a gain of momentum in the opposite direction which results in the negative effective mass.

The concept of effective mass has a physical significance. It provides a satisfactory description of the charge carriers in crystals. In crystals such as alkali metals, which have partially filled energy band, the conduction takes place mainly through electrons. However, in crystals for which the energy band is nearly full, except for a few electron vacancies near the top of the band, these negative charge and negative mass vacancies may be considered as positive charge and positive mass particles called *holes* which act as positive charge carriers to produce conduction.

6.5 DISTINCTION BETWEEN METALS, INSULATORS AND SEMICONDUCTORS

To distinguish between metals, insulators and semiconductors on the basis of the band theory, we consider an energy band which is filled with electrons up to a certain value k_1 ($k_1 < \pi/a$) as shown in Fig. 6.7. It is of interest to know the effective number of free electrons present in this band which gives information about the conductivity of the band. In the expression (6.56), we introduced the term f_k that represents the extent up to which an electron in the k-state behaves as a free electron and hence participates in electrical conduction. Thus, in order to determine the total effective number of free electrons in the energy band, we have to sum up over all the possible f_k's corresponding to the various occupied energy states present in the band. Therefore, we can write

$$N_{eff} = \Sigma f_k \tag{6.57}$$

The Eq. (6.47) gives the number of possible states in the interval dk for a one-dimensional lattice of length L as

$$dn = \frac{L}{2\pi} dk \tag{6.47}$$

Since each state is occupied by a maximum of two electrons, the effective number of free electrons present in the band within the limits $-k_1$ and $+k_1$ be-

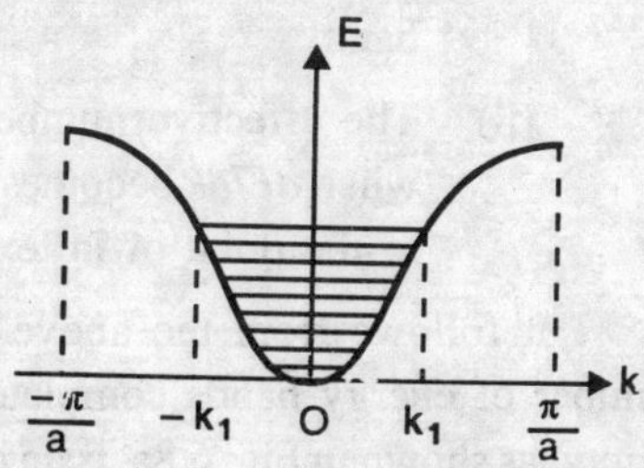

Fig. 6.7. *Energy band (first Brillouin zone) filled up to the state k_1.*

comes

$$N_{eff} = 2\frac{L}{2\pi}\int_{-k_1}^{k_1} f_k dk = \frac{L}{\pi}\int_{-k_1}^{k_1} f_k dk \tag{6.58}$$

Using Eq. (6.56), we obtain

$$N_{eff} = \frac{2L}{\pi}\frac{m}{\hbar^2}\int_0^{k_1}\frac{d^2E}{dk^2}dk$$

$$= \frac{2Lm}{\pi\hbar^2}\left(\frac{dE}{dk}\right)_{k=k_1} \tag{6.59}$$

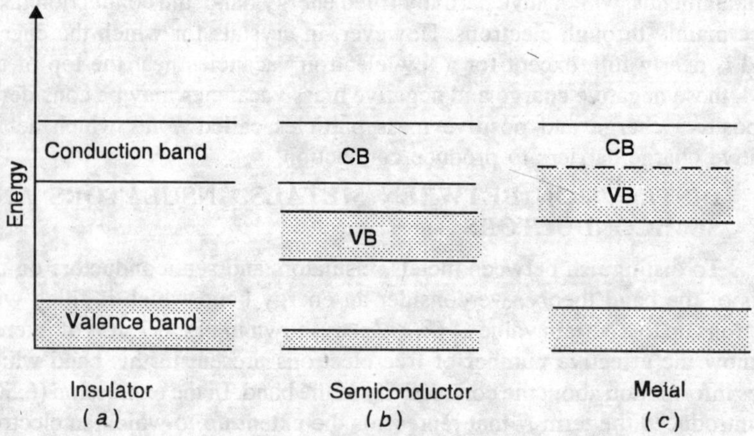

Fig. 6.8. *Electron distribution in an insulator, intrinsic semiconductor and metal at 0 K. The shaded regions are occupied by electrons.*

This yields the following results :

(i) The effective number of free electrons in a completely filled band is zero. This is because *dE/dk* vanishes at the top of the band.

(ii) The effective number of electrons attains a maximum value when *dE/dk* becomes maximum, i.e., when the band is filled up to the point of inflexion.

It follows from the above discussion that a solid having a certain number of energy bands completely filled and all other bands completely empty, as shown in Fig. 6.8a, is an insulator. On the other hand, a solid having a partly filled energy band has a metallic character (Fig. 6.8c). The situation depicted in Fig. 6.8a can strictly occur only at 0 K. At temperatures greater than 0 K, some electrons from the topmost filled energy band (valence band) may get excited to the next higher empty band (conduction band) where they participate in the conduction process. The number of such free or conduction electrons depends on the value of the forbidden energy gap between the valence band and the conduction band. The larger the band gap, the smaller the number of free electrons and hence the larger the tendency of the material to behave as an insulator for all practical purposes. An example of this type of material is diamond which has the band gap of about 6 eV. If the band gap is small (≅1eV) the number of thermally excited electrons becomes appreciable even at ordinary temperatures and the material behaves as an intrinsic semiconductor (Fig. 6.8b). The examples are germanium and silicon. At 0 K, even these materials behave as insulators because electrons from the valence band cannot be thermally excited to the conduction band to cause conduction. Thus it is evident that all the intrinsic semiconductors are insulators at 0 K and all the insulators may behave as semiconductors at temperatures sufficiently higher than 0 K.

It is important to note that the conductivity of semiconductors increases with temperature owing to the increase in the number of conduction band electrons. The conductivity of metals, however, decreases with increase in temperature. This is because more and more phonons are excited at higher temperatures which may scatter conduction electrons and reduce their mobility.

SOLVED EXAMPLES

Example 6.1. Using the Kronig-Penney model, show that for $P << 1$, the energy of the lowest energy band is

$$E = \frac{\hbar^2 P}{ma^2}$$

Solution. Referring to Eq. (6.39), the energy of the lowest band corresponds to $k = \pm \pi/a$, i.e., when

$$P\left(\frac{\sin \alpha a}{\alpha a}\right) + \cos \alpha a = \pm 1$$

Considering only the magnitude on the right hand side, we obtain

$$\frac{P}{\alpha a}(\sin \alpha a) = 1 - \cos \alpha a$$

or
$$\frac{2P}{\alpha a}\sin\left(\frac{\alpha a}{2}\right)\cos\left(\frac{\alpha a}{2}\right) = 2\sin^2\left(\frac{\alpha a}{2}\right)$$

For $P << 1$, we can write

$$\tan\left(\frac{\alpha a}{2}\right) = \frac{P}{\alpha a} = \tan\left(\frac{P}{\alpha a}\right)$$

or
$$\frac{\alpha a}{2} = \frac{P}{\alpha a}$$

or
$$\alpha^2 = \frac{2P}{a^2}$$

Using Eq. (6.23), we get

$$\frac{2mE}{\hbar^2} = \frac{2P}{a^2}$$

or
$$E = \frac{\hbar^2 P}{ma^2}$$

Example 6.2. The energy near the valence band edge of a crystal is given by

$$E = -Ak^2$$

where $A = 10^{-39}$ Jm^2. An electron with wave vector $\mathbf{k} = 10^{10}\ \mathbf{k}_x$ m^{-1} is removed from an orbital in the completely filled valence band. Determine the effective mass, velocity, momentum and energy of the hole.

Solution. We have

$$E = -Ak^2$$

$\therefore$
$$\frac{dE}{dk} = -2Ak,$$

and
$$\frac{d^2E}{dk^2} = -2A$$

The effective mass of the electron is

$$m_e^* = \frac{\hbar^2}{\left(\frac{d^2E}{dk^2}\right)}$$

$$= -\frac{\hbar^2}{2A} = -\frac{(1.05 \times 10^{-34})^2}{2 \times 10^{-39}} = -5.5 \times 10^{-30} \text{ kg.}$$

Since the effective mass of a hole is opposite to that of an electron at the same location in the energy band, the effective mass of hole is

$$m_h^* = -m_e^* = 5.5 \times 10^{-30} \text{ kg}$$

The wave vector of hole is opposite to that of electron, i.e.,

$$\mathbf{k}_h = -\mathbf{k}_e = -10^{10}\ \hat{\mathbf{k}}_x \text{ m}^{-1}$$

The momentum of hole is

$$\mathbf{p}_h = \hbar \mathbf{k}_h$$

$$= -1.05 \times 10^{-34} \times 10^{10}\ \hat{\mathbf{k}}_x = -1.05 \times 10^{-24}\ \hat{\mathbf{k}}_x \text{ Jsm}^{-1}$$

The velocity of hole is

$$\mathbf{v}_h = \frac{\mathbf{p}_h}{m_h^*} = -\frac{1.05 \times 10^{-24}\ \hat{\mathbf{k}}_x}{5.5 \times 10^{-30}} = -1.9 \times 10^5\ \hat{\mathbf{k}}_x \text{ ms}^{-1}$$

The energy of the electron with wave vector $\mathbf{k}_e$ is

$$E_e = -Ak_e^2$$

$$= -10^{-39}(10^{10}\ \hat{\mathbf{k}}_x)^2$$

$$= -10^{-19} \text{ J}$$

Therefore, the energy of the hole referred to zero at the top of the valence band is

$$E_h = -E_e$$

$$= 10^{-19} \text{ J}$$

SUMMARY

1. The band theory assumes that the potential experienced by an electron in a crystal is periodic with period equal to the lattice constant.

2. According to the Bloch theorem, the solution to the Schrodinger wave equation for an electron moving in a periodic potential is of the form of the Bloch function, i.e.,

$$\psi_k(\mathbf{r}) = e^{i\mathbf{k}.\mathbf{r}}\, u_k(\mathbf{r})$$

where $u_k(\mathbf{r})$ is invariant under a crystal lattice translation.

3. The energy spectrum of an electron moving in a periodic potential consists of a set of allowed energy regions or bands which are separated by forbidden energy regions called the band gaps. The forbidden energy regions are those in which $\mathbf{k}$ assumes complex values and no Bloch function type solutions to the wave equation are possible.

4. The width of the allowed energy bands increases with increase in total energy and decrease in binding energy of the electrons. For very large binding energies, the energy bands are transformed into energy levels. If the electrons are completely free, the energy spectrum is quasi-continuous.

5. The *E-k* curve in the extended zone scheme has discontinuities at

$$k = \pm\frac{n\pi}{a}$$

where a is the length of the primitive cell and $n = 1, 2, 3, \ldots$ These *k*-values define the boundaries of the Brillouin zones which produce Bragg reflection of electrons.

6. In a crystal having N primitive cells, the maximum number of energy states per band is N and the maximum number of electrons per band is $2N$.

7. The velocity of an electron is given by the expression

$$v = \frac{1}{\hbar}\frac{dE}{dk}$$

It becomes maximum at the inflextion points of the *E–k* curve where it approaches the free electron velocity. Beyond the inflexion points, the velocity decreases with increase in energy.

8. The effective mass of an electron moving in a crystal depends on the shape of the *E–k* curve and is given by the expression

$$m^* = \frac{\hbar^2}{\left(\dfrac{d^2E}{dk^2}\right)}$$

In the first Brillouin zone, the effective mass is positive in the lower half and negative in the upper half of the energy band. It becomes infinite at the inflexion points of the *E–k* curve.

9. For a one-dimensional lattice of length L, the total effective number of free electrons in an energy band filled with electrons up to a certain value k_1 ($k_1 < \pi/a$) is

$$N_{eff} = \frac{2Lm}{\pi \hbar^2}\left(\frac{dE}{dk}\right)_{k=k_1}$$

where m is the mass of an electron. N_{eff} is zero at the top of the band and attains a maximum value at the points of inflexion.

VERY SHORT QUESTIONS

1. What is the Bloch function?
2. What is forbidden energy gap?
3. Can an electron possess negative effective mass? Justify.
4. Give the order of band gap for a metal, a semiconductor and an insulator.
5. Why a solid whose energy bands is filled can not be a metal?

SHORT QUESTIONS

1. What are the shortcomings of the free electron theory? What is the main cause of its failure?
2. Compare the salient features of the band theory vis-a-vis the free electron theory. Discuss the successes and failures of these theories.
3. What is the Bloch theorem ?
4. Explain the concept of forbidden energy bands.
5. Distinguish between a metal, a semiconductor and an insulator on the basis of their energy band structure.
6. Show that the existence of energy gaps at the boundaries of the Brillouin zone of a one-dimensional lattice is equivalent to the condition of Bragg reflection of electron waves.
7. Show the dependence of velocity of electron on wave vector **k** as predicted by the band theory.
8. Prove that the number of different **k**-states in the first Brillouin zone of a simple cubic lattice is equal to the number of lattice sites.
9. Prove that the total number of possible states in an energy band of a finite crystal is equal to the number of primitive cells in it.
10. How does the band theory lead to the concept of negative effective mass ?
11. The conductivity of metals decreases while that of semiconductors increases with rise in temperature. Explain.

LONG QUESTIONS

1. State and prove the Bloch theorem. Discuss its importance in the band theory.
2. Discuss the formation of allowed and forbidden energy bands on the basis of the Kronig-Penney model. Discuss the extreme conditions when energy levels are either discrete or continuous. What is the effect of changing the binding energy of electron on the energy bands ?
3. Prove that the motion of electrons through the periodic potential in solids gives rise to the band structure.
4. What are Bloch functions? Explain the origin of allowed and forbidden bands for electrons in solids. What is the number of orbitals in an energy band ?
5. Describe the periodic zone scheme, extended zone scheme and reduced zone scheme for representing *E-k* relationships.
6. Discuss the motion of electrons in one-dimension according to the band theory and show the variation of energy, velocity and effective mass as a function of wave-vector.
7. What is meant by the effective mass of an electron? What is its significance? Show that the effective mass of an electron in a crystal is inversely proportional to the second derivative of the *E-k* curve. Discuss the conditions when the effective mass of an electron becomes positive, negative and infinity.

PROBLEMS

1. The potential of an electron in a one-dimensional lattice is of the same type as that used in the Kronig-Penney model. Assuming

$$V_0\, ab << \hbar^2/m,$$

prove that the energy band gap at $k = \pi/a$ is $2V_0 b/a$.

2. A one-dimensional lattice of spacing a has a potential distribution of the type as considered in the Kronig-Penney model. The value of the potential is $-V$ at each lattice point and abruptly changes to zero at a distance of $0.1a$ on either side of the lattice point. Determine the width of the first energy gap in the electron energy spectrum.

(0.37 V)

CHAPTER - VII

SEMICONDUCTORS

7.1 INTRODUCTION

Semiconductors are materials which have electrical conductivities lying between those of good conductors and insulators. The resistivity of semiconductors varies from 10^{-5} to 10^{+4} ohm-m as compared to the values ranging from 10^{-8} to 10^{-6} ohm-m for conductors and from 10^7 to 10^8 ohm-m for insulators. There are elemental semiconductors such as germanium and silicon which belong to Group IV of the periodic table and have resistivity of about 0.6 and 1.5×10^3 ohm-m respectively. Besides these, there are certain compound semiconductors such as gallium arsenide (GaAs), indium phosphide (InP), cadmium sulphide (CdS), etc. which are formed from the combinations of the elements of Groups III and V, or Groups II and VI. Another important characteristic of the semiconductors is that they have small band gap. The band gap of semiconductors varies from 0.2 to 2.5 eV which is quite small as compared to that of insulators. The band gap of a typical insulator such as diamond is about 6 eV. This property determines the wavelength of radiation which can be emitted or absorbed by the semiconductor and hence helps to construct devices such as light emitting diodes (LEDs) and lasers. All the semiconductors have negative temperature coefficient of resistance. The band gap energies for the elements of Group IV at 0 K are as follows :

C (diamond)	5.51 eV
Ge	0.75 eV
Si	1.16 eV
Sn (grey)	0.08 eV
Pd	$\cong 0$

At room temperature, diamond behaves as an insulator, Ge and Si as semiconductors and Sn and Pd as conductors.

The importance of semiconuuctors is further increased due to the fact that the conductivity and the effective band gaps of these materials can be modified by the introduction of impurities which strongly affect their electronic and optical properties. The process of introduction of impurities in semiconductors in a precisely controlled manner is called *doping*. Depending on the nature of impurities added, the semiconductors are classified as follows :

(a) Pure or intrinsic semiconductors

(b) Impurity or extrinsic semiconductors

The intrinsic semiconductors are pure semiconductors in which no impurity atoms are added, whereas the extrinsic semiconductors are doped semiconductors in which suitable impurity atoms are added to modify the properties. In the present chapter, the effects of impurities and charge carrier concentrations in semiconductors are discussed.

7.2 PURE OR INTRINSIC SEMICONDUCTORS

As stated above, the intrinsic semiconductors such as pure Ge or Si are undoped semiconductors. The electrical conductivity of this type of semicondutor is solely determined by thermally generated carriers. To understand the mechanism of conduction, we consider the bonding between

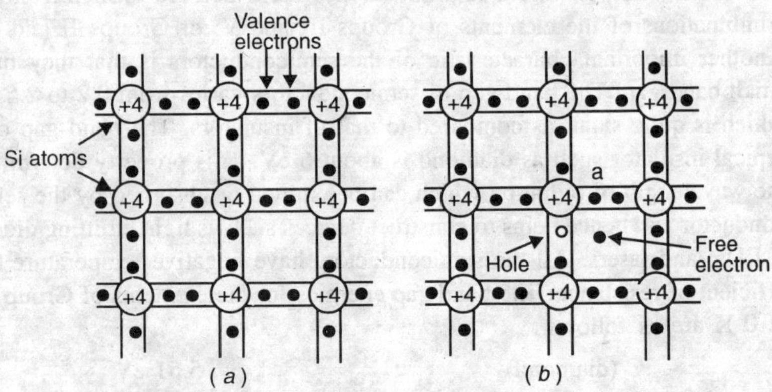

Fig. 7.1. *(a) A two-dimensional representation of Si crystal (b) Silicon crystal containing an electron-hole pair.*

atoms in these semiconductors. Consider, for example, the case of silicon with atomic number 14. Each silicon atom has four valence electrons and can form four covalent bonds with four neighbouring silicon atoms which are directed along the corners of a regular tetrahedron. The silicon crystal, therefore, exhibits a three-dimensional regular network type structure which, for simplicity, is represented by a two-dimensional network as shown in Fig. 7.1a. Apparently, all the valence electrons in a silicon crystal participate in the formation of covalent bonds and no electron is free to cause conduction particularly at 0 K. Hence pure silicon behaves as an insulator at 0 K. As the temperature increases above 0 K, some of the valence electrons may acquire sufficient thermal energy to break their covalent bonds and become free from the influence of cores of the atoms. These electrons move randomly in the crystal and are referred to as the *conduction electrons.* Each escaped electron leaves behind an empty space called a *hole* which also acts as a current carrier. Thus when a valence electron breaks away from a covalent bond, an electorn-hole pair is generated and two carriers of electricity are produced as shown in Fig. 7.1b. When a valence electron located adjacent to a hole acquires sufficient thermal energy, it may jump into the hole position to reconstruct the broken covalent bond and a hole is created at the initial position of the electron. Thus the motion of an electron may also be regarded as the motion of a hole in the opposite direction. These electrons and holes move in opposite directions under the effect of an external electric field and constitute the current.

The energy band diagram of the intrinsic semiconductor is shown in Fig. 7.2. At 0 K, the valence band is completely filled and the conduction band is completely empty. The semiconductor, therefore, behaves as an insulator. The electrons present in the valence band do not conduct as these are bound to their respective cores. As temperature increases, some of the valence band electrons acquire sufficient thermal energy to jump to the conduction band leaving behind an equal number of holes in the valence band. The electrons in the conduction band and holes in the valence band behave as free carriers and increase the conductivity of the material. The conditions for the movement of electrons are, however, different from the conditions for movement of holes; the electrons move when the conduction band is nearly empty and the holes move when the valence band is nearly full. Thus the electrons move mainly under the influence of the applied field while the holes move under the combined effect of the applied electric field and the ionic field of the lattice. Thus the properties such as effective mass, mobility[1], etc. of a hole are quite different from the corresponding properties of electrons. For example, a hole has larger effective mass and lower mobility than the corresponding values for an electron. The charge of a hole is equal and opposite to that of an electron. These properties have been described in the previous chapter. It is now apparent that, in an intrinsic semiconductor, the number of electrons, n_i, in the conduction band is always equal to the number of holes, p_i, in the valence band, i.e., $n_i = p_i$, and either one of these is called the *intrinsic carrier concentration.*

Besides the generation of free electron-hole pairs, there is another process called *recombination* of carriers in semiconductors. A free electron

1. Mobility of a carrier is the velocity acquired by it in a unit electric field.

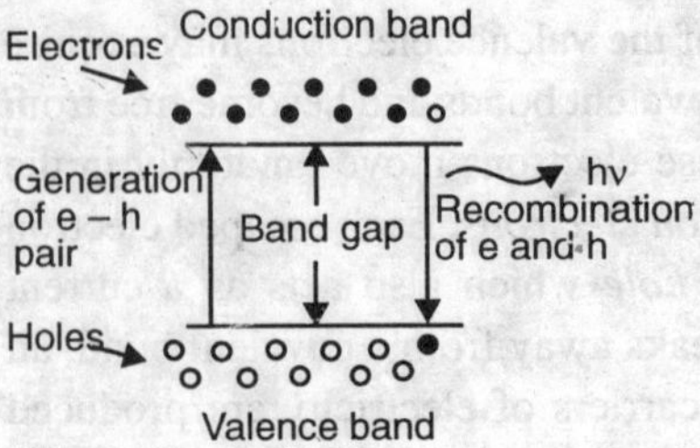

Fig. 7.2. *Energy band diagram of an intrinsic semiconductor showing generation and recombination of an electron and a hole.*

moving randomly in a semiconductor may encounter a hole and combine with it so as to reconstruct the broken covalent bond. Thus the electron-hole pair is destroyed and the free electron is converted into the bound electron. This recombination process is equivalent to an electron jumping from the conduction band to the valence band and occurs with the release of energy equal to the band gap energy in the form of electromagnetic radiation as shown in Fig. 7.2. In an intrinsic semiconductor, the rate of generation of carriers, g, depends on the temperature and nature of the material. The recombination rate, R, on the other hand, depends on the concentration of electrons and holes in a material at that temperature. In equilibrium, the generation rate just equals the recombination rate, i.e.,

$$g = R = C\, n_i p_i$$

where C is a proportionality constant which depends on the nature of the material. Since in an intrinsic semiconductor, $n_i = p_i$, we have

$$g = R = C\, n_i^2 \tag{7.1}$$

For a given temperature, the quantity n_i is a constant and depends only on the nature of the semiconductor. It will be realized later that the constant C is related to the densities of states of electrons and holes at the conduction band and valence band edges respectively.

7.3 IMPURITY OR EXTRINSIC SEMICONDUCTORS

The intrinsic semiconductors, as such, are of little importance owing to their very small and fixed conductivity. The introduction of impurity atoms, i.e. doping, is the most efficient and convenient method of increasing and altering the conductivity of an intrinsic semiconductor. Depending on the type of doping, excess electrons or holes are generated in the material which are free to conduct electricity. The impurity atoms frequently employed to dope pure silicon or germanium are the elements of Group III and Group V of the periodic table. These impurity atoms are referred to as acceptor or *p*-type and donor or *n*-type impurities as they contribute excess holes and electrons respectively to the semiconducting material. The semiconductor is accordingly known as *p*-type or *n*-type semiconductor. The dopants are added in the ratio of about 1 in 10^6 to 10^8 atoms of the semiconducting material. Such a small quantity of dopants does not bring about any structural changes in the semiconductor as the impurity atoms replace the regular atoms in the crystal. However, the conductivity of the semiconductors is greatly affected by such substitutions.

7.3.1 Donor or n-type Semiconductor

When a pentavalent impurity atom of Group V, such as phosphorus, arsenic or antimony, is introduced into silicon, four of its five valence electrons form covalent bonds with the neighbouring four silicon atoms while the fifth valence electron remains loosely bound to its nucleus as shown in Fig. 7.3a. A small but definite amount of energy is required to detach this fifth electron from its nucleus and make it free to conduct. The energy required is, however, quite small as compared to the energy required for breaking a covalent bond and can be easily provided by thermal agitation inside the crystal.

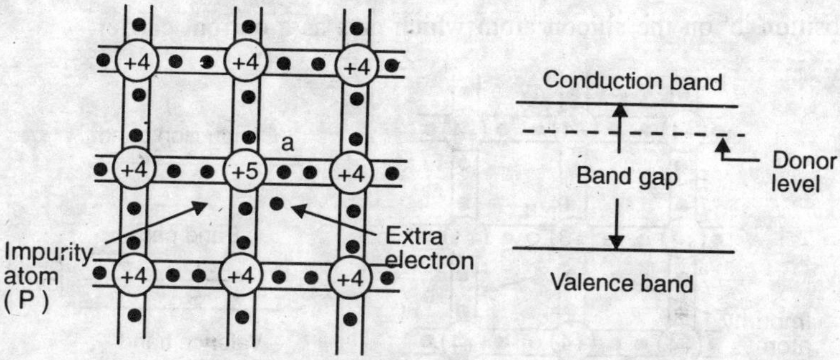

(*a*) (*b*)

Fig. 7.3.*(a) A pentavalent impurity atom (P) in a silicon crystal. (b) Energy level diagram of an n-type semiconductor.*

The energy level corresponding to the fifth valence electron lies in the band gap just below the conduction band edge as shown in Fig. 7.3b. This level is called the *donor level.* The depth of the donor level below the conduction band is merely about 0.01 eV for Ge and 0.03 eV for Si. The electrons are, therefore, easily transferred to the conduction band leaving behind positively charged immobile impurity ions. Thus each pentavalent impurity atom donates one free electron to the semiconductor. Such impurities are, therefore, known as *donors or n-type impurities* and the semiconductor containing such impurity atoms is known as an *n-type semiconductor.* In these semiconductors, the current is carried mainly by electrons which are called *majority carriers.* The thermally generated holes are called *minority carriers.* The electron concentration, n, is obviously quite large as compared to hole concentration, p, but their product always remains constant.

i.e., $$np = n_i p_i = n_i^2 \qquad (7.2)$$

where n_i and p_i are the intrinsic values of the carrier concentration. This relationship is called the *law of mass-action* and will be derived later.

7.3.2 Acceptor or p-type Semiconductor

If a trivalent impurity atom of Group III, such as boron, aluminium, gallium or indium, is introduced into silicon, it forms three covalent bonds with the neighbouring three silicon atoms while the fourth bond is not completed due to the deficiency of one electron. This incomplete bond is shown by broken line in Fig. 7.4a where the small circle (marked 'a') represents the electron deficiency. Thus the trivalent impurity atom has a tendency to accept one electron (say 'b') from a neighbouring silicon atom to complete the fourth covalent bond. This process requires a small amount of energy which is easily provided by the thermal agitation in the crystal. The transferred electron leaves behind a broken covalent bond, i.e., a hole, at position 'b' on the silicon atom which acts as a current carrier.

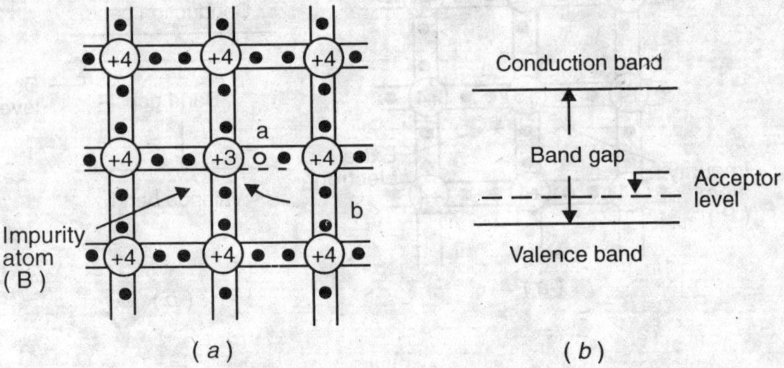

Fig. 7.4. *(a) A trivalent impurity atom (B) in a silicon crystal. (b) Energy level diagram of a p-type semiconductor.*

The energy level corresponding to the electron deficiency of the type 'a' is located just above the valence band and is called the *acceptor level.* The acceptor levels are located at a distance of about 0.01 eV above the top of the valence band in Ge and at about 0.046 to 0.16 eV in Si. An electron can be easily transferred from the valence band to the acceptor level by providing this small amount of energy. This creates a hole in the valence band which acts as a mobile current carrier. The negatively charged impurity atom, however, remains immobile and does not contribute to conduction. Thus each trivalent impurity atom can accept an electron from a neighbouring silicon atom to produce a hole in the semiconductor. Such impurities are, therefore, known as *acceptors or p-type impurities* and the semiconductor containing such impurity atoms is known as a *p-type semiconductor.* In these semiconductors, holes are the *majority carriers* and thermally generated electrons are the *minority carriers.* In this case, p is quite large as compared to n but the law of mass-action still holds, i.e., $np = n_i p_i = n_i^2$. It may also be noted that in either type of semiconductors, the overall charge neutrality is maintained as no charge is added to or removed from the material.

7.4 DRIFT VELOCITY, MOBILITY AND CONDUCTIVITY OF INTRINSIC SEMICONDUCTORS

At finite temperatures, due to thermal agitation and lattice vibrations, some of the valence band electrons are always present in the conduction band, i.e., at ordinary temperatures, an intrinsic semiconductor always contains some free electrons in the conduction band and an equal number of holes in the valence band. In the absence of any applied electric field, these electrons and holes move in random directions and constitute no current. When an electric field is applied, these electrons and holes get accelerated towards the opposite ends of the field and their velocity begins to increase. This increase in velocity, however, does not continue indefinitely because of the collisions of these carriers with the various types of obstacles, such as atomic nuclei, phonons, etc., present in the semiconductor. Depending on the mean free path, the carriers acquire an average increment in velocity which is lost during the subsequent collision. This extra velocity acquired by the carriers in the presence of an applied electric field is called the *drift velocity* and is denoted by v_d. It is proportional to the strength $\mathcal{E}$ of the applied electric field, i.e.,

$$v_d \propto \mathcal{E}$$

or

$$v_d = \mu E \tag{7.3}$$

where the constant μ is called the *mobility* of the charge carrier and is defined as the drift velocity acquired by a carrier per unit electric field strength.

In an intrinsic semiconductor, since the electrons move in nearly empty conduction band while holes move in nearly full valence band, the properties such as mobility, conductivity, etc. of electrons are, in general, different from those of holes. Let v_{dn}, μ_n and n denote the drift velocity, mobility and concentration of electrons respectively in the conduction band. Then current density due to electrons is given by

$$J_n = nev_{dn} \tag{7.4}$$

where e is the electronic charge. We can write (7.3) for the electrons as

$$v_{dn} = \mu_n \mathcal{E}$$

Therefore, from Eq. (7.4), we obtain

$$J_n = ne\mu_n \mathcal{E} \tag{7.5}$$

Comparing it with Ohm's law, i.e.,

$$J_n = \sigma_n \mathcal{E},$$

where σ_n represents the electronic conductivity of the material, we get

$$\sigma_n = ne\mu_n \tag{7.6}$$

Similarly, we can write the expression for the conductivity due to holes in the valence band as

$$\sigma_p = pe\mu_p \tag{7.7}$$

where p and μ_p represent the concentration and mobility of holes respectively. Thus the total conductivity of the material is

$$\sigma = \sigma_n + \sigma_p$$
$$= e\,(n\mu_n + p\mu_p) \tag{7.8}$$

For an intrinsic semiconductor,

$$n = p = n_i$$

Therefore, Eq. (7.8) becomes

$$\sigma = en_i\,(\mu_n + \mu_p) \tag{7.9}$$

It is important to note that, in semiconductors, the movement of carriers or the flow of current is, in fact, the consequence of the following two processes :

(i) drift of carriers under the effect of an applied field; the resulting current is called the *drift current.*

(ii) diffusion of carriers under the effect of concentration gradient of dopants present inside the semiconductor ; the corresponding current is called the *diffusion current.*

In the above treatment, we have considered only the drift current contribution. The diffusion current contribution is absent in semiconductors having a uniform distribution of impurities.

Variation of Conductivity with Temperature

Assuming mobilities to be independent of temperature, the temperature dependence of conductivity arises because of the variation of intrinsic carrier concentration, n_i, with temperature. It will be proved that n_i is given by

$$n_i = \frac{2(2\pi kT)^{3/2}}{h^3} \left(m_n^* m_p^*\right)^{3/4} \exp\left(-\frac{E_g}{2kT}\right) \tag{7.10}$$

where m_n^* and m_p^* represent the effective masses of an electron and a hole respectively, E_g is the band gap, k is the Boltzmann's constant and T is the absolute temperature. Substituting n_i from Eq. (7.10) into (7.9), we get

$$\sigma = e(\mu_n + \mu_p)\,\frac{2(2\pi kT)^{3/2}}{h^3}\left(m_n^* m_p^*\right)^{3/4}\exp\left(-\frac{E_g}{2kT}\right) \tag{7.11}$$

which gives

$$\ln \sigma = -\left(\frac{E_g}{2k}\right)\frac{1}{T} + \frac{3}{2}\ln T + \text{constant} \tag{7.12}$$

The first term on the right hand side is the dominant term. The plot of $\ln\sigma$ versus $1/T$ is a straight line as shown in Fig. 7.5 The slope of the line gives an estimate of the band gap of the semiconductor.

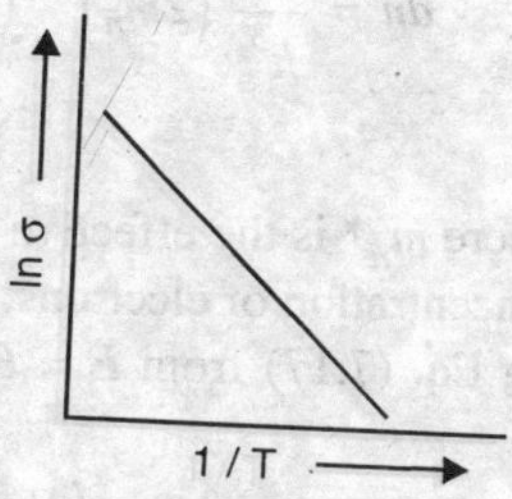

Fig. 7.5. *Plot of lnσ versus 1/T.*

7.5 CARRIER CONCENTRATION AND FERMI LEVEL FOR INTRINSIC SEMICONDUCTOR

The concentration of electrons and holes in a semiconductor can be obtained from the knowledge of the densities of available states in the valence band and the conduction band as well as the Fermi-Dirac distribution function. The expression for the Fermi energy is then obtained from these carrier concentrations.

7.5.1 Electron Concentration in the Conduction Band

Referring to Eq. (5.44) the number of free electrons per unit volume in an energy range E and $E + dE$ can be written as

$$dn = D(E)\, f(E)\, dE \tag{7.13}$$

where $D(E)$ is the density of states defined as the total number of allowed electronic states per unit volume in a semiconductor and $f(E)$ is the Fermi distribution function representing the probability of occupation of a state with energy E. The expression for $f(E)$ is given by Eq. (5.28) as

$$f(E) = \frac{1}{\exp\left(\dfrac{E - E_F}{kT}\right) + 1} \tag{7.14}$$

whereas that for $D(E)$, which is strictly valid for free electrons, is obtained from Eq. (5.40) as

$$D(E) = \frac{4\pi}{h^3}(2m)^{3/2}E^{1/2} \tag{7.15}$$

Using Eqs. (7.15) and (7.13), we obtain

$$dn = \frac{4\pi}{h^3}(2m)^{3/2}E^{1/2}f(E)dE \tag{7.16}$$

It is apparent from Fig. 7.6 that an electron occupying an energy state E in the conduction band, in fact, possesses the kinetic energy $(E–E_c)$. Therefore, in Eq. (7.15), E must be replaced by $(E–E_c)$. Thus Eq. (7.16) becomes

$$dn = \frac{4\pi}{h^3}\left(2m_n^*\right)^{3/2}(E-E_c)^{1/2}\frac{1}{\exp\left(\frac{E-E_F}{kT}\right)+1}dE \tag{7.17}$$

where m_n^* is the effective mass of the electron in the conduction band. The concentration of electrons, n, in the conduction band is obtained by integrating Eq. (7.17) from $E = E_c$ to $E = \infty$, i.e.,

$$n = \frac{4\pi}{h^3}\left(2m_n^*\right)^{3/2}\int_{E_c}^{\infty}\frac{(E-E_c)^{1/2}dE}{\exp\left(\frac{E-E_F}{kT}\right)+1} \tag{7.18}$$

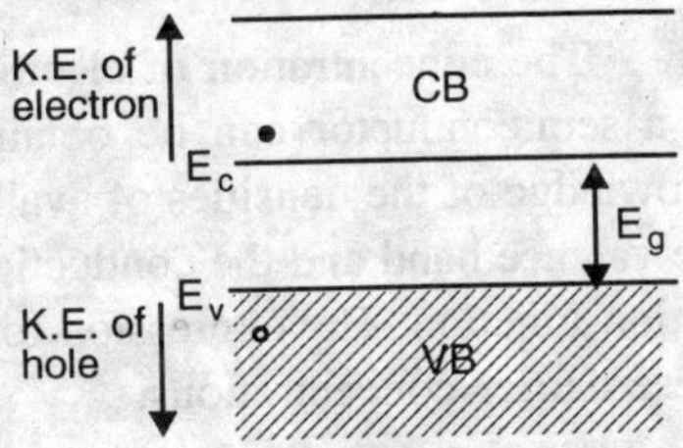

Fig. 7.6. *Band model of an intrinsic semiconductor.*

Now, near room temperature, $kT \cong 0.026$ eV. Therefore, for energies greater than E_c, we have

$$1 + \exp\left(\frac{E-E_F}{kT}\right) \cong \exp\left(\frac{E-E_F}{kT}\right)$$

$$\therefore\ n = \frac{4\pi}{h^3}\left(2m_n^*\right)^{3/2}\int_{E_c}^{\infty}(E-E_c)^{1/2}\exp\left[-\left(\frac{E-E_F}{kT}\right)\right]dE$$

$$= \frac{4\pi}{h^3}\left(2m_n^*\right)^{3/2}\exp\left(\frac{E_F-E_c}{kT}\right)\int_{E_c}^{\infty}(E-E_c)^{1/2}\exp\left[-\left(\frac{E-E_c}{kT}\right)\right]dE$$

Let $$\frac{E-E_c}{kT} = x$$

$\therefore$ $$dE = kT\,dx$$

For $E = E_c$, $x = 0$

$$\therefore \quad n = \frac{4\pi}{h^3}\left(2m_n^*\right)^{3/2} \exp\left(\frac{E_F - E_c}{kT}\right) \int_0^\infty x^{1/2} (kT)^{1/2} e^{-x} kT \, dx$$

$$= \frac{4\pi}{h^3}\left(2m_n^* kT\right)^{3/2} \exp\left(\frac{E_F - E_c}{kT}\right) \int_0^\infty x^{1/2} e^{-x} \, dx$$

Now

$$\int_0^\infty x^{1/2} e^{-x} dx = \left(\frac{\pi}{4}\right)^{1/2}$$

$$\therefore \quad n = 2\left(\frac{2\pi m_n^* kT}{h^2}\right)^{3/2} \exp\left[-\left(\frac{E_c - E_F}{kT}\right)\right] \tag{7.19}$$

From Eq. (7.14), the probability of occupancy of level E_c is given by

$$f(E_c) = \frac{1}{1+\exp\left(\frac{E_c - E_F}{kT}\right)} \cong \exp\left[-\left(\frac{E_c - E_F}{kT}\right)\right]$$

Therefore, Eq. (7.19) becomes

$$n = 2\left(\frac{2\pi m_n^* kT}{h^2}\right)^{3/2} f(E_c)$$

The first term on the right hand side must represent the effective density of states of electrons at the conduction band edge. Denoting it by N_c, we have

$$n = N_c \exp\left[-\left(\frac{E_c - E_F}{kT}\right)\right] \tag{7.20}$$

where

$$N_c = 2\left(\frac{2\pi m_n^* kT}{h^2}\right)^{3/2} \tag{7.21}$$

For silicon,

$$N_c = 2.8 \times 10^{25} \left(\frac{T}{300}\right)^{3/2} \text{m}^{-3}$$

7.5.2 Hole Concentration in the Valence Band

An expression similar to (7.13) for the number of holes per unit volume in the energy range E and $E + dE$ can be written as

$$dp = D(E)[1 - f(E)]dE \tag{7.22}$$

where we have replaced $f(E)$ by $[1-f(E)]$ which represents the probability of an energy state E not to be occupied by an electron, i.e., the probability of finding a hole in the energy state E. Now

$$1-f(E) = 1 - \frac{1}{1+\exp\left(\dfrac{E-E_F}{kT}\right)} = \frac{\exp\left(\dfrac{E-E_F}{kT}\right)}{1+\exp\left(\dfrac{E-E_F}{kT}\right)}$$

In the valence band, since $E<E_F$ the exponential term in the denominator may be neglected in comparison to unity. Thus, we get

$$1 - f(E) \cong \exp\left(\frac{E-E_F}{kT}\right) \tag{7.23}$$

It follows that the probability of finding holes decreases exponentially with increase in depth into the valence band. Also, the kinetic energy of a hole in the energy state E in the valence band is (E_v-E). Therefore, the density of states per unit volume in the valence band can be written as

$$D(E) = \frac{4\pi}{h^3}\left(2m_p^*\right)^{3/2}(E_v - E)^{1/2} \tag{7.24}$$

where m_p^* is the effective mass of a hole in the valence band. Using Eqs. (7.23) and (7.24) in Eq. (7.22) and integrating from $E = -\infty$ to $E = E_v$, we obtain the hole concentration in the valence band as

$$p = \frac{4\pi}{h^3}\left(2m_p^*\right)^{3/2}\int_{-\infty}^{E_v}(E_v - E)^{1/2}\exp\left(\frac{E-E_F}{kT}\right)dE$$

$$= \frac{4\pi}{h^3}\left(2m_p^*\right)^{3/2}\exp\left(\frac{E_v-E_F}{kT}\right)\int_{-\infty}^{E_v}(E_v - E)^{1/2}\exp\left(\frac{E-E_v}{kT}\right)dE$$

Let
$$\frac{E_v-E}{kT} = x$$

$$\therefore \qquad dE = -kT\,dx$$

For $E = E_v$, $x = 0$

$$\therefore \quad p = \frac{4\pi}{h^3}\left(2m_p^*\right)^{3/2} \exp\left(\frac{E_v - E_F}{kT}\right) \int_{\infty}^{0} x^{1/2}(kT)^{1/2} e^{-x}(-kT)\,dx$$

$$= \frac{4\pi}{h^3}\left(2m_p^*\right)^{3/2} \exp\left(\frac{E_v - E_F}{kT}\right) (kT)^{3/2} \int_{0}^{\infty} x^{1/2} e^{-x}\,dx$$

$$= \frac{4\pi}{h^3}\left(2m_p^* kT\right)^{3/2} \exp\left(\frac{E_v - E_F}{kT}\right)\left(\frac{\pi}{4}\right)^{1/2}$$

$$= 2\left(\frac{2\pi m_p^* kT}{h^2}\right)^{3/2} \exp\left[-\left(\frac{E_F - E_v}{kT}\right)\right] \tag{7.25}$$

or

$$p = N_v \exp\left[-\left(\frac{E_F - E_v}{kT}\right)\right] \tag{7.26}$$

where

$$N_v = 2\left(\frac{2\pi m_p^* kT}{h^2}\right)^{3/2} \tag{7.27}$$

represents the effective density of holes at the valence band edge. For silicon,

$$N_v = 2.8 \times 10^{25}\left(\frac{T}{300}\right)^{3/2} m^{-3}$$

The electron and hole concentrations given by Eqs. (7.20) and (7.26) respectively are valid for both intrinsic and extrinsic materials. For intrinsic materials, these equations can also be written as

$$n_i = N_c \exp\left[-\left(\frac{E_c - E_i}{kT}\right)\right], \quad p_i = N_v \exp\left[-\left(\frac{E_i - E_v}{kT}\right)\right] \tag{7.28}$$

where the Fermi level E_F has been replaced by the intrinsic level E_i.

7.5.3 Fermi Level

For an intrinsic semiconductor,

$$n = p = n_i$$

Therefore, from Eqs. (7.20) and (7.26), we get

$$N_c \exp\left[-\left(\frac{E_c - E_F}{kT}\right)\right] = N_v \exp\left[-\left(\frac{E_F - E_v}{kT}\right)\right]$$

or $$\exp\left(\frac{2E_F - E_c - E_v}{kT}\right) = \frac{N_v}{N_c}$$

or $$\frac{2E_F - E_c - E_v}{kT} = \ln\left(\frac{N_v}{N_c}\right)$$

or $$E_F = E_i = \frac{E_c + E_v}{2} + \frac{kT}{2}\ln\left(\frac{N_v}{N_c}\right) \tag{7.29}$$

Using Eqs. (7.21) and (7.27) in Eq. (7.29), we obtain

$$E_F = \frac{E_c + E_v}{2} + \frac{3}{4}kT\ln\left(\frac{m_p^*}{m_n^*}\right) \tag{7.30}$$

At 0 K,

$$E_F = \frac{E_c + E_v}{2} \tag{7.31}$$

i.e., the Fermi level lies in the middle of the conduction band and valence band. This is also true at all other temperatures provided $m_p^* = m_n^*$. However, in general, $m_p^* > m_n^*$ and the Fermi level is raised slightly as T exceeds 0 K. For Si at 300 K, the increase in Fermi energy is about 0.01 eV only which may be neglected for all practical purposes.

7.5.4 Law of Mass Action and Intrinsic Carrier Concentration

Since for an intrinsic semiconductor,

$$n = p = n_i,$$

the Eqs. (7.20) and (7.26) yield

$$np = n_i^2 = N_c N_v \exp\left(-\frac{E_c - E_v}{kT}\right)$$

$$= N_c N_v \exp\left(-\frac{E_g}{kT}\right) \tag{7.32}$$

Using Eqs. (7.21) and (7.27) , the above expression becomes

$$np = n_i^2 = 4\left(\frac{2\pi kT}{h^2}\right)^3 \left(m_n^* m_p^*\right)^{3/2} e^{-E_g/kT} \tag{7.33}$$

This shows that, for a given semiconductor, the product of electron and hole concentrations is a constant at a given temperature and is equal to the square of the intrinsic carrier concentration. This is called the *law of mass action* and holds for both intrinsic and extrinsic semiconductors. If impurity atoms are added to a semiconductor to increase n, there will be a corresponding decrease in p such that the product np remains constant. Thus we always have

$$np = n_i^2 \tag{7.34}$$

The intrinsic carrier concentration can be directly obtained from Eq. (7.32) or (7.33) as

$$n_i = (N_c N_v)^{1/2} \exp\left(-\frac{E_g}{2kT}\right) \tag{7.35}$$

$$= 2\left(\frac{2\pi kT}{h^2}\right)^{3/2} \left(m_n^* m_p^*\right)^{3/4} \exp\left(-\frac{E_g}{2kT}\right) \tag{7.36}$$

For pure Ge at 300 K, the intrinsic electron concentration is about 2.4×10^{19} m^{-3} when the concentration of germanium atoms is 4.4×10^{28} m^{-3}. This shows that, at ordinary temperatures, only about five covalent bonds per 10^{10} atoms of germanium are broken and contribute to intrinsic conduction. On the other hand, in metals such as copper, about 10^{28} electrons per cubic metre are available for conduction.

7.6 CARRIER CONCENTRATION, FERMI LEVEL AND CONDUCTIVITY FOR EXTRINSIC SEMICONDUCTOR

As described earlier, the donor or acceptor levels are present in an extrinsic semiconductor depending on the type of impurity present. The concentration of donors or acceptors in a semiconductor affects the Fermi energy, the carrier concentration and the conductivity of the semiconductor. We consider the following cases :

(a) N-type Semiconductor

The energy level diagram for an n-type semiconductor is shown in Fig. 7.7. At 0 K, all the donors are in the unionized state, i.e., all the donor levels are occupied with electrons. As the temperature increases slightly, some of the donors get ionized and contribute electrons to the conduction band. Also, some of the valence band electrons may jump to the conduction band leaving behind holes in the valence band. The number of holes produced in this process is, however, quite small. Therefore, the Fermi level must lie somewhere near the middle of the donor level and the bottom of the conduction band. We determine the equilibrium carrier concentration at a temperature T. Let there be N_d donors per unit volume occupying the donor levels with energy E_d. Assuming that E_F lies more than few kT below the conduction band, the electron concentration in the conduction band, as given by Eq. (7.20), is

$$n = N_c \exp\left(-\frac{E_c - E_F}{kT}\right) \tag{7.20}$$

The electron concentration must be equal to the sum of the concentration of ionized donors, N_d, in the donor levels and the concentration of thermally generated holes in the valence band, i.e.,

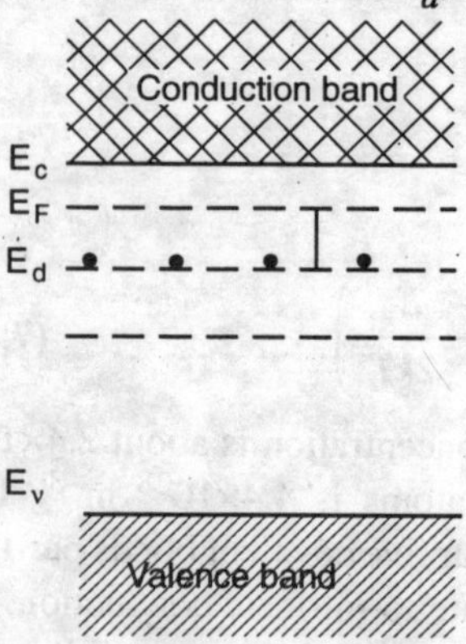

Fig. 7.7. *Band model of an n-type semiconductor.*

$$n = N_d^+ + p \tag{7.37}$$

If a sufficient number of donors are present to produce electrons in the conduction band, the concentration of thermally generated holes gets suppressed as a consequence of the law of mass action [Eq. (7.34)]. Thus p may be neglected in Eq. (7.37) which, therefore, becomes

$$n = N_d^+ \tag{7.38}$$

Also, from Eq.(7.34), the concentration of minority carriers (holes) is given by

$$p = \frac{n_i^2}{N_d^+} \tag{7.39}$$

and the concentration of ionized donors is calculated as

$$N_d^+ = N_d\,[1 - f(E_d)]$$

$$= N_d\left[1 - \frac{1}{1+\exp\left\{-\left(\frac{E_F - E_d}{kT}\right)\right\}}\right]$$

$$= N_d \left[\frac{\exp\left\{-\left(\frac{E_F - E_d}{kT}\right)\right\}}{1+\exp\left\{-\left(\frac{E_F - E_d}{kT}\right)\right\}} \right]$$

$$= N_d \exp\left[-\left(\frac{E_F - E_d}{kT}\right)\right] \tag{7.40}$$

where we have neglected the exponent term in the denominator assuming that E_F lies more than a few kT above E_d. Using Eqs. (7.20) and (7.40) in Eq. (7.38), we obtain

$$N_c \exp\left[-\left(\frac{E_c - E_F}{kT}\right)\right] = N_d \exp\left[-\left(\frac{E_F - E_d}{kT}\right)\right]$$

or
$$\ln N_c - \left(\frac{E_c - E_F}{kT}\right) = \ln N_d - \left(\frac{E_F - E_d}{kT}\right)$$

or
$$-\frac{E_c - E_F}{kT} + \frac{E_F - E_d}{kT} = \ln N_d - \ln N_c$$

or
$$E_F = \frac{E_d + E_c}{2} + \frac{kT}{2} \ln\left(\frac{N_d}{N_c}\right) \tag{7.41}$$

This gives the position of the Fermi level at moderate temperatures. Since N_c varies as $T^{3/2}$ as given by Eq. (7.21), this equation is not valid for T = 0 K, Also, it is not valid for $T \to \infty$ because, at very high temperatures, the assumption of suppressing the holes does not hold good. The only valid information obtainable from this equation is that the Fermi level lies somewhere near the middle of the donor level and the conduction band edge at moderate temperatures. This is particularly true for those values of T and N_d for which the second term on the right hand side is negligible. As T increases, the Fermi level moves downwards and crosses the donor level. For sufficiently large temperatures, it drops to $E_g/2$, i.e., coincides with the intrinsic level E_i. This is, however, not apparent from Eq. (7.41). In such a case, the extrinsic semiconductor behaves like an intrinsic one. The variations of $(E_F - E_i)$ with temperature for both n and p type silicon are shown in Fig. 7.8 for different impurity concentrations.

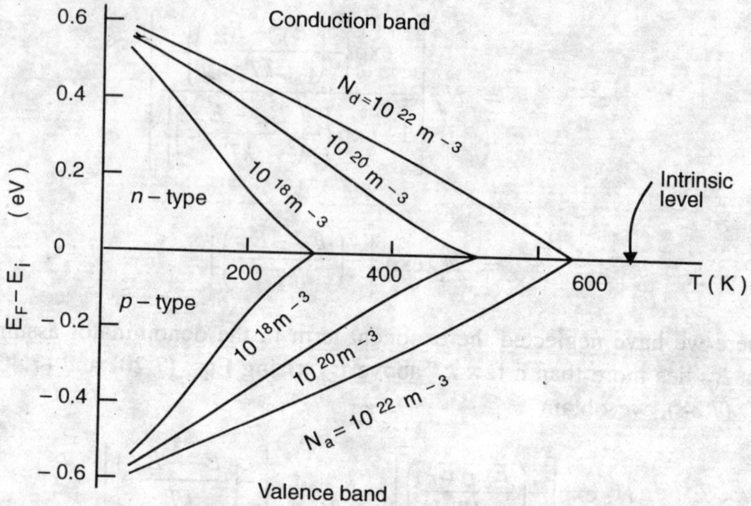

Fig. 7.8. *Variations of Fermi level with temperature and impurity concentration for Si.*

A direct relationship between the equilibrium carrier concentration and the position of the Fermi level relative to the intrinsic level, as obtained from Eqs. (7.20) and (7.28), is

$$n = n_i \exp\left(\frac{E_F - E_i}{kT}\right) \tag{7.42}$$

It is obvious from this equation that the equilibrium electron concentration approaches the intrinsic carrier concentration as E_F approaches E_i. Also, the electron concentration increases exponentially as the Fermi level moves away from E_i towards the conduction band. A similar relation for the hole concentration will be obtained later.

The free electron concentration in the conduction band is obtained by substituting the value of E_F from Eq. (7.41) into (7.20), i.e.,

$$n = N_c \exp\left[\frac{E_d - E_c}{2kT} + \frac{1}{2}\ln\left(\frac{N_d}{N_c}\right)\right]$$

$$= N_c \left(\frac{N_d}{N_c}\right)^{1/2} \exp\left(\frac{E_d - E_c}{2kT}\right)$$

$$= (N_d N_c)^{1/2} \exp\left(\frac{E_d - E_c}{2kT}\right) \tag{7.43}$$

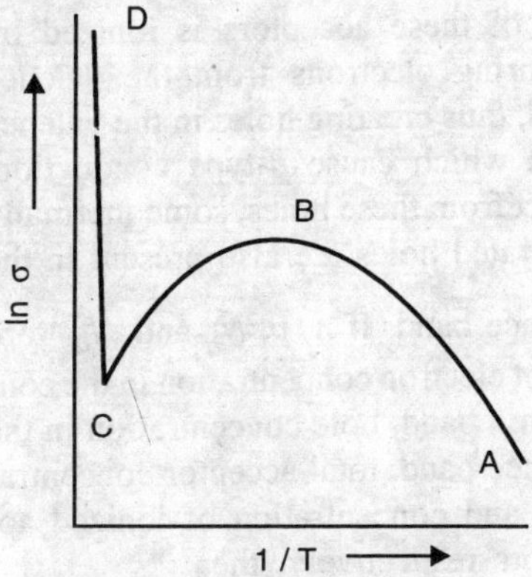

Fig. 7.9. *Conductivity versus temperature for a typical n-type germanium sample.*

$$= (N_d N_c)^{1/2} \exp\left(-\frac{\Delta E}{2kT}\right) \qquad (7.44)$$

where $\Delta E = E_c - E_d$ represents the ionization energy of the donors. This shows that the carrier concentration at moderate temperatures varies as $\sqrt{N_d}$.

The electrical conductivity of an *n*-type semiconductor can be calculated from

$$\sigma = en\mu_n \cong e\, N_d^+\, \mu_n \qquad (7.45)$$

where it has been assumed that the conduction is mainly due to electrons. Knowing the electron concentration, *n*, and the electron mobility, μ_n, at any given temperature, the electrical conductivity can be calculated by using Eq. (7.45). The temperature dependence of electrical conductivity for an *n*-type germanium is shown in Fig. 7.9. The following conclusions can be drawn from this curve :

(i) Starting from the very low temperature (about 10 K) corresponding to the point A, the conductivity increases with rise in temperature. This is due to the increase in the number of conduction electrons as a result of ionization of the donors. The conductivity attains a maximum value (point B) when all the donors are ionized. The temperature corresponding to this is about 50 K for a moderately doped *n*-type germanium.

(ii) The conductivity decreases with further increase in temperature up to about room temperature (point C) and is attributed to the decrease in the value of mobility with rise in temperature. There is also an increase in the intrinsic conductivity but to a lesser extent.

(iii) The sharp rise in conductivity from C to D is due to the large increase in intrinsic conductivity which offsets the decrease in mobility.

(b) P-type Semiconductor

The case of a *p*-type semiconductor can be treated in the same way as that of an *n*-type semiconductor. The energy level diagram for a *p*-type semiconductor is shown in Fig. 7.10. The acceptor impurity atoms occupy the acceptor levels, E_a, which lie above the valence band. For T > 0 K, a

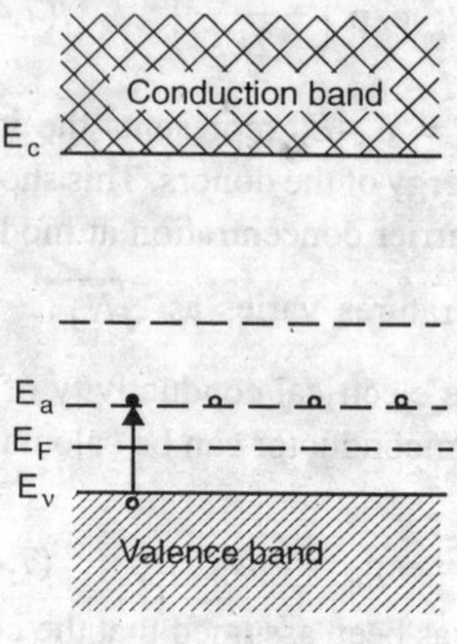

Fig. 7.10. *Band model for a p-type semiconductor.*

part of these acceptors is ionized by acquiring electrons from the valence band, thus creating holes in the valence band which cause *p*-type conduction. Apart from these holes, some thermally generated holes are also present in the valence band. If n, p, N_a and N_a^- represent electron concentration in the conduction band, hole concentration in the valence band, total acceptor concentration, and concentration of ionized acceptors respectively, then

$$p = n + N_a^-$$

Neglecting n in comparison with N_a^- for a *p*-type semiconductor, we get

$$p \cong N_a^- \tag{7.46}$$

The concentration of ionized acceptors is given by

$$N_a^- = N_a f(E_a)$$

$$= \frac{N_a}{1+\exp\left(\dfrac{E_a - E_F}{kT}\right)} \tag{7.47}$$

Assuming $(E_a - E_F)$ to be large as compared to kT, Eq. (7.47) can be written as

$$N_a^- \cong N_a \exp\left(\frac{E_F - E_a}{kT}\right) \tag{7.48}$$

The hole concentration in the valence band is given by Eq. (7.26) as

$$p = N_v \exp\left(-\frac{E_F - E_v}{kT}\right) \tag{7.26}$$

Using Eqs. (7.48) and (7.26) in Eq. (7.46), we obtain

$$N_v \exp\left(-\frac{E_F - E_v}{kT}\right) = N_a \exp\left(\frac{E_F - E_a}{kT}\right)$$

which, on simplification, gives

$$E_F = \frac{E_a + E_v}{2} - \frac{kT}{2} \ln\left(\frac{N_a}{N_v}\right) \tag{7.49}$$

As described earlier, the Fermi level at moderate temperatures lies near the middle of the acceptor level and the top of the valence band. It moves upwards with increase in temperature and finally coincides with the intrinsic level as shown in Fig. 7.8. The expression (7.49) is not valid at very high temperatures where n cannot be neglected.

Using Eqs. (7.26) and (7.28), we obtain an expression identical to Eq. (7.42), i.e.,

$$p = n_i \exp\left(\frac{E_i - E_F}{kT}\right) \tag{7.50}$$

It indicates that the equilibrium hole concentration approaches the intrinsic carrier concentration as E_F approaches E_i. Also, p increases exponentially as E_F moves away from E_i towards the valence band. An expression for the hole concentration in the valence band can be obtained by using Eq. (7.49) into (7.26) as in the previous case.

The electrical conductivity of a p-type semiconductor is given by

$$\sigma_p = ep\mu_p \cong eN_a^- \mu_p \tag{7.51}$$

where the conduction due to minority carriers has been ignored. The variation of conductivity with temperature is similar to that for the n-type semiconductor.

It also follows from Eqs. (7.45) and (7.51) that the increase in concentration of either type of impurity atoms increases the conductivity of a semiconductor. For impurity concentration ranging from 10^{20} to 10^{22} m^{-3}, the resistivity of Si and Ge varies from 10^{-3} to 10^{-1} ohm-m. If doping is heavy (10^{23} to 10^{24} atoms/m^3), the conductivity of semiconductors becomes comparable to metals. Such semiconductors are called *degenerate semiconductors* and find applications in high power and high frequency devices.

(c) Mixed Semiconductor

In a semiconductor containing both n and p-type impurities, the law of electrical neutrality is written as

$$N_d^+ + p = N_a^- + n \tag{7.52}$$

Taking $N_d^+ = N_a^-$ and using Eq. (7.34), we obtain

$$p = n = n_i$$

This shows that, for equal concentrations of donors and acceptors, the semiconductor behaves as pure or intrinsic semiconductor. All the ionized acceptors combine with free electrons of the donors and all the ionized donors combine with free holes of the acceptors to produce no net free carriers. For $N_d^+ \neq N_a^-$ or simply $N_d \neq N_a$ (donors and acceptors are assumed to be ionized), the semiconductor behaves as *n*-type or *p*-type depending on the relative magnitudes of N_d and N_a, and the cases described above become applicable.

SOLVED EXAMPLES

Example 7.1. The electron and hole mobilities in a Si sample are 0.135 and 0.048 m²/V-s respectively. Determine the conductivity of intrinsic Si at 300 K if the intrinsic carrier concentration is 1.5×10^{16} atoms/m³. The sample is then doped with 10^{23} phosphorus atoms/m³. Determine the equilibrium hole concentration, conductivity and position of the Fermi level relative to the intrinsic level.

Solution. Given

$$\mu_n = 0.135 \text{ m}^2/\text{V–s}$$

$$\mu_p = 0.048 \text{ m}^2/\text{V–s}$$

$$n_i = 1.5 \times 10^{16} \text{ m}^{-3}$$

In the case of intrinsic semiconductors,

$$n = p = n_i$$

Therefore, the conductivity is given by

$$\sigma = en_i\,(\mu_n + \mu_p)$$

$$= 1.6 \times 10^{-19} \times 1.5 \times 10^{16} \times (0.135 + 0.048)$$

$$= 4.39 \times 10^{-4}\ (\Omega\text{-m})^{-1}$$

In the extrinsic case, since $N_d >> n_i$, and assuming all the donors to be ionized, we have

$$n \cong N_d^+ = 10^{23} \text{ atoms/m}^3$$

Therefore, the equilibrium hole concentration is

$$p = \frac{n_i^2}{n} = \frac{\left(1.5\times10^{16}\right)^2}{10^{23}} = 2.25 \times 10^9\ m^{-3}$$

The conductivity is given by

$$\sigma = en\mu_n$$

$$= 1.6 \times 10^{-19} \times 10^{23} \times 0.135$$

$$= 21.6 \times 10^2\ (\Omega\text{-m})^{-1}$$

From Eq. (7.42),

$$E_F - E_i = kT \ln \left(\frac{n}{n_i} \right)$$

$$= 8.62 \times 10^{-5} \times 300 \ln \left(\frac{10^{23}}{1.5 \times 10^{16}} \right)$$

$$= 0.406 \text{ eV}$$

The position of the Fermi level in the band diagram is shown in Fig. 7.11.

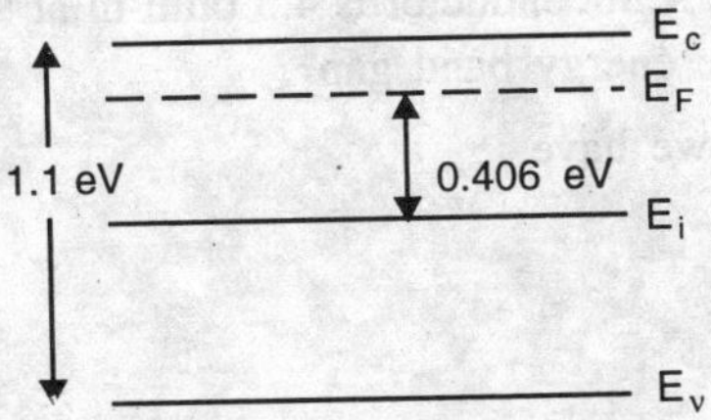

Fig. 7.11

Example 7.2. In intrinsic GaAs, the electron and hole mobilities are 0.85 and 0.04 m^2/V-s respectively and the corresponding effective masses are 0.068 m_o and 0.5 m_o respectively where m_o is the rest mass of an electron. Given the energy band gap at 300 K as 1.43 eV, determine the intrinsic carrier concentration and conductivity.

Solution. Given,

$$\mu_n = 0.85 \text{ m}^2/\text{V-s}$$
$$\mu_p = 0.04 \text{ m}^2/\text{V-s}$$
$$m_n^* = 0.068\, m_o$$
$$m_p^* = 0.5\, m_o$$
$$E_g = 1.43 \text{ eV}$$
$$T = 300 \text{ K}$$

From Eq. (7.36), the intrinsic carrier concentration is given by

$$n_i = 2 \left(\frac{2\pi kT}{h^2} \right)^{3/2} \left(m_n^* m_p^* \right)^{3/4} \exp\left(-\frac{E_g}{2kT} \right)$$

$$= 2 \left[\frac{2\pi \times 1.38 \times 10^{-23} \times 300}{(6.63 \times 10^{-34})^2} \right]^{3/2} \times$$

$$\times \left[0.068 \times 0.5 \times \left(9.1 \times 10^{-31} \right)^2 \right]^{3/4} \exp\left(-\frac{1.43}{2 \times 8.62 \times 10^{-5} \times 300} \right)$$

$$= 1.94 \times 10^{12} \text{ m}^{-3}$$

$$= 1.94 \times 10^{12} \text{ m}^{-3}$$

The conductivity is given by

$$\sigma = en_i (\mu_n + \mu_p)$$
$$= 1.6\times10^{-19}\times1.94\times10^{12}\times(0.85+0.04)$$
$$= 2.76\times10^{-7} \ (\Omega\text{-m})^{-1}$$

Example 7.3. The resistivity of an intrinsic semiconductor is 4.5 ohm-m at 20°C and 2.0 ohm-m at 32°C. What is the energy band gap ?

Solution. For an intrinsic semiconductor, we have

$$\sigma = en_i\mu$$

Using Eq. (7.36), we get

$$\sigma = 2e\mu \left(\frac{2\pi kT}{h^2}\right)^{3/2} \left(m_n^* m_p^*\right)^{3/4} \exp\left(-\frac{E_g}{2kT}\right)$$

$$= CT^{3/2} \exp\left(-\frac{E_g}{2kT}\right)$$

where C is a constant given by

$$C = 2e\mu \left(\frac{2\pi k}{h^2}\right)^{3/2} \left(m_n^* m_p^*\right)^{3/4}$$

If σ_1 and σ_2 are the conductivities at temperatures T_1 and T_2 respectively, then

$$\frac{\sigma_1}{\sigma_2} = \frac{\rho_2}{\rho_1} = \left(\frac{T_1}{T_2}\right)^{3/2} \exp\left\{\frac{E_g}{2k}\left(\frac{1}{T_2} - \frac{1}{T_1}\right)\right\}$$

where ρ represents the resistivity. The above equation can also be expressed as

$$\frac{E_g}{2k}\left(\frac{1}{T_1} - \frac{1}{T_2}\right) = ln\frac{\sigma_2}{\sigma_1} + \frac{3}{2} ln\frac{T_1}{T_2}$$

Now, we have

$T_1 = 20 + 273 = 293$ K, $\rho_1 = 1/\sigma_1 = 4.5$ ohm-m

$T_2 = 32 + 273 = 305$ K, $\rho_2 = 1/\sigma_2 = 2.0$ ohm-m

$$\therefore \quad \frac{E_g}{2 \times 1.38 \times 10^{-23}} \left(\frac{1}{293} - \frac{1}{305} \right) = \ln\left(\frac{4.5}{2.0}\right) + \frac{3}{2} \ln \left(\frac{293}{305}\right)$$

or
$$E_g = 1.54 \times 10^{-19} \text{ J}$$
$$= 0.96 \text{ eV}$$

SUMMARY

1. Semiconductors are materials which have electrical conductivity lying between those of conductors and insulators. They exhibit band gaps of the order of 1 eV and negative temperature coefficient of resistance. Examples of these materials are Si, Ge, GaAs, InP, CdS, etc.

2. In a semiconductor, two types of current carriers are present, viz., electrons and holes. Electrons conduct in the conduction band and holes conduct in the valence band.

3. In a pure or intrinsic semiconductor, electrons and holes are thermally generated and are equal in number.

4. In a doped or extrinsic semiconductor, the conduction is mainly by either electrons (*n*-type conduction) or holes (*p*-type conduction). The electrons in the *n*-type and holes in the *p*-type semiconductors are called majority carriers. The holes in the *n*-type and electrons in the *p*-type material are termed the minority carriers.

5. Mobility is the velocity acquired by a carrier in a unit electric field. Electrons have greater mobility than holes. Hence devices with *n*-type conduction are mostly preferred to those with p-type conduction.

6. In general, two types of currents flow in a semiconductor — the drift current and the diffusion current. The motion of carriers in an electric field constitutes the drift current. The diffusion current arises from the motion of carriers under the effect of concentration gradient of the carriers. In a uniformly doped semiconductor, the latter contribution is absent.

7. The conductivity of a semiconductor, in general, is given by

$$\sigma = e(n\mu_n + p\mu_p)$$

8. The conductivity of an intrinsic semiconductor increases with temperature. The plot of ln σ versus $1/T$ is a straight line and can be used to determine the band gap.

9. The concentration of electrons in the conduction band and holes in the valence band are given by the expressions

$$n = 2\left(\frac{2\pi m_n^* kT}{h^2}\right)^{3/2} exp\left[-\left(\frac{E_c - E_F}{kT}\right)\right]$$

$$p = 2\left(\frac{2\pi m_p^* kT}{h^2}\right)^{3/2} exp\left[-\left(\frac{E_F - E_v}{kT}\right)\right]$$

The intrinsic carrier concentration is

$$n_i = 2\left(\frac{2\pi kT}{h^2}\right)^{3/2}\left(m_n^* m_p^*\right)^{3/4} exp\left(-\frac{E_g}{2kT}\right)$$

10. The law of mass action, i.e.,

$$np = n_i^2$$

holds for both intrinsic and extrinsic semiconductors.

11. In an intrinsic semiconductor, the Fermi level lies in the middle of the conduction band and the valence band edges. At moderate temperatures and impurity concentrations, the Fermi level of an *n*-type semiconductor lies almost in the middle of the donor level and the conduction band edge, whereas in a p-type semiconductor, it lies in the middle of the acceptor level and the valence band edge. As the temperature increases, the Fermi level approaches the intrinsic level.

VERY SHORT QUESTIONS

1. What is an intrinsic semiconductor?
2. What is an extrinsic semiconductor?
3. Explain the concept of hole.
4. What type of carriers is present in a semiconductor?
5. Define mobility of a charge carrier.
6. What is doping?
7. Draw the energy level diagram for an *n*-type semiconductor and label it.
8. Draw the enrgy level diagram for a *p*-type semiconductor and label it

9. What are majority carriers and minority carriers?
10. Define law of mass-action. For what type of semiconductors does it hold?
11. Define drift velocity of a carrier?
12. What are degenerate semiconductors?
13. Which has greater mobility, electron or hole?
14. Can a semiconductor containing both *n*-type and *p*-type impurities behave as intrinsic semiconductor? Give reason.

SHORT QUESTIONS

1. Give the properties of holes vis-a-vis electrons.
2. Explain the conduction mechanism for *n*-type and *p*-type semiconductors.
3. What are donors and acceptors? Give two examples of each.
4. Derive and discuss the law of mass action.
5. Explain the concepts of drift current and diffusion current. How are they different?
6. Obtain an expression for conductivity of an intrinsic semiconductor. How does it vary with temperature?
7. Show a typical variation of conductivity with temperature for an extrinsic semiconductor and explain the different regions.
8. Define the Fermi level. What is its importance in electronic grade materials?

LONG QUESTIONS

1. Discuss the current conduction in semiconductors. How do conductivity of a semiconductor and a metal change with impurity content? Explain the difference in behaviour of these two materials due to change in conductivity.
2. Derive expression for density of free electrons and holes in an intrinsic semiconductor. Show that the Fermi level lies halfway between the valence band and the conduction band.
3. What is an extrinsic semiconductor? Discuss the variation of the Fermi level with temperature for an *n*-type semiconductor.

4. Derive an expression for density of electrons in the conduction band for an n-type semiconductor.
5. What are mobility and conductivity? Obtain an expression for conductivity of doped semiconductors.
6. Derive expressions for electron and hole concentrations for an intrinsic semiconductors. Use these results to obtain intrinsic carrier concentration.
7. Show that the product of electron and hole concentrations in a semiconductor is constant at a given temperature. How is the energy gap determined from the measurement of electrical coductivity of a semiconductor?

PROBLEMS

1. An intrinsic germanium crystal has a hole density of 10^{19} m^{-3} at room temperature. When doped with antimony, the hole density decreases to 10^{17} m^{-3} at the same temperature. Calculate the majority carrier density. (10^{21} m^{-3})
2. The band gaps of diamond and silicon are 5.4 and 1.1 eV respectively. Estimate the temperature at which diamond has the same conductivity as Si at 27°C. (1200°C)
3. The conductivity of intrinsic Si is 4.17×10^{-5} and 4×10^{-4} $(\Omega\text{-m})^{-1}$ at 0°C and 27°C respectively. Determine the average band gap of Si. (1.11 eV)
4. The conductivity of n-type germanium semiconductor is 39 Ω^{-1}m^{-1}. If the mobility of electrons in germanium is 0.39 m^2 V^{-1}s^{-1}, then find the concentration of the donor atoms. (6.25×10^{20} m^{-3})
5. How many donor atoms should be added per cubic metre of pure germanium crystal such that an n-type semiconductor is formed whose conductivity is 500 mho/cm. Given that the mobility of electrons in n-type semiconductor is 0.385 m^{2}V^{-1}s^{-1}. (8.12×10^{24})
6. The conductivity of a semiconductor changes when the concentration of electrons is varied by changing the position of impurity level. Show that it passes through a minimum when the concentration of electrons becomes $n_i\sqrt{\mu_p/\mu_n}$ where n_i is the intrinsic carrier concentration, μ_n and μ_p represent the mobilities of electrons and holes respectively. Dtermine the minimum value of conductivity. ($2ne\mu_n$)

7. The mobilities of electrons and holes in intrinsic Ge are 0.39 and 0.19 m^2/V-s respectively. Determine the intrinsic carrier concentration and conductivity of Ge at 300 K if the band gap of Ge is 0.67 eV and the effective masses of electrons and holes are $0.55m_0$ and $0.37m_0$ respectively, m_0 being the electronic rest mass. How many dopants must be added per cubic metre of Ge to increase its conductivity by a factor of 10^4? ($1.81\times10^{19} m^{-3}$, 1.68 Ω^{-1}-m^{-1}, $2.69\times10^{23} m^{-3}$)

8. In an *n*-type semiconductor, the Fermi level lies 0.5 eV below the conduction band. If the concentration of donor atoms is tripled, find the new position of the Fermi level, given $kT = 0.03$ eV.

(0.48 eV below the conduction band)

Chapter-VIII

MAGNETISM IN SOLIDS

Magnetism was observed as early as 800 BC in a naturally occurring material called load stone which was used for navigation purposes. In the modern concept, all materials, viz., metals, semiconductors and insulators, are said to exhibit magnetism, though of different nature. The present chapter reviews the magnetic terminology in brief and describes the various types of magnetism in detail.

8.1 MAGNETIC TERMINOLOGY

When a solid is placed in a magnetic field, it gets magnetised. The magnetic moment per unit volume developed inside a solid is called *magnetization* and is denoted by **M**. There is another important parameter called the *magnetic susceptibility*, χ, which is a measure of the quality of the magnetic material and is defined as the magnetization produced per unit applied magnetic field, i.e.,

$$\chi = \mathbf{M}/\mathbf{H} \tag{8.1}$$

where **H** is the strength of the applied magnetic field, also referred to as the *magnetic field intensity*. As the vectors **M** and **H** can, in general, have different directions, χ is a tensor. However, in isotropic media, **M** and **H** point in the same direction and χ is a scalar quantity. In the SI system, both **M** and **H** are measured in amperes per metre (Am^{-1}) and hence χ is a dimensionless quantity. If **M** refers to a gram molecule of a substance, the susceptibility is termed the *molar susceptibility* and is designated by χ_m. The magnitude and sign of susceptibility vary with the type of magnetism as will be discussed later.

The *magnetic induction* or *magnetic flux density* **B** produced inside the medium as a consequence of the applied magnetic field **H** is given by

$$\mathbf{B} = \mu_o (\mathbf{H} + \mathbf{M}) \tag{8.2}$$

where μ_o is the *permeability* of the free space or vacuum and is equal to $4\pi \times 10^{-7}$ Henry per metre (Hm^{-1}). The quantity **B** is measured in Weber per square metre (Wbm^{-2}) or Tesla (T). Using Eq. (8.1) in (8.2), we obtain

$$\mathbf{B} = \mu_o (1 + \chi) \mathbf{H} \tag{8.3}$$

or $$\mathbf{B} = \mu \mathbf{H} \tag{8.4}$$

where μ is called the *absolute permeability* of the medium. Like χ, μ is also, in general, a tensor; for the isotropic medium, however, it represents a scalar quantity having dimensions same as that of μ_o. It is more convenient to introduce a dimensionless parameter μ_r which is called the *relative permeability* of the medium and is given by

$$\mu = \mu_o \mu_r \tag{8.5}$$

Thus, Eq. (8.4) becomes

$$\mathbf{B} = \mu_o \mu_r \mathbf{H} \tag{8.6}$$

From Eqs. (8.3) and (8.6), we get

$$\mu_r = 1 + \chi \tag{8.7}$$

For free space, i.e., in the absence of any material medium, $\mathbf{M} = 0$, $\chi = 0$, $\mu = \mu_o$ and $\mu_r = 1$, and from the above relations, we obtain

$$\mathbf{B} = \mu_o \mathbf{H} \tag{8.8}$$

8.2 TYPES OF MAGNETISM

According to the modern theories, the magnetism in solids arises due to orbital and spin motions of electrons as well as spins of the nuclei. The motion of electrons is equivalent to an electric current which produces the magnetic effects. The major contribution comes from the spin of unpaired valence electrons which produces permanent electronic magnetic moments. A number of such magnetic moments may align themselves in different directions to generate a net non-zero magnetic moment. Thus the nature of magnetization produced depends on the number of unpaired valence electrons present in the atoms of the solid and on the relative orientations of the neighbouring magnetic moments. The magnetism in solids has been classified into the following five categories :

(i) Diamagnetism
(ii) Paramagnetism
(iii) Ferromagnetism
(iv) Antiferromagnetism
(v) Ferrimagnetism

Diamagnetism is a very weak effect and is observed in solids which do not contain any permanent magnetic moments. The existence of a small non-zero magnetic moment in these materials is attributed to the orbital motion of electrons. This magnetic moment is always directed opposite to the applied magnetic field. The other types of magnetisms arise due to the presence of permanent atomic or electronic magnetic moments. *Paramag-*

netism is also a weak effect but, unlike diamagnetism, the magnetic moment is aligned in the direction of the field. *Ferromagnetism* is a very strong effect and arises when the adjacent magnetic moments align themselves in the same direction. In *antiferromagnetism,* the adjacent moments are equal and opposite to each other and hence complete cancellation of moments takes place. *Ferrimagnetism* is similar to antiferromagnetism except that adjacent moments are unequal in magnitude and hence complete cancellation of moments does not take place. The magnetic materials are classified according to the nature of the magnetism present. It may, however, be noted that above a certain critical temperature, the distinction between various types of magnetic materials disappears and all the materials become paramagnetic. The various types of magnetisms have been described in detail in the following sections.

8.3 DIAMAGNETISM

As indicated above, diamagnetism in a material arises due to changes in the atomic orbital states induced by the applied magnetic field. An electron revolving in an orbit constitutes an electric current. When a magnetic flux linked with such an electric circuit is changed, an induced current is set up in such a direction as to oppose the change in flux in accordance with the Lenz's law. The magnetic field of the induced current is opposite to the applied field and produces the diamagnetic effect. The occurrence of diamagnetism is manifested by the very small and negative value of the magnetic susceptibility. Diamagnetism exists in all materials but is usually suppressed due to the presence of stronger effects such as paramagnetism, ferromagnetism, etc. The diamagnetism has been described quantitatively by applying the classical and quantum theories.

8.3.1 Langevin's Classical Theory

Consider an electron revolving around the nucleus in a circular orbit with frequency ω_0. Since a moving electron is equivalent to a current, some magnetic flux is linked with such a current loop. When an external magnetic field is applied, the magnetic flux linked with this current loop tends to change which alters the current in the loop in such a way as to oppose the change in flux. This change in current is manifested by the change in the frequency of revolution of the electron. If ω_0 is the frequency of revolution of the electron in the absence of an applied field and ρ is the radius of the orbit, then we have

$$m\omega_0^2\rho = \frac{Ze^2}{4\pi\varepsilon_0\rho^2} \tag{8.9}$$

where *Ze* represents the nuclear charge and *m* is the mass of the electron. It gives the value of ω_o as

$$\omega_o = \sqrt{\frac{Ze^2}{4\pi\varepsilon_o m\rho^3}} \tag{8.10}$$

In the presence of a magnetic induction **B**, the Lorentz force on an electron moving with velocity **v** is given by

$$\mathbf{F} = e(\mathbf{v} \times \mathbf{B}) \tag{8.11}$$

Assuming the magnetic field to be perpendicular to the orbit, the magnitude of this force is

$$F = e\rho\omega \mathrm{B} \tag{8.12}$$

If ω is the freqency of electron in the presence of the field, the force equation can be written as

$$m\omega^2\rho = \frac{Ze^2}{4\pi\varepsilon_o\rho^2} - e\rho\omega B$$

$$= m\omega_o{}^2\rho - e\rho\omega B \qquad \text{[Using Eq. (8.10)]} \tag{8.13}$$

where the minus sign with the Lorentz force represents the negative charge. Thus

$$\omega^2 + \frac{eB}{m}\omega - \omega_o{}^2 = 0$$

or

$$\omega = \frac{1}{2}\left[-\frac{eB}{m} \pm \sqrt{\left(\frac{eB}{m}\right)^2 + 4\omega_o^2}\right]$$

$$= \pm\sqrt{\omega_o^2 + \left(\frac{eB}{2m}\right)^2} - \frac{eB}{2m}$$

For $\frac{eB}{2m} << \omega_o$, we obtain

$$\omega \cong \pm\,\omega_o - \frac{eB}{2m} \tag{8.14}$$

This shows that the frequency of revolution of an electron changes by a factor of $\frac{eB}{2m}$ in the presence of the magnetic induction **B**. The ± sign in (8.14) signifies that the electrons with orbital moments parallel to the field are slowed down whereas those with moments antiparallel to the field are speeded up. This result is called the *Larmor theorem.* The change in frequency produces an additional current which can be written as

$$I = \text{charge} \times \text{revolutions per unit time}$$

$$= (-e)\left(\frac{1}{2\pi}\frac{eB}{2m}\right)$$

$$= -\frac{e^2B}{4\pi m} \tag{8.15}$$

The magnetic moment of the current loop containing a single electron is given by the product of the current and the area of the loop, i.e.,

$$\mu_e = -\frac{e^2B}{4m}\rho^2 \tag{8.16}$$

The negative sign signifies that the induced magnetic moment points in a direction opposite to the applied field. If **B** is not perpendicular to the plane of the orbit, then ρ in Eq. (8.16) represents the projection of the radius of the orbit on a plane perpendicular to the magnetic field. If an atom contains Z electrons with their orbits randomly oriented, the total induced magnetic moment becomes

$$\mu_a = -\frac{Ze^2B}{4m}<\rho^2> \tag{8.17}$$

If the field points along z-axis, then

$$<\rho^2> = <x^2> + <y^2>$$

is the mean of the squares of perpendicular distances of electrons from the axis of the field. The mean square distance of electrons from the nucleus is

$$<r^2> = <x^2> + <y^2> + <z^2>$$

For a spherically symmetric charge distribution, we have

$$<x^2> = <y^2> = <z^2>$$

$$<r^2> = \frac{3}{2}<\rho^2> \tag{8.18}$$

Equation (8.17) now gives

$$\mu_a = -\frac{Ze^2B}{6m}<r^2> = -\frac{\mu_o Ze^2 H}{6m}<r^2> \tag{8.19}$$

For a solid consisting of N atoms per unit volume, each containing Z electrons, the expression for the *diamagnetic susceptibility* becomes

$$\chi_{dia} = \frac{M}{H} = -\frac{N\mu_0 Ze^2}{6m}<r^2> \tag{8.20}$$

This is the classical Langevin result and shows that the diamagnetic susceptibility is independent of temperature.

8.3.2 Quantum Theory

Using quantum mechanics, it can be shown that, in the presence of a magnetic induction **B**, the Hamiltonian $\mathcal{H}$ of an atomic electron changes to $\mathcal{H} + \mathcal{H}'$ where the term $\mathcal{H}'$ may be treated as a small perturbation and is given by

$$\mathcal{H}' = -\frac{ie\hbar}{2m}(\nabla.\mathbf{A}+2\mathbf{A}.\nabla) + \frac{e^2}{2m}A^2 \tag{8.21}$$

where A is the magnetic vector potential expressed by

$$\mathbf{B} = \nabla \times \mathbf{A}$$

If the magnetic induction is uniform, it can be described by the vector potential **A** as

$$\mathbf{A} = \frac{1}{2}\mathbf{B}\times\mathbf{r}$$

$$= \frac{1}{2}\left[\hat{\mathbf{i}}(B_y z - B_z y) + \hat{\mathbf{j}}(B_z x - B_x z) + \hat{\mathbf{k}}(B_x y - B_y x)\right] \tag{8.22}$$

For the magnetic field pointing along the z-direction, we have

$$B_x = B_y = 0,\ B_z = B,$$

and Eq. (8.22) gives

$$A_x = -\frac{1}{2}yB,\ A_y = \frac{1}{2}xB,\ A_z = 0 \tag{8.23}$$

Also,

$$\nabla.\mathbf{A} = \frac{\partial A_x}{\partial x} + \frac{\partial A_y}{\partial y} + \frac{\partial A_z}{\partial z} = 0$$

and

$$\mathbf{A}.\nabla = A_x \frac{\partial}{\partial x} + A_y \frac{\partial}{\partial y} + A_z \frac{\partial}{\partial z}$$

$$= \frac{1}{2} B \left(x \frac{\partial}{\partial y} - y \frac{\partial}{\partial x} \right)$$

Therefore, Eq. (8.21) becomes

$$\mathcal{H}' = -\frac{ie\hbar B}{2m} \left(x \frac{\partial}{\partial y} - y \frac{\partial}{\partial x} \right) + \frac{e^2 B^2}{8m} \left(x^2 + y^2 \right) \qquad (8.24)$$

In quantum mechanics, the angular momentum operator is given by

$$\mathbf{L} = -i\hbar \mathbf{r} \times \nabla$$

$$= -i\hbar \left[\hat{\mathbf{i}} \left(y \frac{\partial}{\partial z} - z \frac{\partial}{\partial y} \right) + \hat{\mathbf{j}} \left(z \frac{\partial}{\partial x} - x \frac{\partial}{\partial z} \right) + \hat{\mathbf{k}} \left(x \frac{\partial}{\partial y} - y \frac{\partial}{\partial x} \right) \right] \qquad (8.25)$$

Thus, it follows that the first term on the right hand side of (8.24) is proportional to L_z. The distance r is measured from the nucleus.

If we consider a unit volume of a substance containing N atoms each having Z electrons, the second term in the expression (8.24) becomes

$$\frac{Ne^2 B^2}{8m} \sum_{i=1}^{Z} \left(x_i^2 + y_i^2 \right) = \frac{NZe^2 B^2}{8m} < \rho^2 > \qquad (8.26)$$

where $<\rho^2>$ represents the mean of the squares of radii of the projections of orbits on a plane perpendicular to the magnetic field. Now if the magnetic field induces a dipole moment in the material, the corresponding energy term should be of the second power in B. Hence the term (8.26) may be regarded as the energy term associated with diamagnetism of the material. Comparing (8.26) with the diamagnetic energy, $\frac{1}{2}\mathbf{M}.\mathbf{B}$ or $-\frac{1}{2\mu_o}\chi_{dia} B^2$ we get,

$$\chi_{dia} = -\frac{N\mu_o Ze^2}{4m} < \rho^2 > \qquad (8.27)$$

Using Eq. (8.18), it becomes

$$\chi_{dia} = -\frac{N\mu_o Ze^2}{6m} < r^2 > \qquad (8.28)$$

where $<r^2>$ represents the mean square distance of the electrons relative to the nucleus. The result is identical to the one obtained from the classical theory. Thus, in order to calculate χ_{dia}, we must determine $<r^2>$ which requires the determination of electronic charge distribution within the atom. It may, however, be noted that in the calculations involving quantum mechanics, the concept of orbits is replaced by the wave function and the quantity $<r^2>$ has a different significance. The experimental results agree fairly well with the theoretically calculated values particularly for light elements and rare gases, whereas large deviations exist for ions containing a very large number of electrons. The typical value for helium is about -1.9×10^{-6} which shows that the diamagnetism is a very weak effect.

8.4 PARAMAGNETISM

Paramagnetism occurs in those atoms, ions and molecules which have permanent magnetic moments. In the absence of a magnetic field, these magnetic moments are oriented randomly and no net magnetization is produced. When a magnetic field is applied, these moments orient themselves in the direction of the field which results in some net magnetization parallel to the applied field. The paramagnetic materials have small, positive and temperature-dependent susceptibility. The permanent magnetic moments of ions result from the following contributions :

(i) The spin or intrinsic moments of the electrons,

(ii) The orbital motion of the electrons,

(iii) The spin magnetic moment of the nucleus.

Paramagnetism is observed in :

(i) Metals

(ii) Atoms and molecules possessing an odd number of electrons, e.g., free sodium atoms, gaseous nitric oxide, etc.

(iii) A few compounds having an even number of electrons, e.g., oxygen molecule.

(iv) Free atoms and ions having a partially filled inner shell, e.g., rare earth and actinide elements, ions of some transition elements such as Mn^{2+}, etc.

8.4.1 Langevin's Classical Theory

Langevin considered a paramagnetic gas containing N atoms per unit volume each having a permanent magnetic moment μ. The mutual interaction between the magnetic dipoles was as-

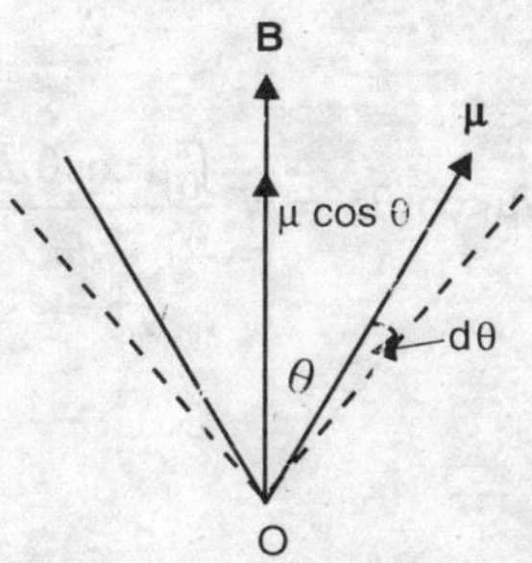

Fig. 8.1. *Magnetic dipole with moment* μ *oriented at an angle* θ *to the applied magnetic field.*

sumed to be negligible. In the presence of a magnetic induction **B**, these dipoles tend to orient themselves in the direction of the field in order to minimize their energy. The thermal energy at ordinary temperatures, however, resists such an alignment of dipoles. In thermal equilibrium, the dipoles orient themselves at an angle θ with the direction of the applied field as shown in Fig. 8.1. The potential energy of each dipole in this position is given by

$$E = -\mu.\mathbf{B} = -\mu B\cos\theta$$

Using Maxwell-Boltzmann distribution law, the number of magnetic dipoles having this particular orientation is proportional to

$$\exp\left(-\frac{E}{kT}\right) \quad \text{or} \quad \exp\left(\frac{\mu B\cos\theta}{kT}\right)$$

Also, according to statistical mechanics, the probability for a magnetic dipole to make an angle between θ and θ + dθ with the magnetic field, or the number of dipoles, *dn,* having axes within the solid angle dω lying between two hollow cones of semiangles θ and θ + dθ (Fig. 8.1) is given by

$$dn \propto \exp\left(\frac{\mu B\cos\theta}{kT}\right) d\omega$$

$$= k\exp\left(\frac{\mu B\cos\theta}{kT}\right) 2\pi\sin\theta\, d\theta \tag{8.29}$$

where *k* is a constant. Each one of these dipoles contributes a component of magnetic moment μ cos θ to the magnetization, whereas the components perpendicular to the field direction cancel each other. Hence the average component of the magnetic moment of each atom along the field direction multiplied by the number of atoms per unit volume, *N*, gives the magnetization *M*, i.e.,

$$M = N\mu\langle\cos\theta\rangle = N\frac{\int_0^\pi \mu\cos\theta\, dn}{\int_0^\pi dn} = N\frac{\mu\int_0^\pi \cos\theta\sin\theta\exp\left(\frac{\mu B\cos\theta}{kT}\right)d\theta}{\int_0^\pi \sin\theta\exp\left(\frac{\mu B\cos\theta}{kT}\right)d\theta} \tag{8.30}$$

Let $\frac{\mu B}{kT} = x,$

and $\cos\theta = y$

∴ $-\sin\theta\, d\theta = dy$

Therefore, Eq. (8.30) becomes

$$M = N\mu \frac{\int_1^{-1} ye^{xy}\,dy}{\int_1^{-1} e^{xy}\,dy} = N\mu\left[\frac{e^x + e^{-x}}{e^x - e^{-x}} - \frac{1}{x}\right] = N\mu\left[\coth x - \frac{1}{x}\right]$$

$$= N\mu\, L(x) \tag{8.31}$$

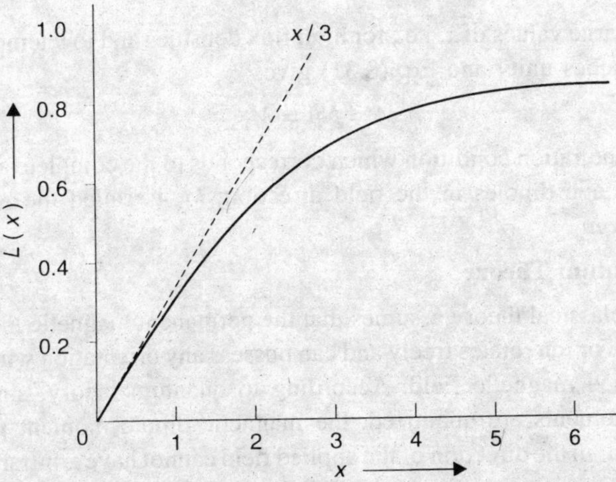

Fig. 8.2. *Variation of L(x) with x.*

where $L(x)$ is called the *Langevin function.* The variation of $L(x)$ with $x = \frac{\mu B}{kT}$ is shown in Fig. 8.2. For $x << 1$, i.e., at normal field strengths and ordinary temperatures, the curve is almost linear and coincides with the tangent to the curve at the origin which is equal to $x/3$. Thus, we have

$$L(x)\big|_{x=0} \cong \frac{x}{3} = \frac{\mu B}{3kT} \tag{8.32}$$

Therefore, Eq. (8.31) gives

$$M \cong \frac{N\mu^2 B}{3kT}$$

The *paramagnetic susceptibility* is given as

$$\chi_{para} = \frac{\mu_o M}{B} = \frac{\mu_o N\mu^2}{3kT} \tag{8.33}$$

which shows that the susceptibility is inversely proportional to temperature. It can be written as

$$\chi_{para} = \frac{C}{T} \tag{8.34}$$

where $C = \dfrac{\mu_0 N \mu^2}{3k}$ is a constant. The expression (8.34) is called the *Curie law* and *C* is called the *Curie constant.* Thus we find that the Curie law holds good for $\mu B << kT$.

For large values of *x*, i.e., for high flux densities and low temperatures, $L(x)$ approaches unity and Eq. (8.31) gives

$$M = N\mu = M_s \tag{8.35}$$

This is the saturation condition which corresponds to the complete alignment of the magnetic dipoles in the field direction; M_s is called the *saturation magnetization.*

8.4.2 Quantum Theory

The classical theory assumes that the permanent magnetic moment of a given atom or ion rotates freely and can possess any orientation with respect to the applied magnetic field. According to quantum theory, since these magnetic moments are quantized, the magnetic dipole moment **μ** and its component μ_z in the direction of the applied field cannot have arbitrary values. We have, in general, a direct relationship between the magnetic dipole moment **μ** of an atom or ion in free space and its angular momentum **J** as

$$\boldsymbol{\mu} = -g\mu_B \mathbf{J} \tag{8.36}$$

The quantity μ_B is called the Bohr magneton and is equal to $\dfrac{e\hbar}{2m}$ in SI system and $\dfrac{e\hbar}{2mc}$ in CGS system of units; *g* is known as the Lande's *g*-factor and is equal to 2 if the net angular momentum of the dipole is due to electron spin and 1 if it is due to orbital motion only. In general, it has mixed origin and is obtained from the expression :

$$g = 1 + \frac{J(J+1) + S(S+1) - L(L+1)}{2J(J+1)} \tag{8.37}$$

where *S* and *L* represent the spin and orbital quantum numbers of the dipole respectively. The orientations of the magnetic moment **μ** with respect to the direction of the applied magnetic field are specified by the rule that the possible components of **μ** along the field direction are given by

$$\mu_z = -g\mu_B m_J \tag{8.38}$$

where $m_J = -J, -J+1, \ldots, J-1, J$ is the magnetic quantum number associated with J. For each value of J, m_J can have $(2J+1)$ values which means that the magnetic moment of the atom can have $(2J+1)$ different orientations relative to the field. The potential energy of such a magnetic dipole in the presence of a magnetic field B is, therefore, given by

$$E = m_J g \mu_B B \tag{8.39}$$

According to the Maxwell-Boltzmann distribution, the number of atoms having a particular value of m_J is thus proportional to

$$e^{-m_J g \mu_B B/kT}$$

Considering a unit volume of a paramagnetic material containing a total of N atoms, the magnetization in the direction of the field is given by

$M = N \times$ (Statistical average of the magnetic moment component per atom along B)

$$= N \sum_{m_J=-J}^{m_J=+J} \frac{-m_J g \mu_B e^{-m_J g \mu_B B/kT}}{e^{-m_J g \mu_B B/kT}} \tag{8.40}$$

We consider the following two cases :

(i) At normal flux densities and ordinary temperatures,

$$\frac{m_J g \mu_B B}{kT} << 1$$

Therefore, the expression (8.40) can be approximated as

$$M = N \frac{g\mu_B \sum_{m_J=-J}^{+J} -m_J\left(1 - \frac{m_J g \mu_B B}{kT}\right)}{\sum_{m_J=-J}^{+J}\left(1 - \frac{m_J g \mu_B B}{kT}\right)}$$

Now, $$\sum_{m_J=-J}^{+J} m_J = 0$$

and $$\sum_{m_J=-J}^{+J} m_J^2 = 2\sum_{m_J=0}^{J} m_J^2 = \frac{J(J+1)(2J+1)}{3}$$

$$\therefore \quad M = N \frac{\frac{g^2 \mu_B^2 B}{kT} \frac{J(J+1)(2J+1)}{3}}{(2J+1)}$$

$$= N \frac{g^2\mu_B^2 B}{3kT} J(J+1) \tag{8.41}$$

$$\therefore \qquad \chi_{para} = \frac{\mu_o M}{B} = \frac{\mu_o N \mu_B^2}{3kT} g^2 J(J+1) \tag{8.42}$$

or

$$\chi_{para} = \frac{\mu_o N p_{eff}^2 \mu_B^2}{3kT} \tag{8.43}$$

where p_{eff} is the effective number of Bohr magnetons and is given by

$$p_{eff} = g\sqrt{J(J+1)} \tag{8.44}$$

The expression (8.43) is identical to the classical expression (8.33) with

$$\mu^2 = p_{eff}^2 \mu_B^2 \tag{8.45}$$

The Curie law can also be deduced from Eq. (8.43) as described earlier.

(ii) At low temperatures and strong magnetic fields, $\frac{m_J g \mu_B B}{kT}$ is not smaller than unity and it is not possible to make a series expansion of the exponential terms in the expression (8.40). After some algebraic manipulations, Eq. (8.40) yields

$$M = NgJ\mu_B B_J(x) \tag{8.46}$$

where $x = \frac{gJ\mu_B B}{kT}$ and $B_J(x)$ is the Brillouin function defined as

$$B_J(x) = \frac{2J+1}{2J}\coth\left(\frac{2J+1}{2J}\right)x - \frac{1}{2J}\coth\left(\frac{x}{2J}\right) \tag{8.47}$$

For $x \ll 1$, we have

$$\coth x \cong \frac{1}{x} + \frac{x}{3}$$

$$\therefore \qquad B_J(x) = \frac{x(J+1)}{3J}$$

Thus the susceptibility becomes

$$\chi_{para} = \frac{\mu_o M}{B} = \frac{\mu_o N J(J+1) g^2 \mu_B^2}{3kT} = \frac{\mu_o N p_{eff}^2 \mu_B^2}{3kT} \tag{8.48}$$

which is the same as (8.43). For $x >> 1$. i.e., at low temperatures and strong fields,

$$\coth x \cong 1$$

$$\therefore \qquad B_J(x) \cong 1$$

and Eq. (8.46) becomes

$$M = NgJ\mu_B \tag{8.49}$$

This result implies the state of magnetic saturation, i.e., all the dipoles get aligned along the magnetic induction **B**. Thus, in this limit, the expression (8.46) is analogous to the Langevin expression (8.31) with the difference that the latter is applicable to freely rotating dipoles only. In fact, for $J \to \infty$, i.e., for a large number of allowed orientations of a magnetic dipole, we have

$$\coth \frac{x}{2J} \to \frac{2J}{x}$$

and

$$\coth \left(1 + \frac{1}{2J}\right) x \to \coth x$$

$$\therefore \qquad B_J(x) \to \coth x - \frac{1}{x} \text{ or } L(x)$$

Thus the quantum results approach the classical ones. This is what is expected as the classical theory allows all conceivable orientations. For other values of J, however, the two results differ considerably.

The order of magnitude of paramagnetic susceptibility of a solid as estimated from Eq. (8.48) at room temperature is about 10^{-7} which is quite small. It increases by more than hundred times at 1 K. Equation (8.48) is successfully employed to predict the values of susceptibility for various paramagnetic crystals particularly rare earth ions. The value of J is determined by applying the Hund's rules [(1)]. However, Eq. (8.48) is unable to account for the experimental observations of susceptibility for the ions of the iron group. This is because of the presence of crystal field due to other ions which

---

(1) Hund's rules state that, for the ground state of atoms with incompletely filled shells,

(i) the electron spins add up to give the maximum possible S consistent with the Pauli's exclusion principle.

(ii) the orbital momenta combine to give the maximum value for L that is consistent with (i)

(iii) the value of J is given by
$J = |L - S|$ if a shell is less than half-filled
$J = L + S$ if a shell is more than half-filled.
If the shell is just half-filled, $L = 0$ and, therefore, $J = S$

cannot be neglected in comparison with the externally applied field. In paramagnetic materials where the crystal field is negligible, Eq. (8.48) holds good. But when it is strong, it may break the rotational symmetry of the dipole and affect its total angular momentum. Also, the average value of L_z may reduce to zero which is known as quenching of the orbital angular momentum. In such a case, Eq. (8.39) should be written as

$$E = 2m_s\mu_B B \qquad [\text{As } g = 2]$$

$$= \left(\frac{e\hbar}{m}\right) m_s B \tag{8.50}$$

For $\left(\frac{e\hbar}{m}\right) m_s B << kT$, it yields

$$\chi = \frac{\mu_0 N}{3kT}\left(\frac{e\hbar}{m}\right)^2 S(S+1) \tag{8.51}$$

For example, for Mn^{3+} ions, the value of χ as obtained from the above expression with $S = 2$ is in conformity with the experimental observations where Eq. (8.48) predicts a zero susceptibility.

8.5 FERROMAGNETISM

Like paramagnetism, ferromagnetism is also associated with the presence of permanent magnetic dipoles, but unlike paramagnetism, the magnetic moments of adjacent atoms in this case are aligned in a particular direction even in the absence of the applied magnetic field. Thus a ferromagnetic material exhibits a magnetic moment in the absence of a magnetic field. The magnetization existing in a ferromagnetic material in the absence of an applied magnetic field is called the *spontaneous magnetization.* It exists below a certain critical temperature called the *Curie temperature, T_C*. The alignment of magnetic moments below the Curie temperature is due to the exchange interaction between the magnetic ions and will be described later. Above the Curie temperature, the thermal effects offset the spin alignment and the ferromagnetic substance becomes paramagnetic.

The ferromagnetic substances acquire a large magnetization in the presence of even a weak external magnetic field. They possess a large and positive value of susceptibility which, in general, is not constant but varies with field strength. The variation of magnetization with field strength exhibits the well-known hysteresis curve. The examples of ferromagnetic materials are the elements such as Fe, Ni, Co, Gd and Dy, and a number of alloys and oxides such as MnBi, MnAs, CrO_2, etc.

8.5.1 Weiss Theory of Ferromagnetism

The theory of ferromagnetism put forward by Weiss is centred about the following two hypotheses :

(i) A specimen of ferromagnetic material contains a number of small regions called *domains* which are spontaneously magnetized. The magnitude of spontaneous magnetization of the specimen as a whole is determined by the vector sum of the magnetic moments of individual domains.

(ii) The spontaneous magnetization of each domain is due to the presence of an exchange field, B_E, which tends to produce a parallel alignment of the atomic dipoles. The field B_E is assumed to be proportional to the magnetization M of each domain, i.e.,

$$B_E = \lambda M \tag{8.52}$$

where λ is a constant called the *Weiss-field constant* and is independent of temperature.

It may be noted that, in the above expression, we always use the average values of B_E and M for a domain. The field B_E is also called the *molecular field* or the *Weiss field* and is generally quite strong as compared to the applied field B. For iron, for example, $B_E \approx 1000\ T$. Thus the effective mangetic field on an atom or ion becomes

$$B_{eff} = B + B_E = B + \lambda M \tag{8.53}$$

In order to develop the theory of ferromagnetism, we shall use the quantum theory of magnetization rather than the classical Langevin theory used by Weiss originally.

Consider a ferromagnetic solid containing N atoms per unit volume each having a total angular momentum quantum number J. By analogy to Eq. (8.46), we can write the expression for magnetization as

$$M = NgJ\mu_B B_J(x) \tag{8.54}$$

where

$$B_J(x) = \frac{2J+1}{2J}\coth\left(\frac{2J+1}{2J}\right)x - \frac{1}{2J}\coth\left(\frac{x}{2J}\right) \tag{8.55}$$

and

$$x = \frac{gJ\mu_B B_{eff}}{kT} = \frac{gJ\mu_B}{kT}(B + \lambda M) \tag{8.56}$$

In case of spontaneous magnetization, $B = 0$ and Eq. (8.56) becomes

$$x = \frac{gJ\mu_B\lambda M}{kT} \tag{8.57}$$

$$\therefore \qquad M(T) = \frac{xkT}{\lambda gJ\mu_B} \tag{8.58}$$

As $T \to 0$ or $x \to \infty$, $B_J(x) \to 1$; the magnetic moments align themselves parallel to the field and the magnetization M becomes the *saturation magnetization, $M_s(0)$*. Thus, from Eq. (8.59) give

$$M_s(0) = NgJ\mu_B \tag{8.59}$$

From Eqs. (8.58) and (8.59), we obtain

$$\frac{M(T)}{M_s(0)} = \frac{xkT}{\lambda Ng^2J^2\mu_B^2} \tag{8.60}$$

Also, Eqs. (8.54) and (8.58) give

$$\frac{M(T)}{M_s(0)} = B_J(x) \tag{8.61}$$

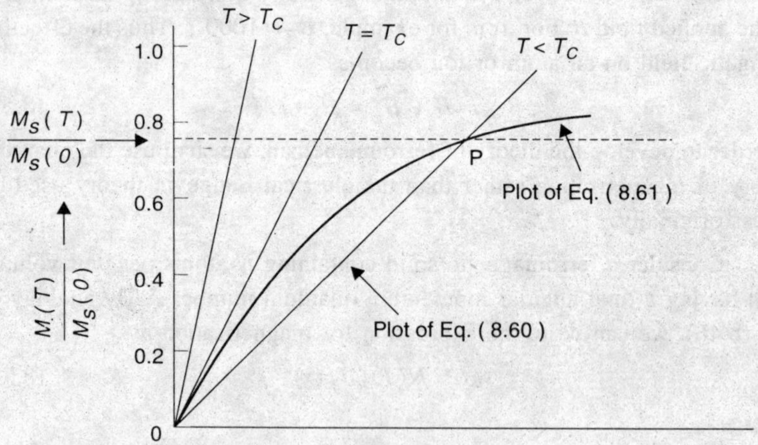

Fig. 8.3. *Graphical solution of the simultaneous equations (8.60) and (8.61). A point of intersection P determines the spontaneous magnetization $M_s(T)$ at a given temperature. No solution exists for $T > T_C$*

The magnetization $M(T)$ at a given temperature can be obtained by solving Eqs. (8.60) and (8.61) simultaneously. The plots of $M(T)/M_s(0)$ versus

x representing Eqs. (8.60) and (8.61) are shown in Fig. 8.3. Note that the Eq. (8.60) represents a straight line passing through the origin and having slope proportional to T. At temperature equal to the critical temperature T_C, this line is tangent to the Brillouin function at the origin. The intersection of the two plots at the point O represents a positive solution, but the magnetization corresponding to this point is unstable. Another point of intersection appears for $T < T_C$ which indicates a non-zero value of M even for zero external field and hence corresponds to spontaneous magnetization. It also follows that the spontaneous magnetization decreases with increase in temperature and vanishes beyond the temperature T_C which is known as the *ferromagnetic Curie temperature.* The variation of spontaneous magnetization with temperature is shown in Fig. 8.4. It becomes maximum at 0 K when all the moments are lined up in a particular direction under the influence of the exchange field. This variation of M_s enables one to classify the paramagnetic-ferromagnetic transition as a second order phase transition, i.e., a transition characterized by an order parameter (e.g. M_s in this case) which is non-zero only below T_C.

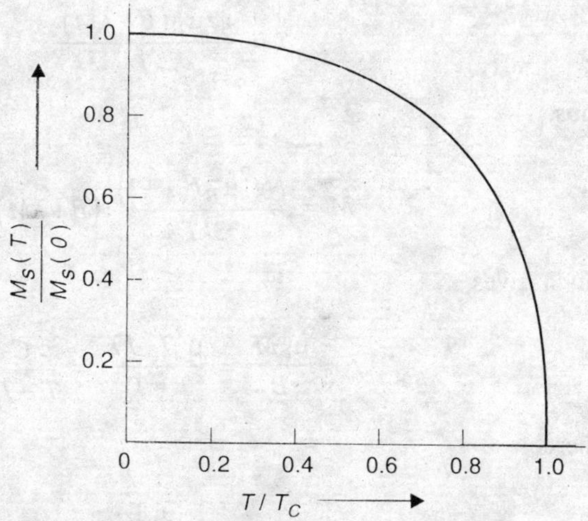

Fig. 8.4. *Spontaneous magnetization versus temperature for* $T < T_C$.

The Paramagnetic Region

Consider the magnetization in the region well above the Curie temperature. For $T > T_C$, the spontaneous magnetization is zero and an external field will have to be applied to produce some magnetization. This field should, however, be weak enough to avoid the saturation state. In such a state, we find from Eq. (8.56) that $x << 1$ and we can write

$$B_J(x) \cong \left(\frac{J+1}{3J}\right)x$$

Therefore, the expression (8.54) becomes

$$M = Ng\mu_B\,(J+1)x/3$$

$$x = \frac{gJ\mu_B(B+\lambda M)}{kT} \tag{8.62}$$

Thus,

$$M = \frac{Ng^2\mu_B^2 J(J+1)}{3kT}(B+\lambda M) \tag{8.63}$$

which gives

$$\chi = \frac{\mu_o M}{B} = \frac{\mu_o T_C/\lambda}{T-T_C} = \frac{C}{T-T_C} \tag{8.64}$$

where

$$C = \frac{\mu_o T_C}{\lambda} \tag{8.65}$$

and
$$T_C = \frac{\lambda N g^2 \mu_B^2 J(J+1)}{3k} \tag{8.66}$$

The expression (8.64) is known as the *Curie-Weiss law.* It satisfactorily describes the temperature dependence of susceptibility in the paramagnetic region provided the temperature is well above the Curie temperature. The Curie temperature determined from the theory of spontaneous magnetization differs by a few degrees from the experimentally determined value for the paramagnetic region.

8.5.2 Nature and Origin of Weiss Molecular Field : Exchange Interactions

The Weiss theory of ferromagnetism is based on the concept of ferromagnetic domains which are spontaneously magnetized due to the presence of an internal molecular field called the *Weiss field* or the *exchange field,* B_E. The theory, however, does not explain the origin and nature of this field. The Weiss field cannot be simply due to magnetic dipole-dipole interaction between the neighbouring dipoles as this would generate fields of the order of 10^3G only whereas the actual field strengths are observed to be quite high. For example, the Weiss field for iron is of the order of 10^7 G. It was Heisenberg who first proposed in 1928 that the Weiss field was the consequence of the quantum-mechanical exchange interaction between the atoms. This interaction arises due to the Pauli's exclusion principle according to which any change in the relative orientation of the two spins would disturb the spatial distribution of charge, thus producing interaction between the two atoms. Apparently, the strength of exchange interaction between the adjacent atoms depends on the extent of overlap of their wave functions as well as the relative orientation of the electron spins but not on the spin magnetic moments. Thus this is an electrostatic and non-magnetic type of interaction.

Using Heitler-London theory of chemical bonding, it can be shown that the total energy of a system of two atoms contains an exchange energy term given by

$$U_{ij} = -2J_e \mathbf{S_i}.\mathbf{S_J} \tag{8.67}$$

where $\mathbf{S_i}$ and $\mathbf{S_j}$ represent the spins of the two atoms and J_e is the exchange integral which is assumed to be the same for any pair of atoms. Its value depends on the overlap of the charge distributions of the two atoms, i.e., on the interatomic distance. In general, J_e is positive for large interatomic distances and negative for smaller ones. The expression (8.67) is known as the *Heisenberg model* of exchange energy. It also follows from (8.67) that if J_e is positive, the parallel arrangement of spins exhibits lower energy and hence is more stable as compared to the antiparallel arrangement, thereby producing magnetization. In a similar way, it can be concluded that the negative value of J_e does not favour magnetism.

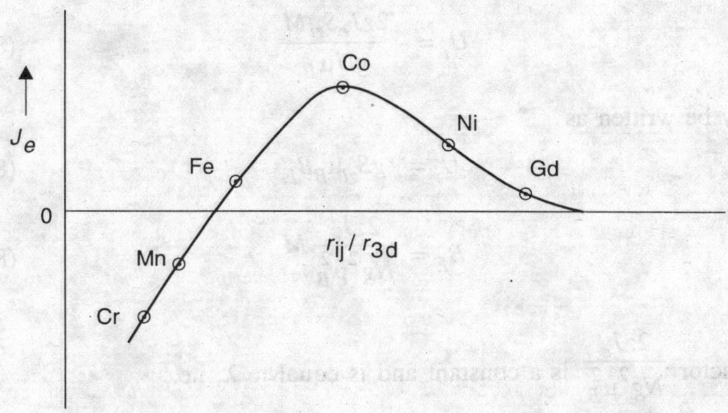

Fig. 8.5. *Plot of exchange integral, J_e versus ratio of interatomic separation to the radius of the 3d orbit, r_{ij}/r_{3d}*

The exchange integral for iron group of atoms is found to be positive inspite of the fact that the interatomic distance for this group is not large. This follows from the observations of Bathe and Slater which indicate that J_e is positive if the ratio r_{ij}/r_{3d} is greater than 3 but is not much larger than 3, r_{ij} being the distance between the atoms *i* and *j* and r_{3d} is the radius of the unfilled

3*d*-shell. The variation of J_e as a function of r_{ij}/r_{3d} is shown in Fig. 8.5. It is apparent that the ratio r_{ij}/r_{3d} favours ferromagnetism for Fe, Co, Ni and Gd whereas for Mn and Cr it does not favour ferromagnetism. Thus elements like Mn and Cr do not exhibit ferromagnetism.

The expression for the exchange field, B_E, and the relationship between J_e and χ can be obtained by following the procedure suggested by Stoner. Assuming J_e to be constant for all the neighbouring pairs and considering contributions from the nearest neighbours only, we may write the exchange energy of an *i*th atom as

$$U_i = -2J_e \sum_j \mathbf{S_i} \cdot \mathbf{S_j} \tag{8.68}$$

where the summation is over the nearest neighbours of the *i*th atom. Stoner replaced the instantaneous values of the neighbouring spins by their time averages. For *z* nearest neighbours, (8.68) can be written as

$$U_i = -2zJ_e\,(S_{xi}\langle S_{xj}\rangle + S_{yi}\langle S_{yj}\rangle + S_{zi}\langle S_{zj}\rangle) \tag{8.69}$$

For magnetization, *M*, along the *z*-axis, we have

$$\langle S_{xj}\rangle = \langle S_{yj}\rangle = 0 \text{ and } \langle S_{zj}\rangle = \frac{M}{g\mu_B N} \tag{8.70}$$

$$\therefore \qquad U_i = -\frac{2zJ_e S_{zi} M}{gN\mu_B} \tag{8.71}$$

It may be written as

$$U_i = -gS_{zi}\mu_B B_E \tag{8.72}$$

with

$$B_E = \frac{2zJ_e}{Ng^2\mu_B^2} M \tag{8.73}$$

The factor $\frac{2zJ_e}{Ng^2\mu_B^2}$ is a constant and is equal to λ, i.e.,

$$\lambda = \frac{2zJ_e}{Ng^2\mu_B^2} \tag{8.74}$$

and Eq. (8.73) gives

$$B_E = \lambda M$$

which is the same as (8.52).

8.5.3 Concept of Domains and Hysteresis

According to the Weiss theory described above, the exchange interactions between the neighbouring dipoles in a ferromagnetic material generate an internal exchange field B_E which aligns them in a particular direction. It is, however, observed that a ferromagnetic material such as iron does not exhibit a net magnetization unless it is placed in an external field. Weiss explained this by introducing the concept of ferromagnetic domains. According to this concept, a single crystal of a ferromagnetic solid is divided into a number of small regions called *domains* each one of which is spontaneously magnetized by the exchange field. The magnetization vectors of different domains are, however, randomly oriented so that no net magnetization is produced in the material as a whole. In the presence of an external magnetic field, the domains pointing in the direction of the field grow at the expense of those pointing in other directions, thereby resulting in some non-zero magnetization in the material.

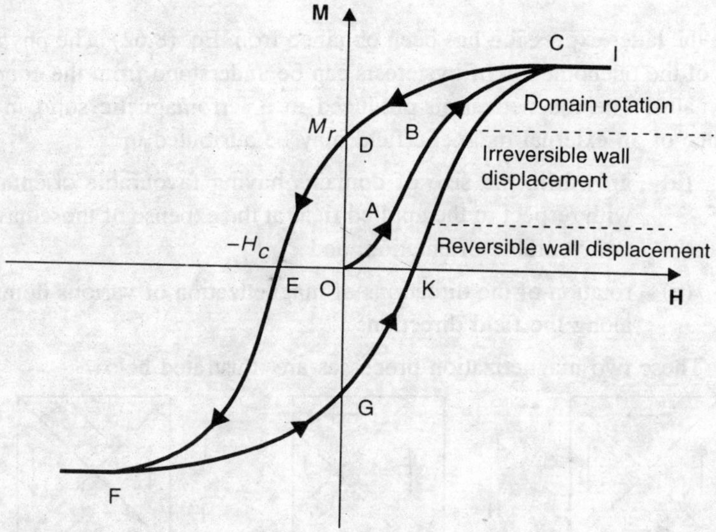

Fig. 8.6. *Typical hysteresis curve for a ferromagnetic material indicating the predominant magnetization processes in different regions.*

According to Neel, the origin of domain structure of a ferromagnetic solid rests in the principle of minimization of the total energy of the material which, apart from exchange energy, comprises the magnetic field energy, anisotropy energy, domain wall energy and magnetostrictive exchange energy. As an example, the presence of free magnetic poles at the ends of a magnetized specimen generates an external field **H** which gives rise to the magnetic field energy equal to ½ **H.B**. This energy can be lowered by reducing the volume of the domain by subdividing it into neighbouring subdomains, thereby reducing the field **H** and hence the magnetic field energy. However, such a subdivision of domains cannot continue indefinitely and is restrained by the fact that the formation of domain walls requires additional energy. Thus, it can be concluded that, whereas the presence of large domains is energetically unfavourable, a ferromagnetic material must possess a domain structure consisting of a number of smaller domains which corresponds to a state of minimum energy.

All ferromagnetic materials exhibit the well known hysteresis curves; a typical one is shown in Fig. 8.6. It is apparent that for $T<T_C$, there are two solutions for M (and hence for B) which trace the boundary of the hysteresis curve. These solutions can be obtained from the intersection of the plots of Eq. (8.61) and the expression

$$M = \left(\frac{kT}{\mu_B g \lambda J}\right) x - \frac{1}{\lambda} B \tag{8.75}$$

where the latter expression has been obtained from Eq. (8.62). The physical cause of the phenomenon of hysteresis can be understood from the concept of domains. The magnetization produced in a ferromagnetic solid in the presence of an external magnetic field may be attributed to

(a) growth in the size of domains having favourable orientation with respect to the applied field at the expense of those having unfavourable orientation, and

(b) rotation of the directions of magnetization of various domains along the field direction.

These two magnetization processes are illustrated below :

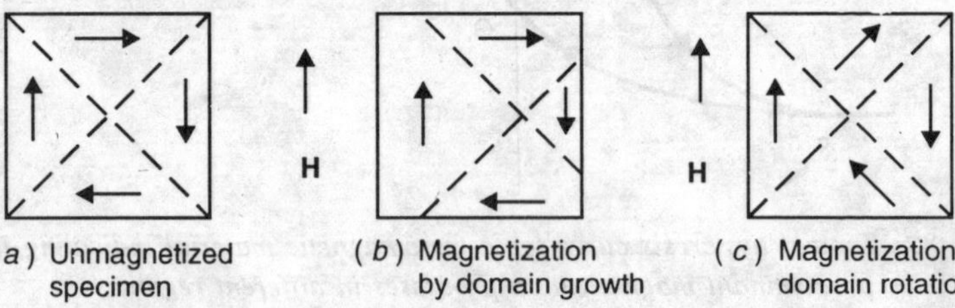

Fig. 8.7. *The two fundamental processes of magnetization.*

When a small magnetic field is applied across a ferromagnetic material, the domains pointing almost along the direction of the field grow at the expense of the domains having opposite orientation, thus resulting in a small magnetization as indicated by the initial portion (OA) of the hysteresis curve. Such displacements of domain boundaries are mostly reversible and hence the portion OA of the curve is also reversible. As the field increases, a greater number of domains grow favourably which results in a large increase in magnetization (portion AB). The boundary displacements in this region are often large and irreversible. The growth of domains continues until the favourable domains grow up to the maximum extent with their magnetization vectors still pointing along the so-called *easy directions* of magnetization. As the field increases further, the domains rotate from their easy directions to the direction of the applied field; the magnetization increases slowly (portion BC) and finally attains a saturation value M_s when all the domains point along the direction of the field. On decreasing the field, the magnetization does not follow the same path because the aligned domains do not regain their random state of orientation easily. There exists some non-zero magnetization even after removing the field altogether. This magnetization is called the *remanent magnetization* or *remanence,* M_r. The magnetization can be reduced to zero by applying a reverse magnetic field known as the *coercive field* or *coercivity,* H_c. A similar variation in the reverse magnetization is observed as the reverse field is first increased and then decreased. The closed loop CEFKC is called the *hysteresis loop.* A similar loop is obtained by plotting B versus H except that the line CI in this curve is never parallel to the field axis. The hysteresis curves are quite important in determining the quality of a magnetic material and selecting the material for a particular application.

8.6 ANTIFERROMAGNETISM

Antiferromagnetism originates when the spin moments of the neighbouring atoms are ordered in an antiparallel arrangement as shown in Fig. 8.8 or when the exchange integral is negative. A crystal exhibiting antiferromagnetism may be considered to be consisting of two interpenetrating sublattices A and B, one of which is spontaneously magnetized in one direction and the other is spontaneously magnetized in the opposite direction. This type of magnetism was first observed in the crystals of MnO. In the absence of an external magnetic field, the neighbouring magnetic moments cancel out each other and the material as a whole exhibits no magnetization. However, when a field is applied, a small magnetization appears in the direction of the field which increases further with temperature. Such a behaviour is typical of an antiferromagnetic material. The magnetization becomes the maximum at a critical temperature T_N, called the *Neel temperature,* which is analogous to the Curie temperature in the paramagnetic or ferromagnetic substances. Above this temperature, the magnetization decreases continuously which is indicative of the paramagnetic state of the material. The variation of susceptibility with temperature is shown in Fig. 8.9 and is compared with the corresponding variations for the paramagnetic and ferromagnetic substances. Unlike the ferromagnetic case, the susceptibility of an antiferromagnetic substance is not infinite for T equal to T_N, but has a weak cusp. The antiferromagnetic behaviour can be explained with the help of the molecular field theory which was initially developed by Neel. A brief description of this theory is given below.

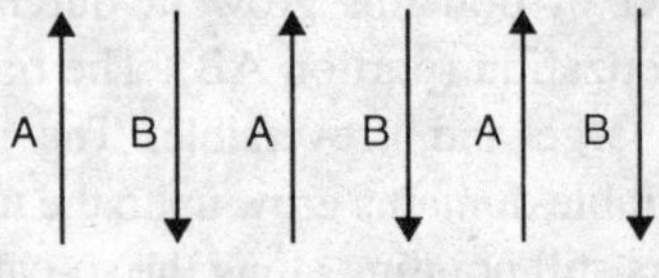

Fig. 8.8. *Neighbouring antiparallel spin moments of A and B sublattices producing antiferromagnetism.*

Consider an antiferromagnetic substance consisting of two sublattices A and B such that an atom of one type has all the nearest neighbours of the opposite type. The internal or *molecular field* acting on an atom can be obtained by using the molecular field approximation developed by Weiss in which the interaction between the various magnetic moments is considered. In an antiferromagnetic substance, besides the nearest neighbour negative AB interaction, there also exists the next nearest neighbour negative AA (or AB) interaction which is quite weak as compared to the former. If α and β represent the interaction parameters, i.e., the Weiss field constants for AA (or BB) and AB interactions respectively, the internal fields present at A and B sites, expressed in terms of flux densities, can be written as

$$\left.\begin{aligned} B_{mA} &= -\alpha M_A - \beta M_B \\ B_{mB} &= -\beta M_A - \alpha M_B \end{aligned}\right] \qquad (8.76)$$

where M_A and M_B are the magnetizations of the corresponding sites. In the presence of an external field B, the effective fields at A and B sites become

$$\left.\begin{aligned} B_A &= B - \alpha M_A - \beta M_B \\ B_B &= B - \beta M_A - \alpha M_B \end{aligned}\right] \qquad (8.77)$$

To determine the total magnetization, $M_A + M_B$, of the specimen, we consider the temperature regions below and above the Neel temperature.

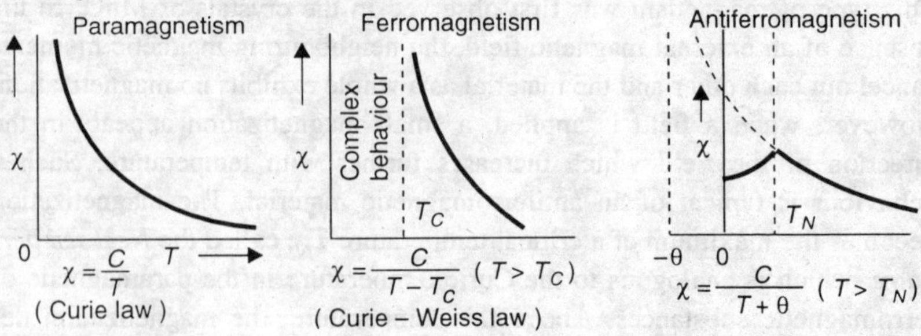

Fig. 8.9.

(*i*) $T > T_N$

In this region, the specimen behaves as paramagnetic; the magnetizations of the two sublattices can be written in accordance with Eq. (8.33) as

$$\left.\begin{aligned} M_A &= \frac{N\mu^2}{3kT} B_A \\ M_B &= \frac{N\mu^2}{3kT} B_B \end{aligned}\right] \tag{8.78}$$

where

$$\mu^2 = \mu_B^2 g^2 J(J+1).$$

In these expressions, it has been assumed that the A-type atoms are identical with B-type atoms and their densities are equal, i.e., $N_A = N_B = N$. The total magnetization M is, therefore, obtained as

$$\begin{aligned} M &= M_A + M_B \\ &= \frac{N\mu^2}{3kT}(B_A + B_B) \\ &= \frac{N\mu^2}{3kT}\left[2B - (\alpha+\beta)(M_A + M_B)\right] \qquad \text{[Using Eq. (8.77)]} \\ &= \frac{N\mu^2}{3kT}\,[2B - (\alpha+\beta)M] \end{aligned} \tag{8.79}$$

This yields the expression for susceptibility as

$$\chi = \frac{\mu_0 M}{B} = \frac{\dfrac{2\mu_0 N\mu^2}{3k}}{T + \dfrac{N\mu^2(\alpha+\beta)}{3k}} = \frac{C}{T+\theta} \tag{8.80}$$

where

$$C = \frac{2\mu_0 N\mu^2}{3k}$$

and

$$\theta = \frac{N\mu^2(\alpha+\beta)}{3k} \tag{8.81}$$

The expression (8.80) satisfies the experimental data on susceptibility quite well. It is analogous to the Curie-Weiss law given by Eq (8.64) for ferromagnetic materials above the critical temperature except that it contains the term $T + \theta$ instead of $T - \theta$ (or $T - T_C$) in the denominator. The $1/\chi$ versus T curves are shown in Fig. 8.10 to compare the antiferromagnetic behaviour with the paramagnetic and ferromagnetic ones. It can be shown that the Neel temperature, T_N, is not identical with θ and the two are related as

$$\frac{T_N}{\theta} = \frac{\beta - \alpha}{\beta + \alpha} \qquad (8.82)$$

Thus the value of T_N depends on the strength of AB and AA (or BB) interactions. For $\alpha = 0$, $T_N = \theta$. If α is positive, $T_N < \theta$. Also, T_N increases with β. These results have been verified experimentally.

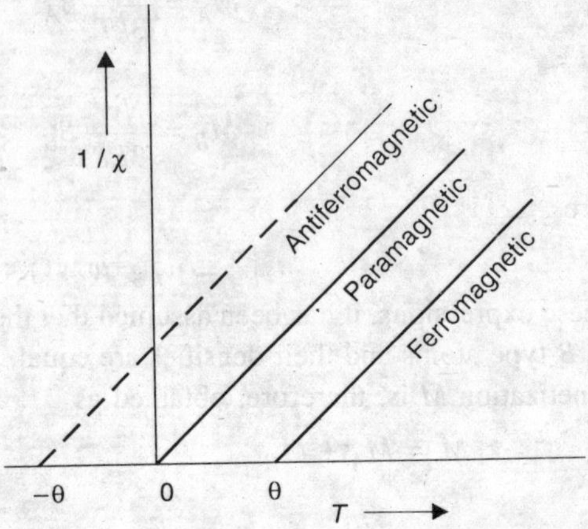

Fig. 8.10. *Plot of 1/χ versus T for para-, ferro- and antiferromagnetic substances above their critical temperatures.*

(ii) $T < T_N$

In this temperature region, each of the two sublattices A and B is spontaneously magnetized along its preferred direction of magnetization, but the net magnetization is still zero. The application of an external field, however, results in some net megnetization. There are two directions of the applied field which are of particular interest — one perpendicular and the other parallel to the preferred direction. The susceptibility in these two directions are determined separately and the average is taken to calculate the net susceptibility It can be shown that the susceptibility in the perpendicular direction, $\chi_\perp$, is independent of temperature, i.e., remains almost constant

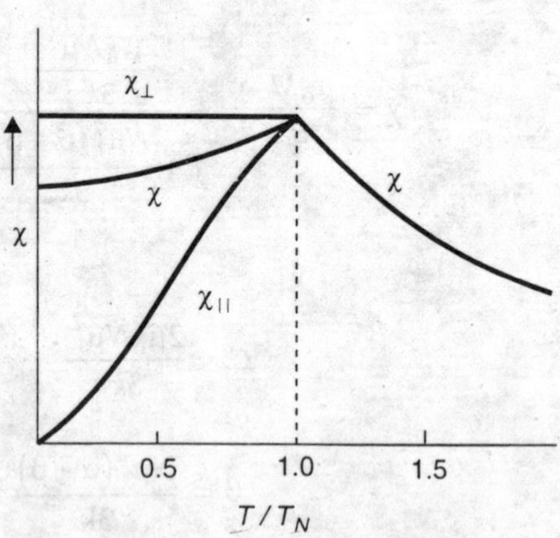

Fig. 8.11 Plot of χ versus T/T_N for an antiferromagnetic single crystal. For $T<T_N$, χ is also a function of direction of external field relative to the preferred direction of magnetization.

below the Neel temperature. The susceptibility in the parallel direction, $\chi_{||}$, decreases continuously with decrease in temperature and approaches zero as T→ 0 K. This means that at 0 K, the internal field is so strong that it does not allow the spins of one sublattice to change their orientation along the spins of the other sublattice even when an external field is present. At $T=T_N$, $\chi_{\perp} = \chi_{||} = \chi$. The variations of $\chi_{\perp}$, $\chi_{||}$ and χ with temperature are shown in Fig. 8.11.

8.7 FERRIMAGNETISM

Ferrimagnetism is identical to antiferromagnetism except that the magnetization of the two sublattices have different magnitudes which result in a non-zero value of net magnetization. This type of magnetism occurs in materials such as ferrites which are basically the oxides of various metal elements. The most common example is the one of magnetite or ferrous ferrite, Fe_3O_4 or FeO. Fe_2O_3. It has been found that for this particular ferrite, one Fe^{3+} ion occupies tetrahedral or A-site, i.e., it is tetrahedrally coordinated to four oxygen anions while the other Fe^{3+} ion and the Fe^{2+} ion occupy octahedral or B-sites, each being octahedrally coordinated to six oxygen anions. The ions present at A and B sites constitute A and B sublattices respectively and have opposite types of magnetizations. The net magnetization of a formula unit of Fe_3O_4 is, therefore, equal to the magnetization of a single Fe^{2+} ion and is represented in Fig. 8.12. However, one unit cell of magnetite having the spinel structure contains eight formula units of Fe_3O_4 in it and accordingly the net magnetization of a unit cell is equal to the sum of the magnetizations of eight Fe^{2+} ions.

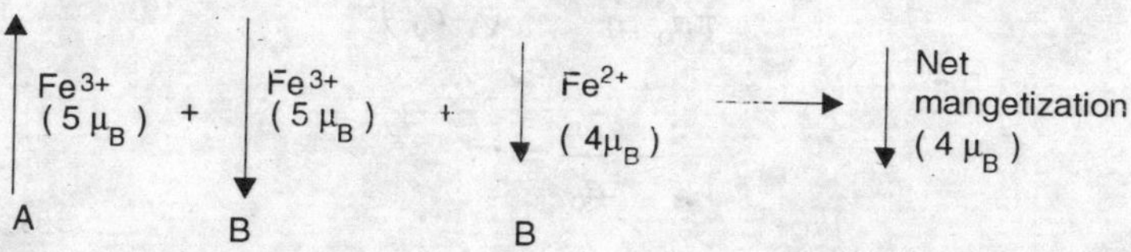

Fig. 8.12. *Magnetizations of the individual sites and the net magnetization in one formula unit of Fe_3O_4.*

It is thus easy to note that a ferrimagnetic solid resembles a ferromagnetic one in the sense that both possess spontaneous magnetization. Also, both exhibit hysteresis and almost identical magnetic properties. Neel attributed the antiparallel arrangement of spins of A and B sites to the negative AB interaction. Besides this, there also exist the negative AA and BB interactions which are much weaker than AB interaction. The magnetic behaviour of ferrimagnetic solids is explained on the basis of Neel's molecular field theory; the detailed description is beyond the scope of this book. Above the Curie temperature, the magnetization at both the sites obeys the Curie-Weiss law, whereas below this temperature the saturation effects come into existence and the magnetizations are obtained using Eq. (8.46).

SOLVED EXAMPLES

Example 8.1. Calculate the diamagnetic susceptibility of atomic hydrogen in the ground state at S.T.P. using the wave function

$$\psi(r) = \frac{1}{\left(\pi a_o^3\right)^{1/2}} \exp\left(-\frac{r}{a_o}\right)$$

where a_o (= 0.46Å) is the atomic radius.

Solution. The wave function for the ground state of hydrogen atom is

$$\psi(r) = \frac{1}{\left(\pi a_o^3\right)^{½}} \exp\left(-\frac{r}{a_o}\right)$$

The mean square distance of electronic charge distribution from the nucleus is calculated as

$$<r^2> = \int \Psi^* r^2 \Psi \mathbf{dr}$$

$$= 4\pi \int_0^\infty \psi^* r^2 \psi r^2 dr$$

$$= \frac{4\pi}{\pi a_o^3} \int_0^\infty r^4 \exp\left(-\frac{2r}{a_o}\right) dr$$

Put $$-\frac{2r}{a_o} = t$$

$\therefore$ $$r = -\frac{a_o}{2} t \quad \text{and} \quad dr = -\frac{a_o}{2} dt$$

$\therefore$ $$<r^2> = \frac{4}{a_o^3}\left(\frac{a_o^4}{16}\right)\left(-\frac{a_o}{2}\right)\int_0^\infty t^4 e^{-t} dt$$

Now

$$\int_0^\infty t^4 e^{-t} dt = 24$$

$\therefore$ $$<r^2> = 3a_o^2 \quad \text{(in magnitude)}$$

From Eq. (8.28), we have

$$\chi_{dia} = -\frac{N\mu_o Ze^2}{6m} \langle r^2 \rangle$$

$$= -\frac{N\mu_o Ze^2}{2m} a_o^2$$

Here $$N = \frac{6.023 \times 10^{26}}{2.24 \times 10^{-2}} = 2.69 \times 10^{28} \text{ m}^{-3}$$

$$Z = 1$$

$$a_o = 0.46 \times 10^{-10} \text{m}$$

$$\therefore \quad \chi_{dia} = -\frac{2.69 \times 10^{28} \times 4\pi \times 10^{-7} \times \left(1.6 \times 10^{-19}\right)^2 \left(0.46 \times 10^{-10}\right)^2}{2 \times 9.1 \times 10^{-31}}$$

$$= -1.01 \times 10^{-6} \text{ (SI units)}$$

Example 8.2. A paramagnetic gas with $L = 0$ and $S = 1/2$ contains N atoms per m^3. Obtain expressions for the atomic densities in the possible energy levels and the resultant magnetization. If the gas contains 10^{28} atoms per m^3, calculate the populations of the levels and the magnetization at 300 K and 5 K for an applied magnetic flux density of 2.5 T.

Solution. Since $L = 0$, $S = 1/2$

$$\therefore \quad J = 1/2, \; m_J = \pm \tfrac{1}{2} \text{ and } g = 2$$

Therefore, there are two energy levels with energies given by

$$E = m_J g \mu_B B$$

Let $N_{1/2}$ and $N_{-1/2}$ represent the populations of the levels with $m_J = 1/2$ and $-1/2$ respectively. Since population of a level E is proportional to $\exp\left(-\frac{E}{kT}\right)$, we have

$$N_{½} = A \exp\left(-\frac{g\mu_B B}{2kT}\right); \; N_{-½} = A \exp\left(\frac{g\mu_B B}{2kT}\right)$$

where A is a proportionality constant determined by the condition :

$$N_{½} + N_{-½} = N$$

It gives

$$A = \frac{N}{2\cosh\left(\frac{g\mu_B B}{2kT}\right)}; \; N_{½} = \frac{N\exp\left(-\frac{g\mu_B B}{2kT}\right)}{2\cosh\left(\frac{g\mu_B B}{2kT}\right)}; \; N_{-½} = \frac{N\exp\left(\frac{g\mu_B B}{2kT}\right)}{2\cosh\left(\frac{g\mu_B B}{2kT}\right)}$$

If μ is the atomic moment, the magnetization is given by

$$M = \mu\,(N_{-½} - N_{½}) = g\mu_B J(N_{-1/2} - N_{1/2})$$

$$= g\mu_B JN \left[\frac{\exp\left(\frac{g\mu_B B}{2kT}\right) - \exp\left(-\frac{g\mu_B B}{2kT}\right)}{2\cosh\left(\frac{g\mu_B B}{2kT}\right)}\right]$$

$$= g\mu_B JN \tanh\left(\frac{g\mu_B B}{2kT}\right) \tag{8.83}$$

Now

$$g = 2,\; J = 1/2$$

$$\mu_B = 9.27 \times 10^{-24}\ \text{Am}^2$$

$$N = 10^{28}\ \text{m}^{-3}$$

$$B = 2.5\ \text{T}$$

$$k = 1.38 \times 10^{-23}\ \text{JK}^{-1}$$

The above expressions become

$$N_{½} = 5 \times 10^{27}\ \frac{\exp\left(\frac{1.68}{T}\right)}{\cosh\left(\frac{1.68}{T}\right)}$$

$$N_{-½} = 5 \times 10^{27}\ \frac{\exp\left(-\frac{1.68}{T}\right)}{\cosh\left(\frac{1.68}{T}\right)}$$

$$M = 9.27 \times 10^{4} \tanh\left(\frac{1.68}{T}\right)$$

At 300 K, we get

$$N_{1/2} = 5 \times 10^{27} \times \frac{0.994}{1.00} = 4.97 \times 10^{27}\ \text{m}^{-3}$$

$$N_{-1/2} = 5 \times 10^{27} \times \frac{1.006}{1.00} = 5.03 \times 10^{27}\ \text{m}^{-3}$$

$$M = 9.27 \times 10^{4} \times 5.6 \times 10^{-3} = 519.1\ \text{Am}^{-1}$$

Similarly, at 5 K, we obtain

$$N_{1/2} = 3.38 \times 10^{27}\ \text{m}^{-3}$$

$$N_{-1/2} = 6.62 \times 10^{27}\ \text{m}^{-3}$$

$$M = 3.0 \times 10^{4}\ \text{Am}^{-1}$$

Example 8.3. Dy^{3+} has outer electronic configuration of $4f^9\ 6s^0$. Calculate the magnetic susceptibility for a salt containing one kg mole of Dy^{3+} ions at 300 K.

Solution. Number of unpaired electrons in $Dy^{3+} = 5$.

Application of Hund's rules yields

$$S = 5(1/2) = 5/2$$

$$L = 3 + 2 + 1 + 0 - 1 = 5$$

$$J = L + S = 5/2 + 5 = 15/2$$

$$\therefore \qquad g = 1 + \frac{J(J+1)+S(S+1)-L(L+1)}{2J(J+1)} = 1.33$$

Now
$$\chi = \frac{\mu_0 N \mu_B^2}{3kT} g^2 J(J+1)$$

For 1 kg mole of salt, $N = 6.023 \times 10^{26}$. Therefore,

$$\chi = \frac{4\pi \times 10^{-7} \times 6.023 \times 10^{26} \times \left(9.27 \times 10^{-24}\right)^2 \times (1.33)^2}{3 \times 1.38 \times 10^{-23} \times 300} \times \frac{15}{2}\left(\frac{15}{2}+1\right)$$

$$= 5.9 \times 10^{-4}$$

Example 8.4. A ferromagnetic material with $J = 3/2$ and $g = 2$ has a Curie temperature of 125 K. Calculate the intrinsic flux density near 0 K. Also, calculate the ratio of the magnetization at 300 K in the presence of an external field of 1 mT to the spontaneous magnetization at 0 K.

Solution. At 0 K, the intrinsic flux density as given by Eqs. (8.52) and (8.59) is

$$B_E = \lambda M = \lambda M_s\,(0) = \lambda N g J \mu_B$$

Using Eq. (8.66), we get

$$B_E = \frac{3kT_C}{g\mu_B(J+1)} = \frac{3\times 1.38\times 10^{-23}\times 125}{2\times 9.27\times 10^{-24}\times (5/2)} = 111.6 \text{ T}$$

The magnetization for $T > T_C$ is obtained by using Eq. (8.64) as

$$M = \frac{CB}{\mu_o(T-T_C)}$$

Using Eqs. (8.65) and (8.66), we obtain

$$M = \frac{Ng^2\mu_B^2 J(J+1)B}{3k(T-T_C)} \tag{8.84}$$

Also,

$$M_s(0) = NgJ\mu_B$$

$$\therefore \qquad \frac{M}{M_s(0)} = \frac{g\mu_B B(J+1)}{3k(T-T_C)}$$

$$= \frac{2\times 9.27\times 10^{-24}\times 10^{-3}\times (5/2)}{3\times 1.38\times 10^{-23}\times (300-125)}$$

$$= 6.4\times 10^{-6}$$

SUMMARY

1. The magnetic moment per unit volume of a solid is called magnetization. The magnetization, **M**, produced per unit applied magnetic field, **H**, is called susceptibility, i.e.,

$$\chi = \mathbf{M}/\mathbf{H}$$

2. The total magnetic flux density, **B**, inside a magnetic material is

$$\mathbf{B} = \mu_o\,(\mathbf{H} + \mathbf{M}) = \mu_o\,(1+\chi)\mathbf{H} = \mu_o\mu_r\,\mathbf{H} = \mu\mathbf{H}$$

where μ_o, μ_r and μ are absolute permeability of free space, relative permeability of medium and absolute permeability of medium respectively.

3. Magnetism in solids arises mainly from the spins of unpaired electrons. The other minor contributions arise due to orbital motion of electrons and spins of nuclei.

4. Diamagnetic materials exhibit a small negative susceptibiiity given by

$$\chi_{dia} = -\frac{N\mu_o Ze^2}{6m} <r^2>$$

where N represents the number of atoms per unit volume of the solid, Z the number of electrons present in each atom, m the electronic mass and $<r^2>$ the mean square distance of electrons relative to the nucleus.

5. Paramagnetic materials possess small permanent magnetic moments which are oriented randomly in the absence of the field. The susceptibility is small but positive.

6. According to the classical theory, the paramagnetic susceptibility at normal field strengths and ordinary temperatures obeys the Curie law and is given by

$$\chi_{para} = \frac{\mu_o N\mu^2}{3kT} = \frac{C}{T}$$

where $C\left(=\frac{\mu_o N\mu^2}{3k}\right)$ is the Curie constant; μ being the magnetic moment of each atom and k the Boltzmann's constant. The quantum theory gives an identical expression with μ^2 replaced by $p^2_{eff}\,\mu^2_B$ where μ_B is the Bohr magneton and p_{eff} the effective number of Bohr magnetons in the magnetic moment of a magnetic dipole. p_{eff} is related to the total angular momentum quantum number J of the dipole as

$$p_{eff} = g\sqrt{J(J+1)}$$

where g is the Lande's g-factor.

6. Ferromagnetic materials consist of spontaneously magnetized regions called domains which are randomly oriented in the absence of an applied magnetic field. The spontaneous magnetization vanishes above a critical temperature called the Curie temperature when the ferromagnetic materials become paramagnetic.

7. Above the Curie temperature, T_C, the susceptibility of a ferromagnetic material obeys the Curie-Weiss law, i.e.,

$$\chi = \frac{C}{T - T_C}$$

where C is the Curie-Weiss constant.

8. An antiferromagnetic material comprises two magnetic sublattices with equal and opposite magnetic moments such that there exists no net magnetization in the material. The magnetization, however, appears in the presence of an applied magnetic field. It increases with temperature and becomes maximum at a critical temperature called the Neel temperature, T_N. For $T > T_N$, the material behaves as paramagnetic.

9. A ferrimagnetic material is identical to an antiferromagnetic material except that the opposite magnetic moments of the neighbouring sublattices have unequal magnitudes. It also resembles the ferromagnetic material in the sense that both possess spontaneous magnetization and exhibit the phenomenon of hysteresis.

VERY SHORT QUESTIONS

1. Define the following terms:

   (*i*) Magnetisation (*ii*) magnetic susceptibility (*iii*) magnetic flux density (*iv*) Permeability (*v*) Relative permeability.

2. Explain the meaning of the following terms in brief:

   (*i*) Diamagnetism (*ii*) Paramagnetism (*iii*) Ferromagnetism (*iv*) Ferrimagnetism (*v*) Antiferromagnetism.

3. Define the Curie law for paramgnetism.
4. What is the basic cause of paramagnetism?
5. What is spontaneous magnetisation?
6. What are domains?
7. What is exchange field?
8. Explain the following terms briefly :

   (*i*) Hysteresis (*ii*) Coercivity (*iii*) Remanence.

9. What is Neel temperature?

SHORT QUESTIONS

1. What is diamagnetism? Why diamagnetic materials have negative susceptibility?
2. What is the essential difference between the classical theory and the quantum theory of paramagnetism?
3. Give the Curie law of paramagnetism. What is the Curie temperature?
4. Discuss the variation of spontaneous magnetisation with temperature for ferromagnetic materials.

5. What is the Curie-Weiss law? Discuss its application for ferromagnetic materials.
6. What is exchange interaction? How does it help to explain magnetism in iron group of atoms?
7. Explain the cause of hysteresis phenomenon in ferromagnetic materials? What does area of the loop signify?
8. Explain the difference between the terms 'Curie temperature' and 'Neel temperature'.
9. Explain the variation of susceptibility with temperature for antiferromagnetic materials.

LONG QUESTIONS

1. Distinguish between the characteristic features of diamagnetism, paramagnetism, ferrimagnetism, antiferromagnetism and ferrimagnetism. Give an example of each type of material. Comment on the temperature variation of susceptibility for all types of materials.
2. Explain the origin of diamagnetism in materials. Obtain an expression for diamagnetic susceptibility using the Langevin's theory. What is the significance of negative susceptibility?
3. Derive an expression for diamagnetic susceptibility using the quantum theory. Discuss the temperature dependence of susceptibility.
4. Describe the Langevin's theory of paramagnetism and obtain an expression for paramagnetic susceptibility. Comment on the temperature dependence of susceptibility.
5. Give an account for the quantum theory of paramagnetism and discuss the low and high temperature cases. Explain how the orbital motion of electrons in transition and rare earth metals is quenched. Write the modified expression for susceptibility for the ions of iron group.
6. Give an account of the Weiss theory of ferromagnetism. Discuss the temperature variation of saturation magnetisation. Explain hysteresis and Curie point on the basis of this theory.
7. Explain how and why are the ferromagnetic domains formed? Draw a typical *B-H* loop and describe the different magnetisation processes, which lead to the formation of a *B-H* loop. What are the advantages and disadvantages of having a *B-H* loop in a material ?
8. Describe the Heisenberg's exchange interaction. How does it explain ferromagnetism? Relate the exchange integral to the Weiss constant and the ferromagnetic Curie temperature.

9. Describe the two-sublattice model to explain antiferromagnetism. How does this model account for the difference between the Neel temperature T_N and the Curie-Weiss temperature, θ?

PROBLEMS

1. Calculate the diamagnetic susceptibility of copper assuming that an atom of copper contributes only one electron and the mean square distance of electronic distribution from the nucleus is equal to the radius of monovalent copper ion. The lattice constant of copper is 3.61 Å and the radius of Cu^+ is 0.96 Å. (-4.62×10^{-6})
2. The electronic configuration of a Cr^{2+} ion is $3d^4 4s^0$. Calculate the magnetic susceptibility for a salt containing one kg mole of Cr^{2+} ions at 300 K. (1.25×10^{-4})

   (**Hint** : Orbital angular momentum of Cr^{2+} is quenched by the presence of the crystal field).
3. Cu^{2+} has nine electrons in the $3d$-shell. What magnetic field must be applied to a salt containing Cu^{2+} ions at 1 K so that 90 per cent of the ions are in the ground state ? (3 T)
4. The paramagnetism in copper sulphate arises mainly from the copper ions with $S = 1/2$. Show that the magnetization is given by the expression

$$M = N\mu_B \tanh\left(\frac{\mu_B B}{kT}\right)$$

5. Iron is a ferromagnet with the Curie temperature of 1043 K and an effective moment of 2.2 μ_B per atom. Compute the internal field for iron. (2100 T)
6. Europium oxide is ferromagnetic with the Curie temperature of 70 K. An europium ion has $J = 7/2$ and $g = 2$. Calculate the internal magnetic flux density. Also, determine the ratio of magnetization at 300 K in a field of 0.05 T to that at 0 K. (34.7 T, 4.38×10^{-4})
7. Nickel (*fcc*) has a Curie temperature of 358°C and a lattice constant of 3.52 Å. A Ni^{2+} ion has 8 electrons in the $3d$-shell. Assuming quenching of the orbital angular momentum and using the Hund's rules and the Weiss theory, determine the saturation magnetization, the Weiss field constant, the internal magnetic field and the Curie constant for nickel. (1.7×10^6 Am^{-1}, 4.14×10^{-4}, 704 T, 1.92)

CHAPTER–IX

DIELECTRIC PROPERTIES OF SOLIDS

Dielectrics are basically electric insulators which ordinarily do not contain any free charge carriers for conduction. They, however, contain positive and negative charges which are bound together and hence could be affected by the applied electric fields. A brief description of the properties of dielectric solids in the presence of an external electric field is given in this chapter. These properties are important to understand the propagation of electromagnetic waves through the material media and fabricate various devices such as capacitors, microphones, etc.

9.1 POLARIZATION AND SUSCEPTIBILITY

When a dielectric is placed in an external electric field $\mathbf{E_o}$, the positive and negative charges are displaced from their equilibrium positions by very small distances (less than an atomic diameter) throughout the volume of the dielectric. This results in the formation of a large number of dipoles each having some dipole moment in the direction of the field. The material is said to be polarized with a *polarization* **P** defined as the dipole moment per unit volume of the material. As shown in Fig. 9.1, the effect of polarization is to reduce the magnitude of the external field $\mathbf{E_o}$. Thus the magnitude of the resultant field is less than the applied field, i.e., $E < E_o$. In vector notation, we may write

$$\mathbf{E} = \mathbf{E_o} + \mathbf{E_p} \tag{9.1}$$

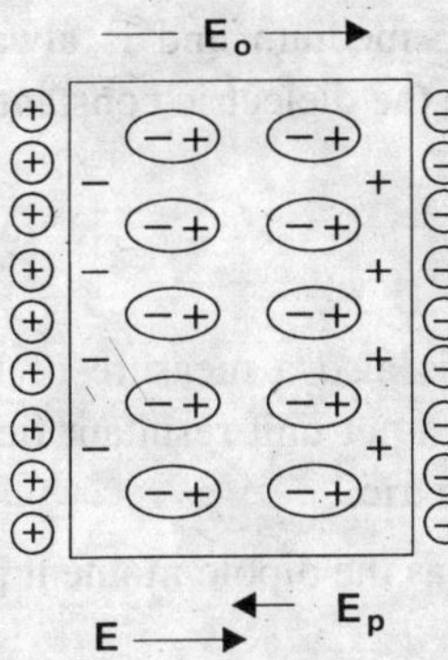

Fig. 9.1. *A dielectric slab placed in an electric field* $\mathbf{E_o}$ *produced by fixed charges (encircled) outside the slab. The internal polarization field* $\mathbf{E_p}$ *is assumed to be due to fictitious bound charges at the surface of the slab and is directed opposite to* $\mathbf{E_o}$.

The field $\mathbf{E_p}$ is called the *polarization field* as it tends to oppose the applied field E_o within the material. For ordinary electric fields, the polarization **P** is proportional to the macroscopic field **E**. In SI units, it is expressed as

$$\mathbf{P} = \varepsilon_o \chi_e \mathbf{E} \tag{9.2}$$

where ε_o is the permittivity of free space and χ_e is the electric susceptibility. Thus, except for a constant factor, the electric susceptibility is a measure of the polarization produced in the material per unit resultant electric field.

9.2 THE LOCAL FIELD

The electric field acting at the site of an atom or molecule is, in general, significantly different from the macroscopic electric field **E** and is called the *local field*. This field is responsible for polarization of each atom or molecule of a solid. For an atomic site with cubic symmetry, the local field is given by the *Lorentz relation*, i.e.,

$$\mathbf{E}_{loc} = \mathbf{E}_o + \mathbf{E}_p + \frac{\mathbf{P}}{3\varepsilon_o} = \mathbf{E} + \frac{\mathbf{P}}{3\varepsilon_0} \tag{9.3}$$

Thus, apart from the macroscopic field, the local field also contains a term which represents the field due to polarization of other atoms in the solid. The expression (9.3) gets modified with shape of the specimen.

9.3 DIELECTRIC CONSTANT AND POLARIZABILITY

The electric displacement vector for an isotropic or cubic medium can be defined as

$$\mathbf{D} = \varepsilon_o\varepsilon_r\mathbf{E} = \varepsilon_o\mathbf{E} + \mathbf{P} \tag{9.4}$$

where ε_r is called the *relative permittivity* or *dielectric constant* of the dielectric. It is a scalar quantity for an isotropic medium and is always dimensionless. The Eq. (9.4) can be used to define the dielectric constant as

$$\varepsilon_r = \frac{\varepsilon_o\,\mathbf{E}+\mathbf{P}}{\varepsilon_o\,\mathbf{E}} = 1 + \chi_e \tag{9.5}$$

Thus, like susceptibility, the dielectric constant is also a measure of the polarization of the material. Larger the polarization per unit resultant field, greater will be the dielectric constant of the dielectric.

The polarizability, α, of an atom is defined as the dipole moment per unit local electric field at the atom, i.e.,

$$p = \alpha E_{loc} \tag{9.6}$$

Thus polarizability is an atomic property whereas dielectric constant is a macroscopic property which depends upon the arrangement of atoms within the crystal. If all the atoms have the same polarizability and there are N number of atoms per unit volume, the polarization can be expressed as

$$P = Np = N\alpha E_{loc} \tag{9.7}$$

The general expression for polarization is, however, given by

$$P = \sum_j N_j\alpha_j E_{loc}(j) \tag{9.8}$$

where the summation is over all the atoms or atomic sites. In case of cubic symmetry, Eqs. (9.3) and (9.7) yield

$$P = N\alpha\left(E + \frac{P}{3\varepsilon_o}\right) \tag{9.9}$$

which, on rearranging the terms, gives

$$\chi_e = \frac{P}{\varepsilon_o E} = \frac{N\alpha / \varepsilon_o}{1 - \dfrac{N\alpha}{3\varepsilon_o}} \tag{9.10}$$

Using Eq. (9.5), we obtain

$$\varepsilon_r = 1 + \frac{N\alpha / \varepsilon_o}{1 - \dfrac{N\alpha}{3\varepsilon_o}}$$

or

$$\varepsilon_r = \frac{1 + \dfrac{2N\alpha}{3\varepsilon_o}}{1 - \dfrac{N\alpha}{3\varepsilon_o}} \tag{9.11}$$

This expression can be put in the form

$$\frac{\varepsilon_r - 1}{\varepsilon_r + 2} = \frac{N\alpha}{3\varepsilon_o} \tag{9.12}$$

This is known as the *Clausius-Mossotti relation*. It relates the dielectric constant to the atomic polarizability provided the condition of cubic symmetry holds. In a more general form, it is expressed as

$$\frac{\varepsilon_r - 1}{\varepsilon_r + 2} = \frac{1}{3\varepsilon_o}\sum_j N_j \alpha_j \tag{9.13}$$

9.4 SOURCES OF POLARIZABILITY

The net polarizability of a dielectric material results mainly from the following three types of contributions :

(*i*) Electronic polarizability

(*ii*) Ionic polarizability

(*iii*) Dipolar or orientational polarizability

The extent to which a particular polarizability contributes depends on the nature of the dielectric and the frequency of the applied electric field.

9.4.1 Electronic Polarizability

The electronic polarizability arises due to the displacement of the electron cloud of an atom relative to its nucleus in the presence of an applied electric field as shown in Fig. 9.2. The polarization as well as the dielectric constant of a material at optical frequencies results mainly from the electronic polarizability. At optical frequencies, Eq. (9.13) may be written as

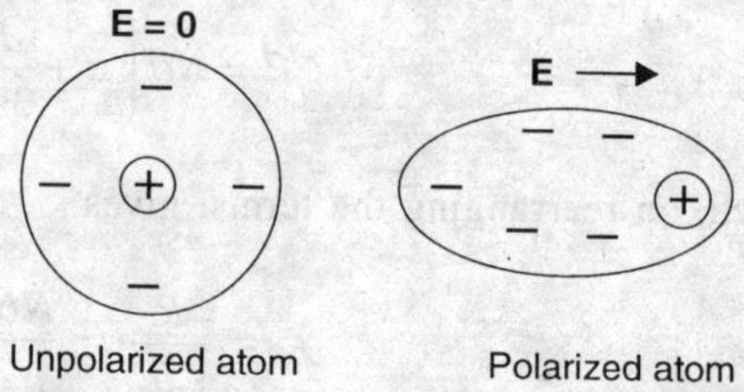

Fig. 9.2. *Electronic polarization in the presence of an external field.*

$$\frac{n^2-1}{n^2+2} = \frac{1}{3\varepsilon_0} \sum_j N_j \alpha_j \text{ (electronic)} \tag{9.14}$$

where ε_r has been replaced by n^2, n being the refractive index.

Classical Theory of Electronic Polarizability

An electron bound harmonically to an atom exhibits resonant absorption at a frequency given by

$$\omega_0 = \sqrt{\beta / m} \tag{9.15}$$

where β represents the force constant and m is the mass of the electron. If x is the displacement of the electron under the effect of the electric field E_{loc}, then we have

$$-eE_{loc} = \beta x = m\omega_0{}^2 x \tag{9.16}$$

The static electronic polarizability is calculated as

$$\alpha_e = \frac{p}{E_{loc}} = -\frac{ex}{E_{loc}} = \frac{e^2}{m\omega_0^2} \tag{9.17}$$

To obtain the frequency dependence of electronic polarizability, we treat the system as a simple harmonic oscillator. If ω is the frequency of the local field, the field at any time t is given by E_{loc} sin ωt and the equation of motion may be written as

$$m\frac{d^2x}{dt^2} + m\omega_0{}^2 x = -eE_{loc} \sin \omega t \tag{9.18}$$

Substituting $x = x_o \sin \omega t$, we obtain

$$m(-\omega^2+\omega_o^2)\,x_o = -e\,E_{loc} \tag{9.19}$$

The amplitude of the dipole is given by

$$p_o = -ex_o = \frac{e^2 E_{loc}}{m\left(\omega_o^2 - \omega^2\right)}$$

which gives

$$\alpha_e = \frac{p_o}{E_{loc}} = \frac{e^2/m}{\omega_o^2 - \omega^2} \tag{9.20}$$

This expression gives the frequency dependence of electronic polarizability. The corresponding expression based on the quantum theory is given by :

$$\alpha_e = \frac{e^2}{m}\sum_j \frac{f_{ij}}{\omega_{ij}^2 - \omega^2} \tag{9.21}$$

where f_{ij} is known as the *oscillator strength* of the electric dipole transition between the atomic states *i* and *j*.

9.4.2 Ionic Polarizability

The ionic polarizability arises due to displacement of a charged ion relative to other ions in a solid. Assuming the forces near equilibrium as simple harmonic, the displacement Δx in the presence of an electric field *E* is given by

$$\beta\Delta x \cong eE \tag{9.22}$$

where β is the force constant. Thus the ionic polarizability is determined as

$$\alpha_i = \frac{p}{E} = \frac{e\Delta x}{E} \cong \frac{e^2}{\beta} \tag{9.23}$$

For $\beta \cong 20$ Nm^{-1}, $\alpha_i \cong 10^{-39}$ Fm^2. The ionic contribution is important at low frequencies. The sodium chloride, for example, has $\varepsilon_r \cong 5.6$ at low frequencies whereas the value reduces to about 2.25 at optical frequencies.

9.4.3 Dipolar Polarizability

A molecule, such as H_2O, having a permanent dipole moment is called a *dipolar* or *polar molecule* and a substance comprising such molecules is called a *dipolar substance.* The dipolar polarizability is the property of dipolar substances. In the absence of an external electric field, the dipoles have random orientations and there is no net polarization. However, when the field is present, the dipoles orient themselves along the field and produce *orientational* or *dipolar polarizability.* The thermal agitation of the molecules tends to counteract the ordering effect of the electric field and an equilibrium state is reached wherein the different dipoles make all possible angles varying from zero to π radians with the field direction. The potential energy of such a molecule of dipole moment **p** oriented at an angle θ with the field direction (Fig. 9.3) is given by

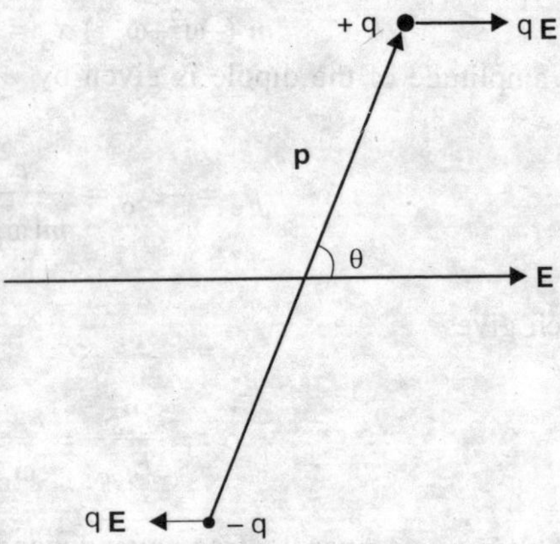

Fig. 9.3. *A molecule of dipole moment* **p** *placed in an electric field* **E**.

$$U = -\mathbf{p}.\mathbf{E} = -pE\cos\theta \tag{9.24}$$

According to the statistical mechanics, the number of dipoles having orientations between θ and $\theta + d\theta$, i.e., which lie within the solid angle $d\Omega$ or $2\pi\sin\theta d\theta$ is proportional to

$$\exp\left(-\frac{U}{kT}\right)d\Omega \quad \text{or} \quad \exp\left(\frac{pE\cos\theta}{kT}\right)2\pi\sin\theta d\theta \tag{9.25}$$

Since a dipole making an angle θ with the field direction contributes a component of dipole moment $p\cos\theta$ parallel to the field, the contribution of the above number of dipoles to the total polarization is

$$2\pi p\exp\left(\frac{pE\cos\theta}{kT}\right)\cos\theta\sin\theta\, d\theta \tag{9.26}$$

The average contribution to polarization is, therefore, given by

$$\overline{p} = \frac{P}{N} = \frac{\text{Total polarization due to all the dipoles}}{\text{Total number of dipoles}}$$

$$= \frac{p\int_0^\pi \exp\left(\dfrac{pE\cos\theta}{kT}\right)\cos\theta\sin\theta\, d\theta}{\int_0^\pi \exp\left(\dfrac{pE\cos\theta}{kT}\right)\sin\theta\, d\theta} \tag{9.27}$$

Putting

$$x = \frac{pE}{kT},\ y = \cos\theta \text{ and } dy = -\sin\theta\, d\theta,$$

Eq. (9.27) becomes

$$\bar{p} = \frac{p\int_{+1}^{-1} ye^{xy}dy}{\int_{+1}^{-1} e^{xy}dy}$$

or

$$\frac{\bar{p}}{p} = \frac{e^x + e^{-x}}{e^x - e^{-x}} - \frac{1}{x}$$

$$= \coth x - \frac{1}{x} \equiv L(x) \qquad (9.28)$$

where $L(x)$ is called the *Langevin function.* The variation of $L(x)$ with x has been shown in Fig. 8.2. For $x << 1$ or $pE << kT$, i.e., for fields not too large and for temperatures not too low, we have

$$L(x) \cong \frac{x}{3} = \frac{pE}{3kT} \qquad (9.29)$$

Using Eqs. (9.28) and (9.29), the polarization P becomes

$$P = N\bar{p} = \frac{Np^2E}{3kT} \qquad (9.30)$$

The dipolar polarizability per molecule is, therefore, given by

$$\alpha_d = \frac{\bar{p}}{E} = \frac{p^2}{3kT} \qquad (9.31)$$

Thus the existence of dipolar polarizability depends on whether the molecules possess a permanent dipole moment. The dipolar polarizability also depends on temperature in accordance with Eq. (9.31). At room temperature, $\alpha_d \sim 10^{-39}$ Fm^2 which is comparable to electronic polarizability.

9.5 FREQUENCY DEPENDENCE OF TOTAL POLARIZABILITY

The total polarizability of a dielectric is given by the expression :

$$\alpha = \alpha_e + \alpha_i + \alpha_d \qquad (9.32)$$

It decreases with increase in frequency as shown in Fig. 9.4. This type of variation of polarizability can be explained on the basis of the relaxation times of the various contributing polarization processes. When the frequency of the applied field is quite large as compared to the inverse of the relaxation time for a particular polarization process, the contribution of that process to polarizability is negligible. As relaxation time is maximum for the dipolar process and minimum for the electronic process, the dipolar contribution disappears first followed by ionic and electronic contributions.

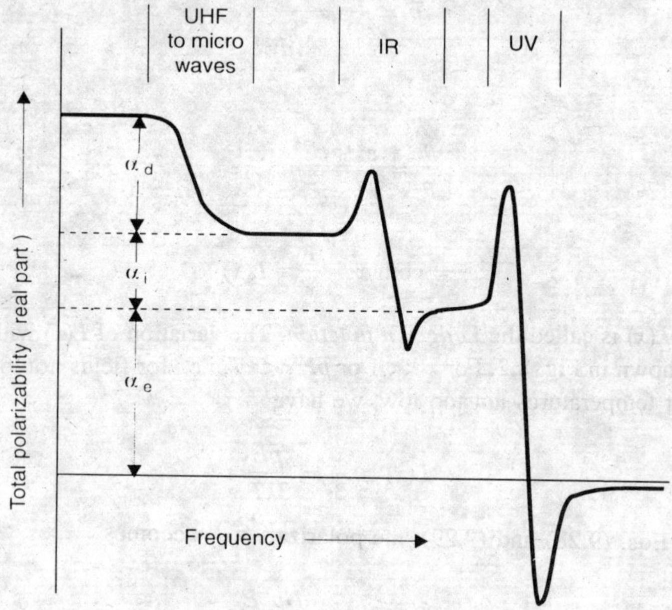

Fig. 9.4. ***Frequency dependence of various contributions to polarizability.***

9.6 FERROELECTRICITY

Ferroelectricity is the phenomenon which refers to the state of *spontaneous polarization,* i.e., polarization of the meterial in the absence of an electric field. It is thus analogous to ferromagnetism which represents the state of spontaneous magnetization of the material. The crystals exhibiting ferroelectricity are called the *ferroelectric crystals.* In such crystals, the centres of positive and negative charges do not coincide with each other even in the absence of the field, thus producing a non-zero value of the dipole moment. The variation of polarization with electric field is not linear for such crystals but forms a closed loop called the *hysteresis loop.* The ferroelectricity disappears above a certain critical temperature called the *transition temperature* or the *Curie point,* T_C when the material gets transformed from ferroelectric to paraelectric state as indicated by a rapid decrease in the dielectric constant with increase in temperature.

Ferroelectricity is observed in a number of substances; the most familiar ones are barium titanate ($BaTiO_3$) and the Rochelle salt ($NaKC_4H_4O_6 \cdot 4H_2O$). The ferroelectric crystals may be classified into two main groups—the *order-disorder group* and the *displacive group*. In the order-disorder class of ferroelectrics, the ferroelectric transition is associated with individual ordering of ions. These are the crystals which contain hydrogen bonds and in which the motion of protons is related to the ferroelectric properties. The examples are potassium dihydrogen phosphate (KH_2PO_4), rubidium hydrogen phosphate (RbH_2PO_4), etc. The displacive group of ferroelectrics is the one in which the ferroelectric transition is associated with the displacement of a whole sublattice of ions of one type relative to a sublattice of another type. The crystals of this class exhibit structures which are closely related

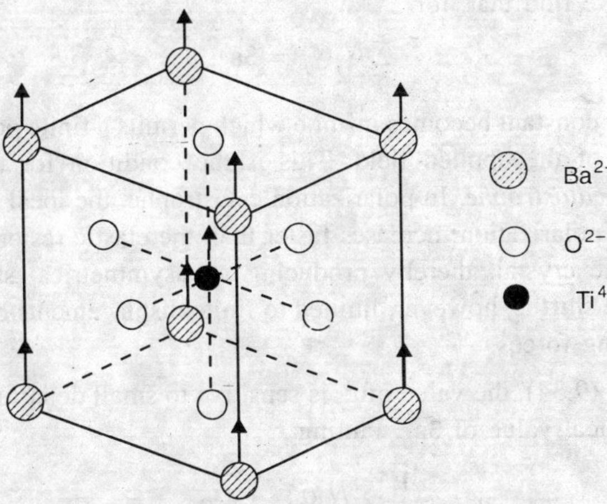

Fig. 9.5. *Structure of $BaTiO_3$ for $T > T_C$.*

to the perovskite and ilmenite structures. The examples are $BaTiO_3$, $KNbO_3$, $LiTaO_3$, etc. Consider the case of $BaTiO_3$ crystal. Below the Curie point (380 K), it exhibits the perovskite structure as shown in Fig. 9.5. The unit cell is cubic with Ba^{2+} ions occupying the corners, O^{2-} ions occupying the face centres and Ti^{4+} ion occupying the body centre of the cube. Thus each Ti^{4+} ion is surrounded by six O^{2-} ions in an octahedral configuration. For $T > T_C$, the centres of gravity of positive and negative charges exactly coincide with each other to produce no net dipole moment. However, for $T<T_C$, the Ti^{4+} and Ba^{2+} ions slightly move upwards while the O^{2-} ions slightly move downwards and the structure becomes tetragonal with centres of positive and negative charges not coinciding with each other. This produces the spontaneous polarization in the crystal.

In fact, the ferroelectricity in $BaTiO_3$ arises from induced electronic and ionic dipole moments. Rewriting Eq. (9.13) for the dielectric constant, we get

$$\varepsilon_r = \frac{1 + \frac{2}{3\varepsilon_o}\sum_j N_j\alpha_j}{1 - \frac{1}{3\varepsilon_o}\sum_j N_j\alpha_j} \tag{9.33}$$

where α_j is the electronic plus ionic polarizability of the jth ion and N_j is the number of jth ions per unit volume. The expression (9.33) has been obtained by assuming that the local field at all the atoms is equal to $\mathbf{E} + \frac{\mathbf{P}}{3\varepsilon_o}$. From Eq. (9.33), we find that, for

$$\sum_j N_j\alpha_j = 3\varepsilon_o, \tag{9.34}$$

the dielectric constant becomes infinite which permits a finite polarization in the absence of the applied field. This is the condition for the so-called *polarization catastrophe*. In polarization catastrophe, the local electric field arising from polarization increases faster than the elastic restoring force on an ion in the crystal, thereby producing an asymmetrical shift in ionic position. The shift is, however, limited to finite displacements by the higher order restoring forces.

In Eq. (9.34), the value of ε_r is sensitive to small departures of $\Sigma N_j\alpha_j$ from the critical value of $3\varepsilon_o$. Putting

$$\frac{1}{3\varepsilon_o}\sum_J N_j\alpha_j = 1-3s, \tag{9.35}$$

where s << 1, in expression (9.33), the dielectric constant becomes

$$\varepsilon_r = \frac{1}{s} - 2 \cong \frac{1}{s} \tag{9.36}$$

We assume that s varies linearly near the critical temperature as

$$s \cong \frac{T - T_C}{\xi} \tag{9.37}$$

where ξ is a constant. This type of variation of s or $\sum_j N_j\alpha_j$ might result from normal thermal expansion of the lattice. From Eq. (9.36), we obtain

$$\varepsilon_r \cong \frac{\xi}{T - T_C} \tag{9.38}$$

Such a variation of dielectric constant with temperature is close to that observed experimentally in the paraelectric state.

9.7 PIEZOELECTRICITY

In cetain crystals, the application of an external stress induces a net dipole moment which produces the electric polarization with the polarization charges appearing on the surfaces of the crystals. Such crystals are called the *piezoelectric crystals* and the phenomenon is known as the *piezoelectricity.* Some examples of such crystals are quartz, the Rochelle salt and tourmaline. The inverse effect is also observed, i.e., the application of an electric field produces strains in the crystal.

In schematic one-dimensional notation, the piezoelectric equations are

$$\left.\begin{aligned} P &= Zd + \varepsilon_o \, E \, \chi \\ e &= Zs + Ed \end{aligned}\right] \quad (9.39)$$

where P represents the polarization, Z the stress, d the piezoelectric strain constant, E the electric field, χ the dielectric susceptibility, e the elastic strain and s the elastic compliance constant. The first one of the equations (9.39) exhibits the development of polarization by an applied stress and the second one shows the development of elastic strain by an applied electric field. Generally, very large electric fields are needed to produce very small strains. In quartz, for example, an electric field of about 10^4 Vm^{-1} produces a strain of about 1 in 10^8 only.

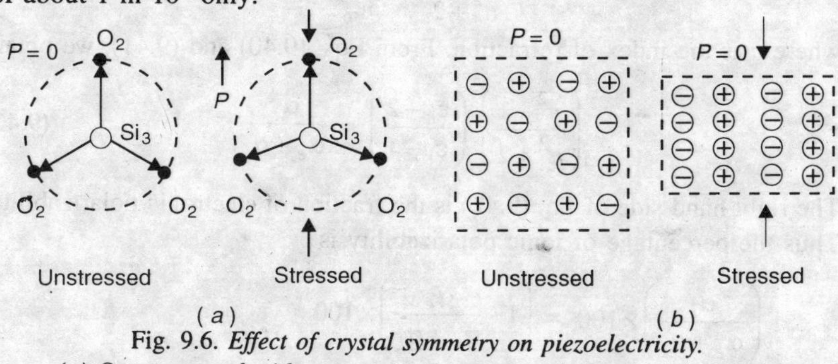

Fig. 9.6. *Effect of crystal symmetry on piezoelectricity.*
(a) Quartz crystal with no centre of inversion shows piezoelectricity.
(b) A crystal with centre of inversion shows no piezoelectricity.

The occurrence of piezoelectricity is the result of displacement of ions in certain crystals under the effect of the applied stresses. In such crystals, the ions are so displaced that their charge distribution loses the original symmetry as shown in Fig. 9.6a. In certain other crystals (Fig. 9.6b), where the symmetry of the charge distribution is not disturbed even after distortion, no piezoelectricity is observed. The latter type of crystals are those which possess the *centre of inversion.* Thus the absence of the centre of inversion is the pre-requisite for the occurrence of piezoelectricity. It may also be noted that all ferroelectric crystals are piezoelectric but all piezoelectric crystals are not necessarily ferroelectric. An example of the former type of crystals is barium titanate and that of the latter type of crystals is quartz.

Piezoelectric materials are used to convert electrical energy into mechanical energy and vice versa, i.e., they act as electro-mechanical transducers. They are used in devices such as gramophone pick-ups, microphones, strain gauges, ultrasonic generators, etc.

SOLVED EXAMPLES

Example 9.1. Determine the percentage of ionic polarizability in the sodium chloride crystal which has the optical index of refraction and the static dielectric constant as 1.5 and 5.6 respectively.

Solution. From the Clausius-Mossotti relation, we have

$$\frac{\varepsilon_r - 1}{\varepsilon_r + 2} = \frac{N(\alpha_e + \alpha_i)}{3\varepsilon_o} \tag{9.40}$$

Here α_e and α_i are the electronic and ionic contributions to polarizability respectively. At optical frequencies, Eq. (9.40) becomes

$$\frac{n^2 - 1}{n^2 + 2} = \frac{N\alpha_e}{3\varepsilon_o}, \tag{9.41}$$

where n is the index of refraction. From Eqs. (9.40) and (9.41), we obtain

$$\left(\frac{n^2 - 1}{n^2 + 2}\right)\left(\frac{\varepsilon_r + 2}{\varepsilon_r - 1}\right) = \frac{\alpha_e}{\alpha_e + \alpha_i} \tag{9.42}$$

The right hand side of Eq. (9.42) is the fraction of electronic polarizability. Thus the percentage of ionic polarizability is

$$\left(\frac{\alpha_i}{\alpha_e + \alpha_i}\right) \times 100 = \left(1 - \frac{\alpha_e}{\alpha_e + \alpha_i}\right) \times 100$$

$$= \left[1 - \left(\frac{n^2 - 1}{n^2 + 2}\right)\left(\frac{\varepsilon_r + 2}{\varepsilon_r - 1}\right)\right] \times 100$$

$$= \left[1 - \left\{\frac{(1.5)^2 - 1}{(1.5)^2 + 2}\right\}\left\{\frac{5.6 + 2}{5.6 - 1}\right\}\right] \times 100$$

$$= 51.4\%$$

SUMMARY

1. A dielectric material gets polarized in the presence of an external electric field **E**. The polarization **P** is defined as the dipole moment per unit volume of the material and is related to the electric susceptibility, χ_e, as

$$\mathbf{P} = \varepsilon_0 \chi_e \mathbf{E}$$

2. The electric displacement vector **D** for an isotropic or cubic medium is defined as

$$\mathbf{D} = \varepsilon_0 \varepsilon_r \mathbf{E} = \varepsilon_0 \mathbf{E} + \mathbf{P}$$

where ε_r is the relative permittivity or dielectric constant of the dielectric and is related to the electric susceptibility as

$$\varepsilon_r = 1 + \chi_e$$

3. The electric field acting at the site of an atom or molecule is called the local field. This is responsible for polarization of the individual atoms or molecules in the solid. For an atomic site with cubic symmetry, it is given by the Lorentz equation :

$$\mathbf{E}_{\mathbf{loc}} = \mathbf{E} + \frac{\mathbf{P}}{3\varepsilon_0}$$

4. The atomic polarizability, α, is defined as the dipole moment, p, per unit local electric field at an atom and is given by the relation :

$$p = \alpha E_{loc}$$

5. The polarizability is an atomic property whereas the dielectric constant is a macroscopic property of the material. The two parameters are related to each other by the Clausius-Mossotti relation given as

$$\frac{\varepsilon_r - 1}{\varepsilon_r + 2} = \frac{N\alpha}{3\varepsilon_0}$$

6. The net polarizability of a dielectric material results from three main contributions — the electronic polarizability, ionic polarizability, and dipolar or orientational polarizability. At low frequencies of the applied electric field, all the three contributions are present. As the frequency increases, the dipolar contribution disappears first followed by ionic and electronic contributions at higher frequencies. At optical frequencies, the polarization is mainly due to electronic contribution.

7. Ferroelectric materials are those which exhibit spontaneous polarization, i.e., polarization in the absence of an applied field. These are analogous to ferromagnetic materials. They possess a Curie point and exhibit hysteresis loop. The examples are barium titanate, the Rochelle salt, etc.

8. Piezoelectric crystals are those which are polarized under the application of an external stress. The examples are quartz, the Rochelle salt, tourmaline, etc. The inverse effect is also observed, i.e., the application of an electric field produces strains in such crystals. The absence of the centre of inversion is the pre-requisite for the occurrence of piezoelectricity.

9. All ferroelectric crystals are piezoelectric, but all piezoelectric crystals are not necessarily ferroelectric. Examples of ferroelectric and piezoelectric crystals are barium titanate and quartz respectively.

VERY SHORT QUESTIONS

1. Define the following terms:

   Polarisation, electric susceptibility, local field, dielectric constant, electric displacement vector, polarisability, spontaneous polarisation.

2. Enlist the various contributions to total polarisability.
3. Explain the following in brief:

   Ionic polarisation, electronic polarisation, orientational polarisation.

4. Which type of polarisation is the most effective in the visible region?
5. What is ferroelectricity? Give an example of ferroelectric crystal.
6. What is piezoelectricity?
7. Give an example of a crystal that is piezoelectric but not ferroelectric.
8. How is dielectric constant related to electric susceptibility ?

SHORT QUESTIONS

1. Derive the Clausius-Mossotti relation expressing the relationship between dielectric constant and atomic polarisability.
2. What is dipolar polarisability? Obtain an expression for dipolar polarisability of a dielectric at moderate temperatures.
3. What is electronic polarisability? Derive an expression for electronic polarisability using the classical theory.
4. Why is ionic polarisability found to be rather insensitive to temperature ? Give a possible explanation.
5. How does the total polarisability depend on frequency?
6. Why do piezoelectric crystals having centre of inversion show no piezoelectricity?
7. Describe the characteristic properties of ferroelectric materials. What is meant by polarisation catastrophe?

8. Explain the meaning and origin of piezoelectricity. Justify the statement " All ferroelectric crystals are piezoelectric, but all piezoelectric crystals are not necessarily ferroelectric.

PROBLEMS

1. The optical index of refraction and the dielectric constant for water are 1.33 and 8.1 respectively. Determine the percentage of ionic polarizability. (71%)
2. Calculate the polarization of a $BaTiO_3$ crystal using the structure as shown in Fig. 9.5. The structure transforms from cubic to tetragonal below the Curie point with lattice parameters c and a equal to 4.03Å and 3.98 Å respectively. The displacement of titanium ion is opposite to that of oxygen ions. The magnitude of displacement is 0.06 Å for titanium, 0.06 Å for oxygen ions on the side faces and 0.08 Å for oxygen ions on the top and bottom faces of the cube. The displacement of barium ions may be neglected. (0.16 cm^{-2})

CHAPTER - X

SUPERCONDUCTIVITY

10.1 INTRODUCTION AND HISTORICAL DEVELOPMENTS

The field of superconductivity has emerged as one of the most exciting fields of solid state physics and solid state chemistry during the last decade. The phenomenon was first discovered in 1911 by Kamerlingh Onnes in Leiden while observing the electrical resistance of mercury at very low temperatures close to 4.2 K, the melting point of helium. It was observed that the electrical resistance of mercury decreased continuously from its melting point (233 K) to 4.2 K and then, within some hundredths of a degree, dropped suddenly to about a millionth of its original value at the melting point as shown in Fig. 10.1. Similar results were obtained by using various other metals such as Pb, Sn and In. The phenomenon of disappearance of electrical resistance of material below a certain temperature was called *superconductivity* by Onnes and the material in this state was called a superconductor.

The discovery of superconductivity aroused considerable interest in this field since the materials with no electrical resistance, and hence negligible heat losses, could be exploited to fabricate powerful and economical devices which consume very little amount of electrical energy. For example, an electromagnet made up of a superconducting material can function for years together even after removal of the sup-

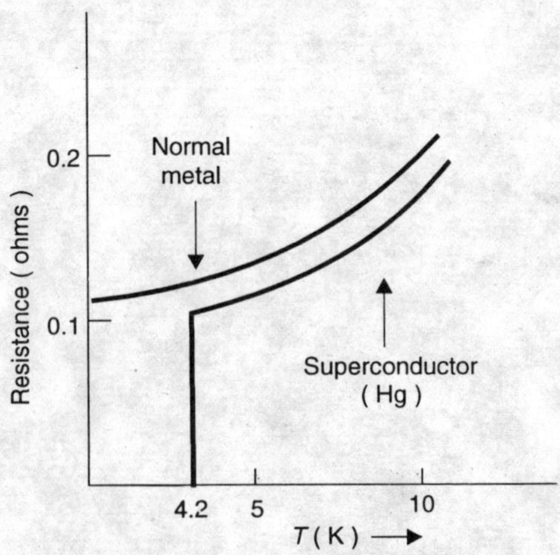

Fig. 10.1. *Temperature dependence of the resistance of a normal metal and a superconductor like Hg.*

ply voltage. However, due to the requirement of very low temperature, it was not feasible to manufacture such devices. It is both difficult and expensive to attain the liquid helium temperature and maintain it for a long time. Thus soon after the discovery of superconductivity, a lot of research work was undertaken to develop a superconducting material having as high critical temperature as possible. A number of materials including various metals, alloys, intermetallic and interstitial compounds, and ceramics were employed for this purpose. Besides all these efforts, the maximum critical temperature (T_c) of only 23 K was achieved in Nb_3Ge, an intermetallic compound of niobium and germanium in the year 1977. Thus the scientists had almost given up the hope of producing superconducting devices for which it was necessary to have a superconductor with the transition temperature equal to or higher than 77 K, the liquid nitrogen temperature, if not the room temperature.

In 1986, Bednorz and Muller reported their discovery on the La-Ba-Cu-O system of ceramic superconductors which showed T_c equal to 34 K. Thus, contrary to the previous findings, a new class of ceramic superconductors was discovered which showed critical temperature considerably greater than that of the metallic superconductors. They named these materials as *high-T_c ceramic superconductors.* They were awarded the Nobel Prize in 1988 for such an important discovery which created an unprecedented world wide interest in the field of oxide ceramic superconductors. In 1987, a ceramic superconductor of the composition $YBa_2Cu_3O_7$ was discovered which showed T_c equal to 90 K. In 1988, the value of T_c further shot up to about 125 K for thallium cuprates. Table 10.1 gives some data on superconductors in chronological order.

It would be apparent from the following discussions that the superconducting state is a distinct phase of matter having characteristic electrical, magnetic, thermodynamic and other physical properties. The most easily observed characteristics of bulk superconductors are the zero electrical resistance and the perfect diamagnetism. We describe below the empirical properties of superconductors; the relevant theory will be presented in a subsequent section.

10.2 ELECTRICAL RESISTIVITY

As described earlier, a superconductor exhibits no electrical resistance. The resistance of a superconductor suddenly drops to an extremely small value near the transition temperature as shown in Fig. 10.1. The careful investigations have shown that the resistivity of a metal in the superconducting state drops to less than one part in 10^{17} of its value in the normal state.

TABLE 10.1. Properties of some selected superconductors in chronological order

Year	T_c (K)	Material	Class	Crystal Structure	Type	H_c^* (MAm^{-1})
1911	4.2	Hg	Metal	Tetragonal	I	0.033
1913	6.2	Pb	Metal	*fcc*	I	0.064
1930	9.25	Nb	Metal	*bcc*	II	0.164
1940	15	NbN	Interstitial compound	NaCl	II	12.2
1950	17	V_3Si	Intermetallic compound	β-tungsten (W_3O)	II	12.4
1954	18	Nb_3Sn	Intermetallic compound	W_3O	II	18.5
1960	10	Nb-Ti	Alloy	*bcc*	II	11.9
1964	0.7	$SrTiO_3$	Ceramic	Perovskite	II	Small
1970	20.7	Nb_3 (Al, Ge)	Intermetallic compound	W_3O	II	34.0
1977	23	Nb_3Ge	Intermetallic compound	W_3O	II	29.6
1986	34	$La_{1.85}Ba_{0.15}CuO_4$	Ceramic	Tetragonal	II	43
1987	90	$YBa_2Cu_3O_7$	Ceramic	Orthorhombic	II	111
1988	108	Bi cuprates	Ceramic	Orthorhombic	II	–
1988	125	Tl cuprates	Ceramic	Orthorhombic	II	–

*extrapolated to 0 K. H_{c2} for type II materials. (1 Oersted = 79.6 Am^{-1})

10.3 PERFECT DIAMAGNETISM OR MEISSNER EFFECT

Meissner and Ochsenfeld discovered in 1933 that a superconductor expelled the magnetic flux as the former was cooled below T_c in an external magnetic field, i.e., it behaved as a perfect diamagnet. This phenomenon is known as the *Meissner effect.* Such a flux exclusion is also observed if the superconductor is first cooled below T_c and then placed in the magnetic field. It thus follows that the diamagnetic behaviour of a superconductor is independent of its history as illustrated by Fig. 10.2. It also follows from this figure that the Meissner effect is a reversible phenomenon. Since **B** = 0 inside the superconductor, we can write

$$\mathbf{B} = \mu_o (\mathbf{H} + \mathbf{M}) = 0$$

i.e.,
$$\mathbf{H} = -\mathbf{M}$$

Therefore, the susceptibility is given by

$$\chi = \mathbf{M}/\mathbf{H} = -1 \tag{10.1}$$

which is true for a perfect diamagnet.

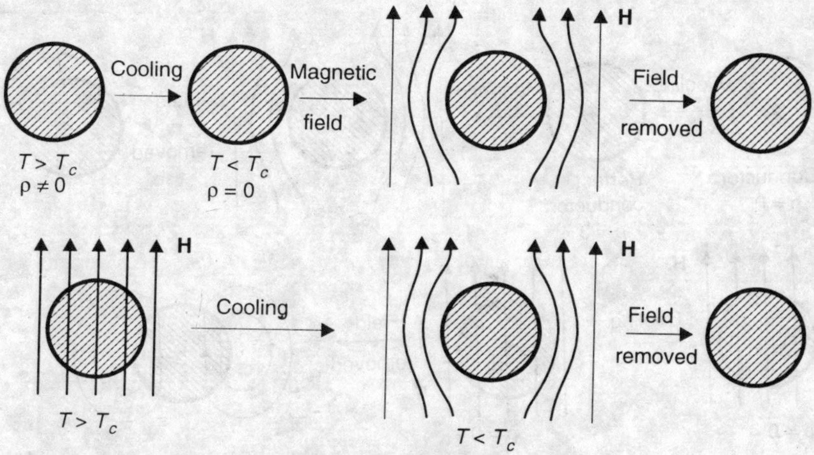

Fig. 10.2. *A superconductor showing a perfect diamagnetism independent of its history.*

It is interesting to note that the perfect diamagnetic behaviour of a superconductor cannot be explained simply by considering its zero resistivity. Such a perfect conductor would behave differently under different conditions as illustrated by Fig. 10.3. Since the resistivity, ρ, is zero for a perfect conductor, the application of Ohm's law ($E = \rho J$) indicates that no electric field can exist inside the perfect conductor. Using one of the Maxwell's equations, i.e.,

$$\nabla \times \mathbf{E} = -\frac{\partial \mathbf{B}}{\partial t}$$

we obtain

$$\mathbf{B} = \text{constant}$$

Thus the magnetic flux density passing through a perfect conductor becomes constant. This means that when a perfect conductor is cooled in the magnetic field until its resistance becomes zero, the magnetic field in the material gets frozen in and cannot change subsequently irrespective of the applied field. This is in contradiction to the Meissner effect.

Thus we conclude that the behaviour of a superconductor is different from that of a perfect conductor and the superconducting state may be considered as a characteristic thermodynamic phase of a substance in which the substance cannot sustain steady electric and magnetic fields. Hence the two mutually independent properties defining the superconducting state are the zero resistivity and perfect diamagnetism, i.e.,

$$E = 0 \quad \text{and} \quad B = 0 \tag{10.2}$$

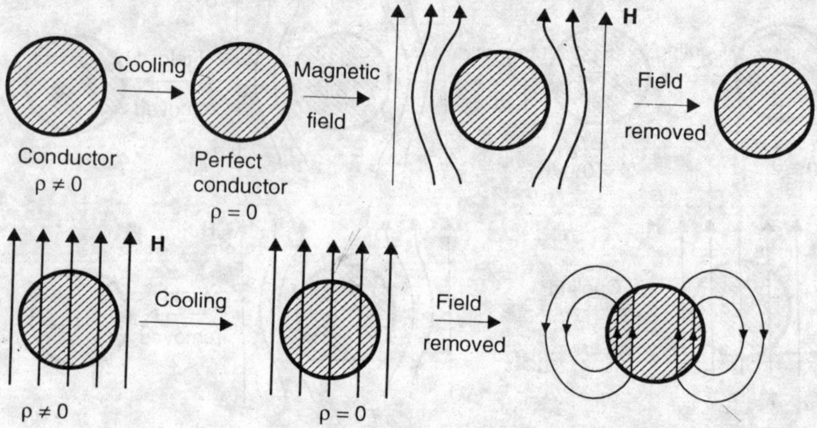

Fig. 10.3. *Magnetic behaviour of a perfect conductor.*

10.4 SUPERCURRENTS AND PENETRATION DEPTH

Consider a superconductor with a plane surface placed in an external magnetic field H_e (or B_e/μ_0) acting parallel to the plane surface as shown in Fig. 10.4. Consider a rectangular loop ABCD with sides AB and CD lying outside and inside the superconducting regions respectively. An application of the Ampere's theorem, i.e.,

$$\oint \mathbf{B}.\mathbf{dl} = \mu_0 I \tag{10.3}$$

to the loop ABCD indicates that a current I must flow perpendicular to the loop pointing inside the plane of the paper. Since $B = 0$ inside a superconductor, Eq. (10.3) gives

$$\mu_0 H_e (AB) = \mu_0 I$$

$\therefore$ Surface current per unit length $= H_e = B_e / \mu_0$ (10.4)

This current flows along the surface of the superconductor and produces a magnetization **M** which exactly cancels $\mathbf{H_e}$ inside the superconductor. Since a superconductor has zero electrical resistance, this current will remain almost constant and can flow indefinitely. Such currents are known as *supercurrents.*

If the width AD or BC of the rectangle ABCD is reduced, the current through the loop remains unchanged as indicated by the Amperes' theorem. This means that the current density must increase as the loop shrinks. For zero loop area, the current density should become infinite which is physically not possible. Thus the supercurrents cannot become abruptly zero as one moves from the surface to the interior of the superconductor. This is possible only if the applied magnetic field penetrates the superconductor up to a small thickness near the surface. This, infact, has been verified experimentally. The penetration of the external magnetic field inside a superconductor is depicted in Fig. 10.5. The field decreases exponentially inside the superconductor and is, according to London, given by the relation

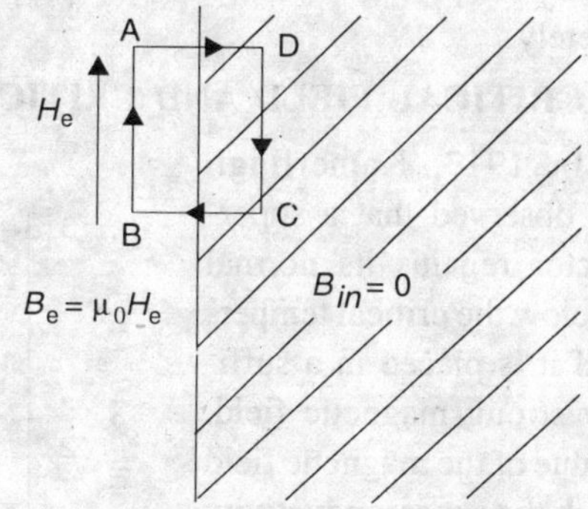

Fig. 10.4. *A superconductor placed in an external magnetic field.*

$$H_e(x) = H_e(0) \exp\left(-\frac{x}{\lambda_L}\right) \qquad (10.5)$$

Here $H_e(0)$ represents the magnetic field at the surface and λ_L is called the *characteristic length* or the *penetration depth* of the field. Thus the penetration depth is defined as the distance inside a superconductor where the magnetic field reduce to $1/e$ of its value at the surface. For superconductors below the critical temperature, λ_L~10^3 to 10^4Å. Thus, with the introduction of the idea of field penetration, it becomes apparent that the supercurrent density remains finite everywhere and the Eq. (10.4) gives the total surface supercurrent.

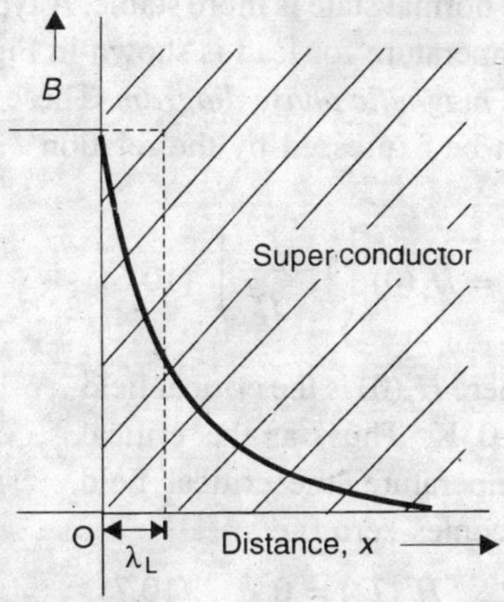

Fig. 10.5. *Penetration of an external field H_e $(=B_e/\mu_0)$ inside a superconductor. The flux density, B, does not fall abruptly to zero inside the superconductor but decreases exponentially.*

The penetration depth depends on temperature; a typical variation is shown in Fig. 10.6. When a specimen of tin is placed in a weak magnetic field, the penetration depth changes only slightly as the temperature rises until the critical temperature is reached where it increases sharply. Thus at and above T_c, the field penetrates the metal completely.

10.5 CRITICAL FIELD AND CRITICAL TEMPERATURE

In 1913, Kamerlingh Onnes observed that a superconductor regains its normal state below the critical temperature if it is placed in a sufficiently strong magnetic field. The value of the magnetic field at which the superconductivity vanishes is called the *threshold* or the *critical field*, H_c, and is of the order of a few hundred oersteds for most of the pure superconductors. This field changes with temperature. Thus we find that the superconducting state is stable only in some definite ranges of magnetic fields and temperatures. For higher fields and temperatures, the normal state is more stable. A typical plot of critical magnetic field versus temperature for lead is shown in Fig. 10.7. Such a plot is also referred to as the *magnetic phase diagram.* These types of curves are almost parabolic and can be expressed by the relation

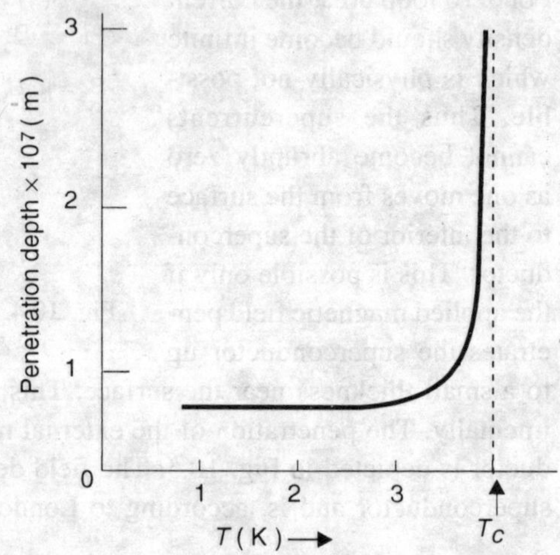

Fig. 10.6. *Variation of penetration depth with temperature for tin.*

$$H_c = H_c(0)\left(1 - \frac{T^2}{T_c^2}\right) \quad (10.6)$$

where $H_c(0)$ is the critical field at 0 K. Thus, at the critical temperature, the critical field becomes zero, i.e.,

$$H_c(T_c) = 0 \quad (10.7)$$

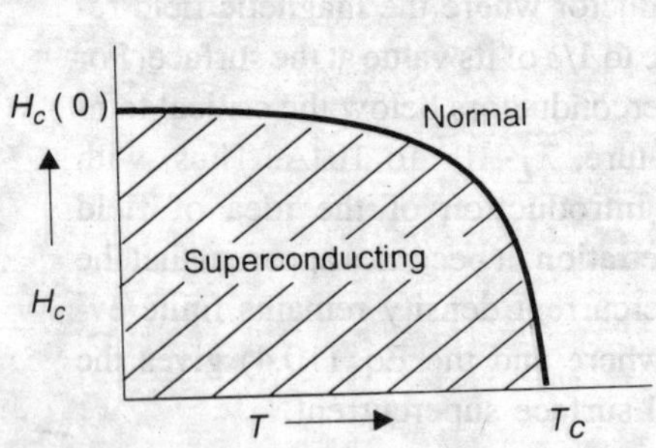

Fig. 10.7. *Variation of H_c with T for Pb.*

10.6 TYPE I AND TYPE II SUPERCONDUCTORS

Superconductors have been classified as type I and type II depending upon their behaviour in an external magnetic field, i.e., how strictly they follow the Meissner effect. We describe below these two types of superconductors.

10.6.1. Type I or Soft Superconductors

The superconductors which strictly follow the Meissner effect are called *type I superconductors.* The typical magnetic behaviour of lead, a type I superconductor, is shown in Fig. 10.8. These superconductors exhibit perfect diamagnetism below a critical field H_c which, for most of the cases, is of the order of 0.1 tesla. As the applied magnetic field is increased beyond H_c, the field penetrates the material completely and the latter abruptly reverts to its normal resistive state. These materials give away their superconductivity at lower field strengths and are referred to as the *soft superconductors.* Pure specimens of various metals exhibit this type of behaviour. These materials have very limited technical applications owing to the very low values of H_c.

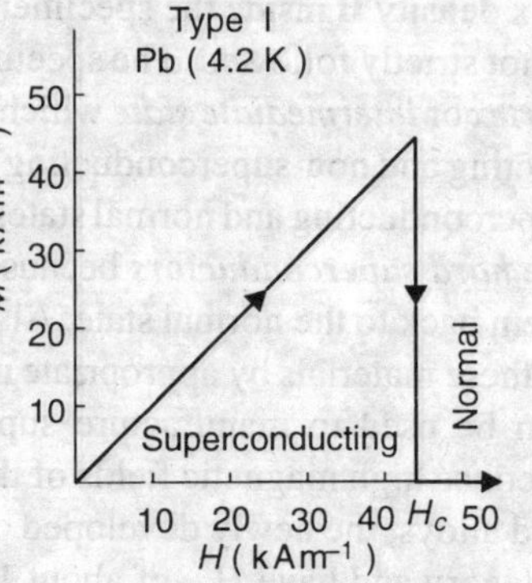

Fig. 10.8. *Magnetization curve of pure lead at 4.2 K.*

10.6.2. Type II or Hard Superconductors

These superconductors do not follow the Meissner effect strictly, i.e., the magnetic field does not penetrate these materials abruptly at the critical field. The typical magnetization curve for Pb-Bi alloy shown in Fig. 10.9 illustrates the magnetic behaviour of such a superconductor. It follows from this curve that for fields less than H_{c1}, the material exhibits perfect diamagnetism and no flux penetration takes place. Thus for $H < H_{c1}$, the material exists in the superconducting state. As the field exceeds H_{c1}, the flux begins to penetrate the specimen and, for $H = H_{c2}$, the complete penetration occurs and the material becomes a normal conductor. The fields H_{c1} and H_{c2} are called the lower and upper critical fields respectively. In the region between the fields H_{c1}

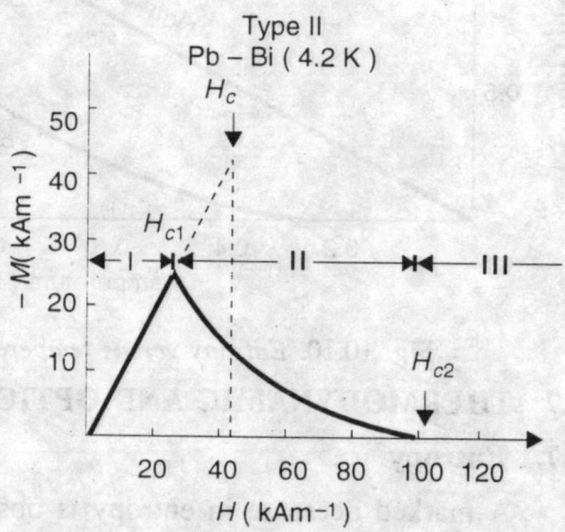

Fig. 10.9. *Magnetization curve of a lead-bismuth alloy at 4.2 K.*

I : Superconducting state

II : Vortex or mixed or intermediate state

III : Normal state

and H_{c2}, the diamagnetic behaviour of the material vanishes gradually and the flux density **B** inside the specimen remains non-zero, i.e., the Meissner effect is not strictly followed. The specimen in this region is said to be existing in the *vortex* or *intermediate state* which has a complicated distribution of superconducting and non-superconducting regions and may be regarded as a mixture of superconducting and normal states. The type II superconductors are also called the *hard superconductors* because relatively large fields are needed to bring them back to the normal state. Also, large magnetic hysteresis can be induced in these materials by appropriate mechanical treatment. Hence these materials can be used to manufacture superconducting wires which can be used to produce high magnetic fields of the order of 10 tesla. Apart from some metals and alloys, the newly developed copper oxide superconductors belong to this category and have H_{c2} of about 150 tesla.

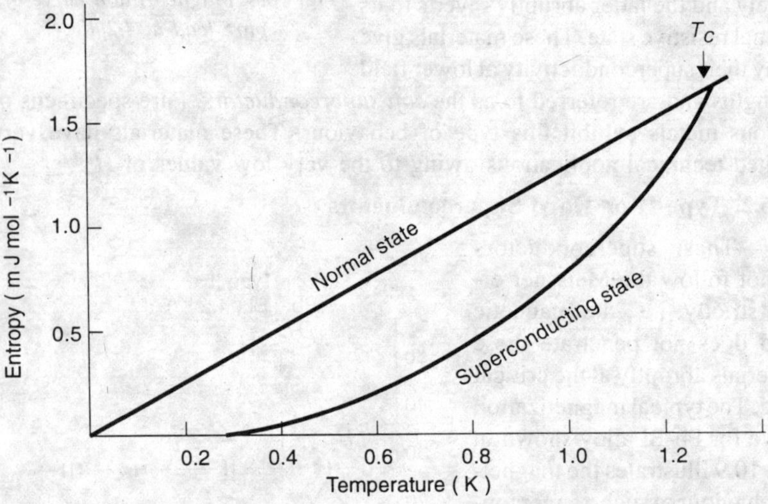

Fig. 10.10. *Entropy versus temperature for aluminium.*

10.7 THERMODYNAMIC AND OPTICAL PROPERTIES

10.7.1 Entropy

A marked decrease in entropy is observed during normal to superconducting transition near the critical temperature which indicates that the superconducting state is more ordered than the normal state. The plots of entropy versus temperature for aluminium in the normal and superconducting states are shown in Fig. 10.10. It has been established that it is the electronic structure of a solid which is mainly affected during the superconducting transition. Some or all the thermally excited electrons in the normal state are ordered in the superconducting state. Such an order may extend up to a distance of the order of 10^{-6} m in type I superconductors.

This range is called the *coherence length.*

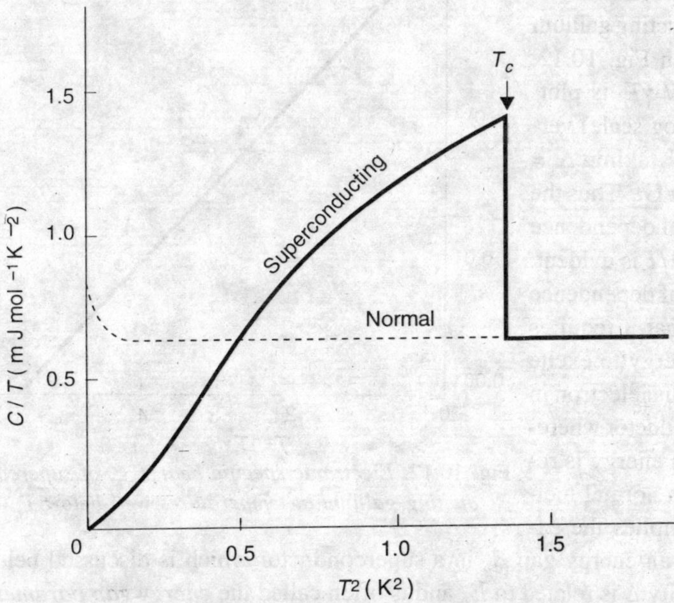

Fig. 10.11. *Specific heat of gallium near T_c. The normal state has been obtained from the superconducting state by applying a magnetic field of 200 gauss which is greater than H_c for Ga.*

10.7.2. Specific Heat

The variations of specific heat with temperature for gallium in the superconducting and normal states are shown in Fig. 10.11. The latter state can be achieved from the former by applying a magnetic field H greater than H_c. Such a small field has no effect on the specific heat of the normal metal. For the normal metal, the specific heat obeys the following relationship :

$$C_n = \gamma T + \beta T^3 \tag{10.8}$$

Thus the plot of C_n/T versus T^2 is a straight line as shown in Fig. 10.11. The first term on the right hand side of (10.8) represents the electronic contribution to the specific heat while the second one represents the contribution of lattice vibrations at low temperatures. Since the latter contribution remains unaffected in the superconducting state, it is obvious that only the electronic specific heat changes in the superconducting state. Also, unlike in the normal state, the electronic specific heat, C_{es}, in the superconducting state does not show linear variation with temperature; rather it varies exponentially as

$$C_{es} \propto \exp\left(-\frac{\Delta}{k_B T}\right) \tag{10.9}$$

i.e., the plot of ln C_{es} versus $1/T$ is a straight line. One such plot for superconducting gallium is shown in Fig. 10.12, where $C_{es} / \gamma T_c$ is plotted (on a log scale) versus T_c / T taking $\Delta = 1.4 k_B T_c$ for Ga. Thus the exponential dependence of C_{es} on $1/T$ is evident. This type of dependence suggests that it requires a finite energy to excite an individual electron in a superconductor whereas no such energy is required in a normal metal. This implies the existence of an energy gap E_g in a superconductor which is discussed below. The quantity Δ is related to E_g and is often called the *energy gap parameter.*

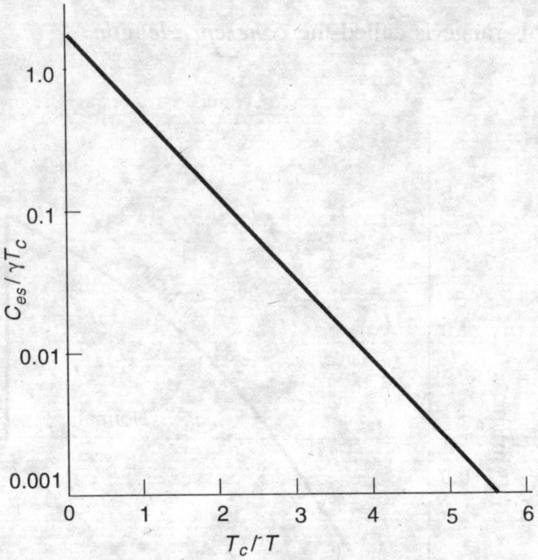

Fig. 10.12. *Electronic specific heat, C_{es}, of superconducting gallium at temperatures well below T_c.*

10.7.3. Energy Gap

As described above, the temperature dependence of electronic specific heat indicates the existence of an energy gap E_g in a superconductor. This energy gap is of entirely different nature compared with the energy gap in insulators. The gap is tied to the lattice in an insulator whereas it is tied to the Fermi gas in a superconductor. The gap separates the lowest excited state in a superconductor from the ground state as shown in Fig. 10.13 and is related to Δ as

$$E_g = 2\Delta \tag{10.10}$$

Therefore, for gallium,

$$E_g = 2 \times 1.4\ k_B T_c = 2.8\ k_B T_c \sim 10^{-4}\ \text{eV}$$

The electrons present in the excited states behave as normal electrons and create resistance whereas those present below it behave as superconducting electrons. The energy gap varies with temperature. It is maximum at 0 K and decreases continuously to zero as the temperature is increased to the critical temperature as shown in Fig. 10.14. Thus, at 0 K, there are no electrons above the gap and at $T = T_c$, all the superconducting electrons become normal electrons. Due to the presence of an energy gap, the superconductors respond to high frequency electromagnetic radiations of a particular frequency. Thus the energy gap is a characteristic feature of all superconductors which determines their thermal properties as well as their response to high frequency electromagnetic fields. The existence of an energy gap is accounted for in the BCS theory.

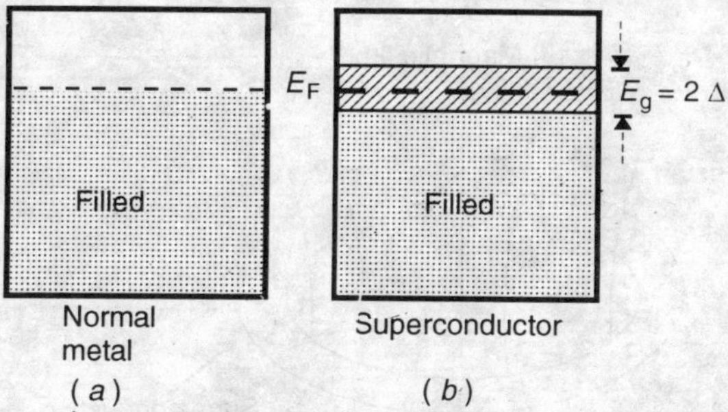

Fig. 10.13. *(a) Conduction band in the normal metal (b) Energy gap at the Fermi level in the superconducting state.*

10.8 ISOTOPE EFFECT

It was observed in the year 1950 that the transition temperatures of a superconductor varies with its isotopic mass M as

$$T_c \propto M^{-1/2}$$

or $$T_c M^{1/2} = \text{constant} \quad (10.11)$$

Thus larger the isotopic mass, lower is the transition temperature. For example, the transiton temperature of mercury changes from 4.185 K to 4.146 K when its isotopic mass is changed from 199.5 to 203.4 amu.

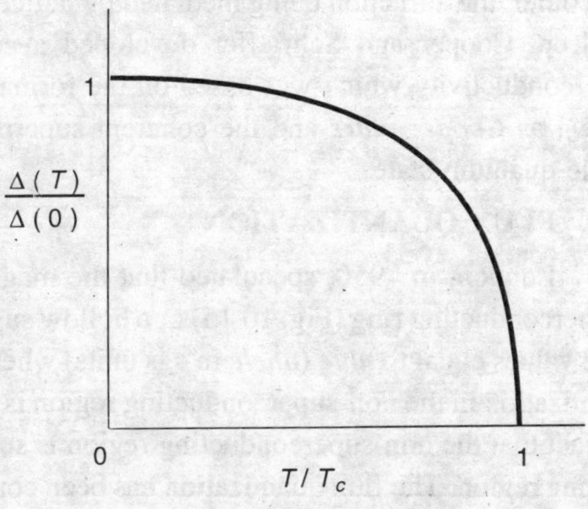

Fig. 10.14. *Variation of energy gap, Δ(T), with temperature for a superconductor.*

Now it is known that the mean square amplitude of atomic or lattice vibrations at low temperatures is proportional to $M^{-1/2}$ and the Debye temperature, θ_D, of the phonon spectrum is related to M as

$$\theta_D M^{1/2} = \text{constant} \quad (10.12)$$

From Eqs. (10.11) and (10.12), we obtain

$$T_c/\theta_D = \text{constant} \quad (10.13)$$

or, in general, it can be written as

$$T_C \propto \theta_D \propto M^{-1/2} \tag{10.14}$$

The Eqs. (10.13) and (10.14) indicate that the lattice vibrations are likely to be involved in causing superconductivity, i.e., the electron-phonon interactions might be playing an important role for the occurrence of superconductivity. This led Frohlich to show that two electrons in a metal can effectively attract each other, the attraction being mediated by lattice vibrations. Later, in 1957, Bardeen, Cooper and Schrieffer developed a complete atomic theory of superconductivity which was based on the formation of such electron pairs known as *Cooper pairs* and the coherent superposition of the pairs into a single quantum state.

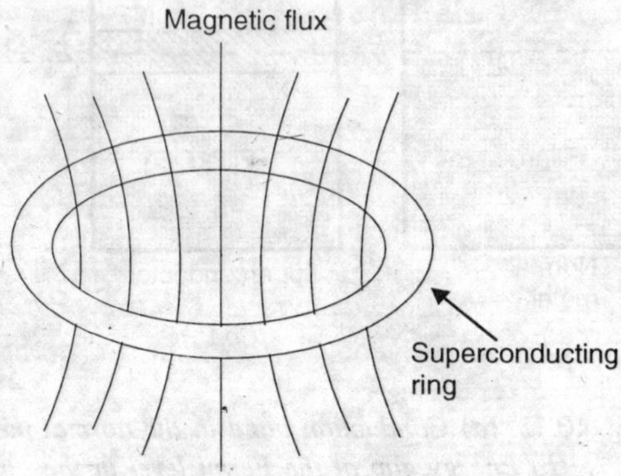

Fig. 10.15. *Magnetic flux through a superconducting ring.*

10.9 FLUX QUANTIZATION

London, in 1950, speculated that the magnetic flux passing through a superconducting ring (Fig. 10.15) or a hollow superconducting cylinder can have values equal to nh/e (nhc/e in cgs units) where n is an integer. This flux quantization in the non-superconducting region is simply the consequence of the fact that the non-superconducting region is surrounded by the superconducting region. The flux quantization has been confirmed experimentally but the quantum of flux has been found to be $h/2e$ rather than h/e. This unit of flux is called a *fluxoid* and is nearly equal to 2.07×10^{-15} Weber.

10.10 THE JOSEPHSON EFFECTS AND TUNNELLING

Josephson observed some remarkable effects associated with the tunnelling of superconducting electrons through a very thin insulator (1–5 nm) sandwiched between two superconductors. Such an insulating layer forms a weak link between the superconductors which is referred to as the *Josephson junction.* The effects observed by Josephson are given as follows :

(*i*) *The dc Josephson effect*

According to this effect, a dc current flows across the junction even when no voltage is applied across it.

(*ii*) *The ac Josephson effect*

If a dc voltage is applied across the junction, *rf* current oscillations of frequency $f = 2eV/h$ are set up across it. For example, a *dc* voltage of 1μV produces a frequency of 483.6 MHz. By measuring the frequency and the voltage, the value of e/h can be determined. Hence this effect has been utilized to measure e/h very precisely and may be used as a means of establishing a voltage standard. Furthermore, an application of *rf* voltage along with the *dc* voltage can result in the flow of direct current through the junction.

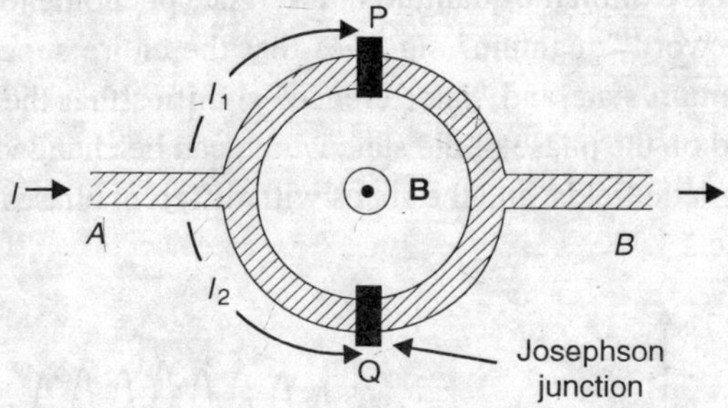

Fig. 10.16. *A SQUID.*

(*iii*) *Macroscopic quantum interference*

This effect describes the influence of the applied magnetic field on the supercurrent flowing through the junction. According to this effect, if a *dc* magnetic field is applied through a superconducting circuit containing two junctions, the maximum supercurrent shows interference effects which depend on the intensity of the magnetic field.

Consider the arrangement shown in Fig. 10.16 which is known as the *superconducting quantum interference device (SQUID)*. It consists of a ring of superconducting material having two side arms A and B which act as an entrance and exit for the supercurrent respectively. The insulating layers P and Q may, in general, have different thicknesses and let the currents through these layers be I_1 and I_2 respectively. The variations of I_1 and I_2 versus the magnetic field as obtained by Jaklevic, Lambe, Mercerean and Silver are shown in Fig. 10.17. Both I_1 and I_2 vary periodically with the magnetic field, the periodicity of I_1 being greater than that of I_2. The variation of I_2 is an interference effect of the two junctions while that of I_1 is a diffraction effect that arises from the finite dimension of each junction. Since the current is sensitive to very small changes in the magnetic field, the SQUID can be used as a very sensitive galvanometer.

The Josephson effects are the consequence of the fact that the superconductor is characterized by a single wave function. The flow of a supercurrent takes place between any two points where the wave function has different phases. The change in phase can also be brought about by the applied electric and magnetic fields. The states having different phases can be superimposed by using an arrangement as shown in Fig. 10.16. Thus the Josephson effects exhibit the quantum interference phenomenon on a macroscopic scale. The word "quantum" signifies that the entire superconductor is in a single quantum state and "interference" signifies that the measured properties depend on the phase of the state which can be changed by applying electric and magnetic fields, and the states with different phases can produce interference effects.

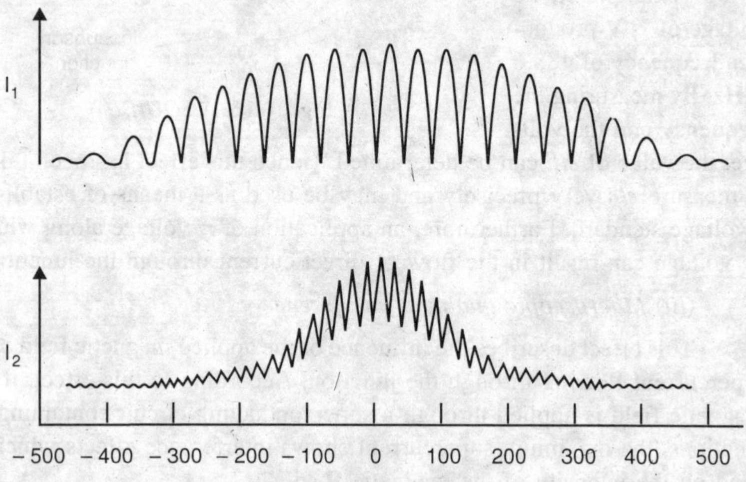

Fig. 10.17. *Dependence of supercurrents on the magnetic flux through the SQUID arrangement.*

10.11 ADDITIONAL CHARACTERISTICS

Besides the above-mentioned properties, some other characteristic features of superconductors are listed as follows :

(i) The crystal lattice remains unchanged during the transition from normal to superconducting state. This follows from the observed positions of x-ray diffraction lines which remain unchanged below and above the transition temperature. Also, the absence of any appreciable change in the intensities of diffraction lines indicates that the change in the electronic structure, if any, is very small.

(ii) Some of the properties of superconductors are modified when the size of the specimen is reduced below 10^{-14} cm approximately. For example, the magnetic permeability of very small specimen is non-zero and increases further as the temperature approaches T_c.

(*iii*) The critical temperature and the critical magnetic field of a superconductor change slightly under the influence of an applied stress. A stress which increases the dimensions of the specimen increases the transition temperature and produces a corresponding change in the critical magnetic field.

(*iv*) The introduction of chemical impurities modifies almost all the superconducting properties particularly the magnetic ones.

(*v*) The elastic properties and the thermal expansion coefficient remain unaffected below and above T_c.

(*vi*) The thermal conductivity of a material changes discontinuously during the transition from normal to superconducting state or vice versa. It is smaller in the superconducting state in case of pure metals but larger in case of some alloys. In each state, however, it increases continuously with temperature up to the critical temperature.

(*vii*) The superconducting state does not exhibit any thermolectric effect.

(*viii*) No changes in photoelectric properties are observed.

(*ix*) No appreciable changes in the reflectivity are observed in the visible and infrared regions.

(*x*) The zero resistance of superconductors changes slightly at very high frequencies (above 10 MHz) of the alternating current.

10.12 THEORETICAL ASPECTS

A number of theories have been proposed to explain the phenomenon of superconductivity with varying success. These are, for example, the phenomenological theory by London and London (1935), semiphenomenological theory of Ginzburg and Landau (1950) and the microscopic theory by Bardeen, Cooper and Schrieffer (1957), also called the BCS theory. Of these theories, the BCS theory is the most successful one and explains all the properties of superconductors except those of high-T_c ceramic superconductors. Bardeen, Cooper and Shrieffer were awarded the Nobel prize in 1972 for this work. The BCS theory is briefly presented in the following section.

10.12.1. The BCS Theory

The superconducting state of a metal may be considered to be resulting from a cooperative behaviour of conduction electrons. Such a cooperation or coherence of electrons takes place when a number of electrons occupy the same quantum state. This, however, appears to be impossible for both statistical and dynamic reasons. Statistically, electrons are fermions and nence occupy the quantum states singly. Secondly, the repulsive force among electrons tends to take them away from one another. In metals, however, the repulsive forces are not very strong owing to screening. According to the BCS theory, both these difficulties can be overcome under certain circumstances. In such a case, the electrons attract each other in a certain energy range and form pairs. A pair of electrons behaves like a boson. Thus a number of pairs can occupy the same quantum state which causes coherence among electrons. The complete BCS theory is too technical to be described here. We present below some physical arguments and ideas underlying this theory.

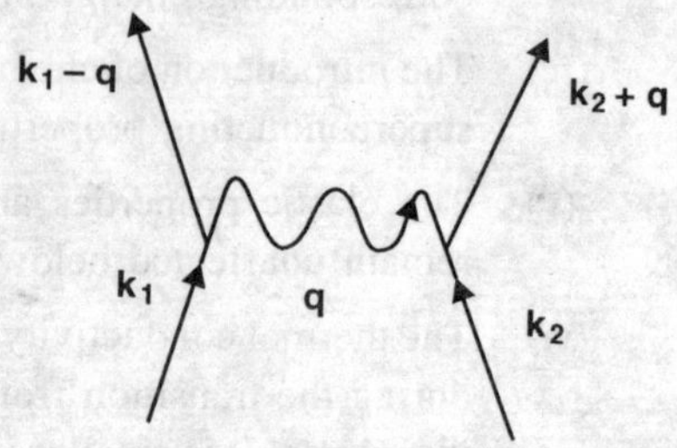

Fig. 10.18. *Electron-phonon-electron interaction.*

(*i*) *Electron-phonon interaction*

Frohlich, in 1950, realized that electrons could attract each other via distortion of the lattice. When an electron moves through a crystal, it produces lattice distortion and sets the heavier ions into slow forced oscillations. Since the electron moves very fast it leaves this region much before the oscillations can die off. Meanwhile, if another electron happens to pass through this distorted region, it experiences a force which is one of attraction and is of the type of polarization force. This attractive force lowers the energy of the second electron. The repulsive force between the electrons is small since the Coulomb's repulsion is instantaneous while the attraction mediated by lattice distortion is highly retarded in time. Therefore, the attraction caused by even a weak lattice distortion can overcome a stronger Coulomb's repulsion. Thus the net effect is the attraction of two electrons via a lattice distortion (or phonon) to form a pair of electrons known as the *Cooper pair.*

In quantum-mechanical terms, the first electron of wave vector $\mathbf{k_1}$ creates a virtual phonon $\mathbf{q}$ and loses momentum while the second electron of wave vector $\mathbf{k_2}$ acquires this momentum during its collision with the virtual phonon so that the overall momentum remains conserved. This is depicted in Fig. 10.18. The phonons involved are called *virtual phonons* due to their very short life time which renders it unnecessary to conserve energy during interaction in accordance with the uncertainty principle. Infact, the nature of the resulting electron-electron interaction depends on the relative magnitudes of the electronic energy change and the phonon energy. If the phonon energy exceeds the electronic energy change, the interaction is attractive. Also the interaction is the strongest when the two electrons have equal and opposite momenta and spins, i.e., $\mathbf{k}_1 = -\mathbf{k}_2$ and $\mathbf{s}_1 = -\mathbf{s}_2$. Such a pair of electrons is called the Cooper pair as L.N. Cooper first discovered that it was energetically favourable for such electrons to enter into an attractive interaction of this type.

It is now obvious that the mass of an ion has an important role in superconductivity. The smaller the mass of ions, the larger is the energy of phonons emitted and hence larger should be the transition temperature. The same conclusion is drawn from the isotope effect discussed earlier. In fact, it was the isotope effect which led Frohlich to consider the electron-phonon interaction as the possible cause of supercoductivity.

(ii) Cooper pair

As described above, a Cooper pair is formed when the phonon mediated attractive interaction between two electrons dominates the usual repulsive Coulombic interaction. The energy of such a pair of electrons in the bound state is less than the energy of two unbound or free electrons. The difference in energy is the binding energy, E_B, of the electron pair and is basically the same as the energy gap parameter, which was discussed in Sec. 10.7.2. Its typical value is of the order of 10^{-3} eV or about 10 K in temperature units. The binding is the strongest when the total momentum of the pair is zero and the pairs are in a spin singlet state with symmetrical spatial wave function. Cooper calculated the size of the Cooper pair as

$$r_o = \frac{\hbar v_F}{E_B} \tag{10.15}$$

wher v_F is the characteristic velocity of an electron in a metal and is called the *Fermi velocity*. In metals, v_F is related to the conduction electron concentration and is typically of the order of 10^6 ms^{-1}. Using this value of v_F along with E_B equal to 10^{-3} eV in Eq. (10.15), we obtain $r_o = 4 \times 10^{-7}$ m. This is rather a large value compared to the typical distance between two electrons which is of the order of 10^{-10} m. Hence the Cooper pairs overlap with each other considerably and the coherence between them becomes important. It was shown by Barden, Cooper and Schrieffer that the energy of the system is the lowest when total momentum of each pair is the same and is zero. This is the single quantum-mechanical state into which the electron pairs condense. The flow of Cooper pairs constitutes the supercurrent.

(iii) Existence of energy gap

The Cooper pairs are bound together by a very small energy, Δ , and form a new ground state which is superconducting and is separated by an energy gap 2Δ from the next lowest excited state above it. The Fermi level lies at the middle of the gap. The normal electron states lie above the energy gap and the superconducting electron states lie below the gap at the Fermi surface.

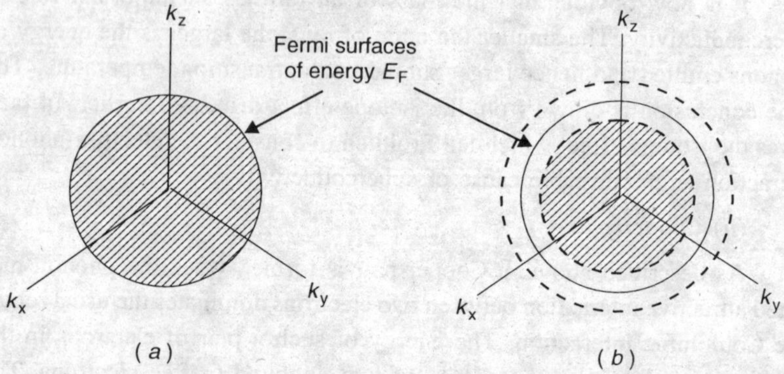

Fig. 10.19. *Ground state of non-interacting Fermi gas (a), and BCS ground state (b) in three dimensions.*

BCS ground state

The BCS ground state differs from the ground state of the non-interacting Fermi gas as shown in Fig. 10.19. In the non-interacting Fermi gas, all the states below the Fermi surface are occupied and all above it are vacant. The lowest excited state is separated from the ground state by an arbitrary low excitation energy, i.e., one can form an excited state by taking an electron from the Fermi surface and raising it just above this surface. As described earlier, the phonon assisted attractive interaction between electrons gives rise to the BCS ground state which is superconducting. This state is separated from the lowest excited state by a finite energy gap E_g. The formation of the BCS ground state can be understood from Fig. 10.20. The BCS state appears to have a higher kinetic energy than the Fermi state, but the attractive potential energy (not shown) of the BCS state acts to decrease the total energy of the BCS state relative to the Fermi state. Thus the BCS state is more stable than the Fermi state and superconductivity persists. The occupancy of one-particle orbitals near E_F in the BCS state resemble somewhat like that obtained from the Fermi-Dirac distribution at a finite temperature. However, in the BCS state, the one particle orbitals are occupied in pairs which are called Cooper pairs. If an orbital with wave vector **k** and spin up is occupied, then the one with wave vector – **k** and spin down is also occupied. Likewise, if **k** (↑) is vacant, then – **k** (↓) is also vacant.

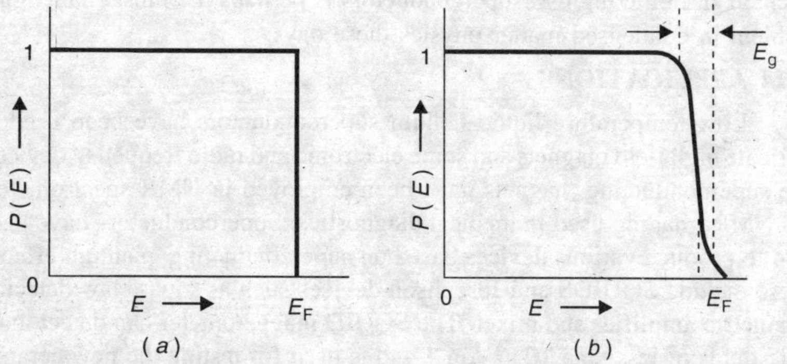

Fig. 10.20. *(a) Probability of occupancy, P(E), of an orbital of kinetic energy E versus the energy E in the ground state of the non-interacting Fermi gas (b) The BCS ground state differing from the Fermi state in a region of width ~ E_g (Both curves are at 0 K).*

10.13 HIGH TEMPERATURE CERAMIC SUPERCONDUCTORS

As described in Sec 10.1, a new class of oxide ceramic superconductors having the critical temperature greater than 30 K was discovered by Bednorz and Muller in 1986 which ushered in a new era in the field of superconductivity. These are called high-T_c superconductors. The first group of such superconductors discovered was $La_{2-x}M_xCuO_4$ (M = Ba, Sr, Ca) with T_c ranging from 25 to 40 K and is usually referred to as *'214' system.* This system possesses K_2NiF_4 structure with an orthorhombic distortion. This discovery was followed by the discovery of another important system having the general formula $LnBa_2Cu_3O_{7-x}$ (Ln = Y, Nd, Sm, Eu, Gd, Dy, Ho, Er, Tm, Yb) with x ≈ 0.2. This is called *'123' system* and has orthorhombic structure. In 1988, several other non-rare earth based copper oxide systems involving Bi and Tl were discovered which showed superconductivity between 60 K and 125 K. Some data on superconductors in chronological order is given in Table 10.1.

Many of the properties of these conventional high-T_c superconductors are identical to those of conventional low-T_c metallic superconductors. These include, for example, the existence of energy gap over the entire Fermi surface below T_c and the Josephson tunnelling. These superconductors, however, possess certain properties which do not match with those of conventional ones. These are, for example, small isotope effect, small coherence lengths (~ a few lattice spacings) and unconventional temperature dependencies of normal state response functions. Also, the pressure is found to increase the transition temperature in high-T_c superconductors, whereas usually an opposite effect is observed in conventional superconductors. Thus there appears to be something essentially new in these high-T_c superconductors which has not yet been clearly understood. The identification of the possible conduction mechanism in the high-T_c superconductors is perhaps the most challenging problem in condensed matter physics these days.

10.14 APPLICATIONS

Low temperature liquid helium superconductors have been used to fabricate high field magnets and some electronic and radio frequency devices. The superconducting magnets have been employed in NMR spectrometers and NMR imaging used in medical diagnostics. Superconductors have been used to produce various devices based on superconducting quantum effects. These include SQUIDS and Josephson devices such as square law detector, parametric amplifier and mixer. The SQUID magnetometer can detect magnetic fields of less than 10^{-4} Am^{-1} and is used for testing the new ceramic superconductors. Besides these, superconductors have also been used to produce electromagnetic shields and magneitc levitating trains. Most of the applications based on liquid helium superconductors make use of Nb alloys, particularly NbTi, owing to their ductile nature and ability to carry moderate currents. The utility of low temperature superconductors is limited because of the requirement of liquid helium temperature which is a great economic disadvantage.

The high-T_c oxide superconductors with $T_c > 77$ K have advantage over low-T_c superconductors in the sense that liquid nitrogen can be used as a coolant which greatly reduces the cost. In addition to this, the liquid nitrogen serves as a better coolant than helium because of its larger heat capacity. However, due to their low current densities and ceramic nature, the utiltity of oxide superconductors is limited. These materials are being used to build prototype liquid nitrogen cooled motors or generators operating at modest currents and magnetic fields. The SQUIDS fabricated using these superconductors find application in medical diagnostics, under sea communications, submarine detection and geophysical prospecting. It is apparent that the discovery of new superconductors with large current densities and T_c near room temperature will bring revolution in the scientific world.

SOLVED EXAMPLES

Example 10.1. Lead in the superconducting state has critical temperature of 6.2 K at zero magnetic field and a critical field of 0.064 MAm^{-1} at 0 K. Determine the critical field at 4 K.

Solution. Using Eq. (10.6), we have

$$H_c = H_c(0)\left[1-\left(\frac{T}{T_c}\right)^2\right]$$

$$= 0.064\left[1-\left(\frac{4}{6.2}\right)^2\right] = 0.037\ \text{MAm}^{-1}$$

Example 10.2. The transition temperature of mercury with an average atomic mass of 200.59 amu is 4.153 K. Determine the transition temperature of one of its isotopes, $_{80}Hg^{204}$

Solution. The transition temperature of a superconductor is related to its isotopic mass as

$$T_c \propto \frac{1}{\sqrt{M}}$$

which gives

$$\frac{T_{c2}}{T_{c1}} = \sqrt{\frac{M_1}{M_2}}$$

or $$T_{c2} = T_{c1}\sqrt{\frac{M_1}{M_2}} = 4.153\sqrt{\frac{200.59}{204}} = 4.118 \text{ K}$$

SUMMARY

1. A superconductor is characterized by two independent properties, viz., zero electrical resistance and perfect diamagnetism.

2. The phenomenon of flux exclusion in a superconductor is called the Meissner effect. It is independent of the history of superconductors and is reversible.

3. An applied magnetic field penetrates a superconductor up to a distance called the penetration depth. The resulting surface currents are known as supercurrents. The field penetration is quite small for $T < T_c$ and is complete for $T > T_c$.

4. The superconducting state is stable only below a particular applied magnetic field called the critical field. This field decreases with increase in temperature so long as $T < T_c$ and becomes zero when T equals T_c.

5. There exist two types of superconductors — type I or soft and type II or hard superconductors. Type I superconductors behave as perfect superconductors below a critical field H_c and as normal cunductors above it. Type II superconductors have two critical fields, H_{c1} and H_{c2}. Supercon-

ducting and normal states exist below and above the fields H_{c1} and H_{c2} respectively whereas a vortex state exists in the range between H_{c1} and H_{c2}.

6. The superconducting state is more ordered than the normal state. This order may extend up to a certain range called the coherence length.

7. Experiments on specific heat, infrared absorption and tunnelling suggest the existence of an energy gap which separates the superconducting electrons from the normal electrons.

8. The flux passing through the non-superconducting region of a superconducting ring is quantized; the quantum of flux is called a fluxoid and is equal to $h/2e$, where h is the Planck's constant and e the electronic charge.

9. The BCS theory accounts for a superconducting state by considering coherence among the Cooper pairs. A Cooper pair is a pair of electrons having equal and opposite momenta and spins.

VERY SHORT QUESTIONS

1. What is superconductivity?
2. What are high-T_C superconductors? Give one example.
3. Give any three main characteristics of superconductors?
4. What is Meissner effect?
5. What is the magnetic susceptibility of a superconductor?
6. What are supercurrents?
7. Define penetration depth for a superconductor. What is its value at the critical temperature?
8. What is critical field and what is its value at the critical temperature?
9. Which type of superconductors does not follow the Meissner effect strictly?
10. What is vortex state of a superconductor?
11. What is the effect of an external magnetic field on the superconducting state of a material?
12. What is coherence length?
13. What is the meaning of SQUID?
14. What is a Cooper pair?
15. What are high-T_C superconductors?

16. Enlist the main applications of superconductors.

17. What is a fluxoid?

SHORT QUESTIONS

1. Show that when a superconductor is placed in an external magnetic field, the field must penetrate up to a ertain depth inside the superconductor. Hence define the penetration depth.
2. What are soft and hard superconductors?
3. How do entropy and specific heat vary with temperature for a superconductor?
4. How does the energy gap in superconductors differ from the energy gap in insulators? How does it vary with temperature for superconductors?
5. Describe the Josephson effect underlying a SQUID. Discuss applications of SQUID.
6. Describe the isotope effect in superconductors.
7. Explain how the electron-phonon interaction helps to produce the Cooper pairs.
8. Explain the concept of the BCS ground state.
9. Compare the main properties of high-T_c superconductors with those of conventional superconductors.

LONG QUESTIONS

1. What are superconductors? What is the difference between a conductor cooled to 0K and a superconductor? Show that the material gets cooled when its superconductivity is destroyed by a magnetic field.
2. How do the electrical, magnetic, thermodynamic and optical properties of superconductors differ from those of normal conductors? Give some potential applications of superconductors.
3. Explain the difference between type I and type II superconductors using the Meissner effect. Prove that the Meissner effect and the disappearance of resistivity in a superconductor are mutually consistent.
4. Give a qualitative description of the BCS theory. How does it account for the superconducting state?

PROBLEMS

1. The critical temperature of a superconductor at zero magnetic field

is T_c. Determine the temperature at which the critical field becomes half of its value at 0 K. $(0.707\ T_c)$

2. The critical fields at 6 K and 8 K for a NbTi alloy are 7.616 and 4.284 MAm^{-1} respectively. Determine the transition temperature and the cirtical field at 0 K. (10 K, 11.9 MAm^{-1})

3. Determine the frequency of the electromagnetic waves radiated by a Josephson junction across which a dc voltage of 0.5 mV is applied. $(2.41\times10^{11}$ Hz)

APPENDIX - I

TABLE OF PHYSICAL CONSTANTS AND CONVERSION FACTORS

Speed of light in vacuum	c	2.998×10^{8} ms^{-1}
Plank's constant	h	6.626×10^{-34} Js
Plank's constant / (2π)	$h = \hbar/(2\pi)$	1.055×10^{-34} Js
Electron rest mass	m_e, m_o	9.1096×10^{-31} kg
Electron charge	e	1.602×10^{-19} C
		$= 4.803 \times 10^{-10}$ e.s.u.
Proton rest mass	m_p	1.6726×10^{-27} kg
Avogadro's number	N, N_A	6.0225×10^{23} mole^{-1}
Atomic mass unit	$a.m.u.$	1.6605×10^{-27} kg
Bohr magneton	$\mu_B = e\hbar\ (2m_e)$	9.274×10^{-24} Am^2
Nuclear magneton	$\mu_N = e\hbar/(2m_p)$	5.509×10^{-27} Am^2
Boltzmann constant	k, k_B	1.3806×10^{-23} JK^{-1}
		$= 8.614 \times 10^{-5}$ eVK^{-1}
Gas constant	R	8.314 J mo1^{-1}K^{-1}
Permeability of free space	μ_o	$4\pi \times 10^{-7}$ Hm^{-1}
		$= 1.257 \times 10^{-6}$ Hm^{-1}
Permittivity of free space	ε_o	8.854×10^{-12} Fm^{-1}

1 Pa = 1 Nm^{-2} = 10^{-5} bar

1 torr = 1 mm of Hg = 133.3 Pa

1 atm = 0.101325 MPa

1 calorie = 4.18 J

1 eV = 1.602×10^{-19} J = 1.602×10^{-12} erg

1 eV/entity = 96.49 kJmo1^{-1} (of entities)

1 cm^2/volt–sec = 10^{-4} m^{2}V^{-1} s^{-1}

1 C = 2.998×10^{9} *e.s.u.* (charge)

1 Am^{-1} = $4\pi \times 10^{-3}$ Oe (magnetic field intensity)

1 Am^{-1} $\cong$ 10^{-3} *e.m.u.* (magnetization)

1 *m.k.s.* unit = $1/(4\pi)$ *e.m.u.* (volume susceptibility)

1 T = 1 Wbm^{-2} = 1 Vsm^{-2} = 10^{4} G

References

1. Azaroff, L.V., *Introduction to solids,* Tata McGraw-Hill, New Delhi, 1971.
2. Bates, L.F., *Modern Magnetism,* 4th ed., Cambridge, London, 1961.
3. Blatt, F.J., *Physics of Electron Conduction in Solids,* McGraw-Hill, New York, 1968.
4. Born, M. and Huang, K., *Dynamical Theory of Crystal Lattices,* Oxford University, Oxford, 1956.
5. Buerger, M.J., *Elementary Crystallography,* Wiley, New York, 1956.
6. Chikazumi, S., *Physics of Magnetism,* Wiley, New York, 1964.
7. Dekker, A.J., *Solid State Physics,* Macmillan India, Madras, 1981.
8. Hutchison, T.S. and Baird, D.C., *The Physics of Engineering Solids,* 2nd ed., Wiley, 1963.
9. Kittel, C., *Introduction to Solid State Physics,* 5th ed., Wiley Eastern, New Delhi, 1979, and 6th ed., Wiley, New York, 1986.
10. Levy, R.A., *Principles of Solid State Physics,* Academic Press, New York, 1968.
11. Long, D., *Energy Bands in Semiconductors,* Wiley, New York, 1968.
12. Lynton, E., *Superconductivity,* Mathuen and Company, London, 1969.
13. Mattics, D.C., *Theory of Magnetism,* Harper and Row, New York, 1965.
14. McKelvey, J.P., *Solid State and Superconductor Physics,* Harper and Row, New York, 1966.
15. Morrish, A.H., *Physical Principles of Magnetism,* Wiley, New York, 1965.
16. Omar, M.A., *Elementary Solid State Physics,* Addison-Wesley, Reading, Massachusetts, 1975.
17. Philips, F.C., *Introduction to Crystallography,* Longman, London, 1949.
18. Ramakrishnan, T.V. and Rao, C.N.R., *Superconductivity Today,* Wiley Eastern, New Delhi, 1972.
19. Rose-Innes, A.C. and Rhoderick, E.H., *Introduction to Superconductivity,,* 2nd ed., Pergamon Press, New York, 1978.
20. Rosenberg, H.M., *The Solid State,* 2nd ed., Clarendon Press, Oxford,

1978.

21. Seitz, F., *The Modern Theory of Solids,* McGraw-Hill, New York, 1940.

22. Streetman, B.G., *Solid State Electronic Devices,* 3rd ed., Prentice-Hall, New Jersey, 1990.

23. Sze, S.M., *Physics of Semiconductor Devices,* Wiley Eastern, New Delhi, 1969.

24. Tareev, B., *Physics of Dielectric Materials,* Mir Publishers, Moscow, 1975.

25. White, R.M., *Quantum Theory of Magnetism,* McGraw-Hill, New York, 1970.

26. Wolf, H.F., *Semiconductors,* Wiley-Interscience, New York, 1971.

27. Ziman, J.M., *Principles of the Theory of Solids,* Cambridge, London, 1964.

Index